LAN Times Guide to Telephony

"Technically detailed yet very readable and understandable . . . Vital and accurate information right at your fingertips!"

—Steven A. Cates, Scripps Medical Center

LAN Times Guide to Telephony

David D. Bezar

Osborne **McGraw-Hill**

Berkeley New York St. Louis San Francisco
Auckland Bogotá Hamburg London Madrid
Mexico City Milan Montreal New Delhi Panama City
Paris São Paulo Singapore Sydney
Tokyo Toronto

Osborne **McGraw-Hill**
2600 Tenth Street
Berkeley, California 94710
U.S.A.

For information on translations or book distributors outside the U.S.A., or to arrange bulk purchase discounts for sales promotions, premiums, or fundraisers, please contact Osborne McGraw-Hill at the above address.

LAN Times Guide to Telephony

1234567890 DOC 998765

ISBN 0-07-882126-6

Publisher
Lawrence Levitsky

Acquisitions Editor
Cindy Brown

Project Editor
Bob Myren

Copy Editor
Kathryn Hashimoto

Proofreader
Bill Cassel

Indexer
Valerie Robbins

Computer Designer
Jani Beckwith

Illustrator
Rhys Elliott

Cover Design
emdesign, cary mc.

Quality Control Specialist
Joe Scuderi

Dedicated
with love
to my wife, Anna
and
son, Daniel
for all the time together
sacrificed for the sake of this project.

About the Author

David D. Bezar received his Bachelor of Science Degree in Computer Information Systems from Cal Poly, Pomona. He has over ten years of diversified telecommunications experience, including network planning and management of medium to large telecommunications systems (500 lines and up.) He is currently a consultant in many areas of telephony including, but not limited to, the Internet, switch procurement, network planning, optimization and troubleshooting. He is a co-founder of Big Byte Ventures, an Internet consulting firm, and a principle in Select Switch Systems, Inc., a PBX and long distance provider. Some of the clients he has worked for in the past and present include Carson productions, Warner Brothers, Sprint Long Distance, Pepperdine University, the United States Air Force, Scripps Medical Clinic and Jet Propulsion Laboratories.

Any comments or questions can be directed to e-mail: info@big-byte.com.

Contributing Writers

Eric Tholomé graduated from Ecole Centrale de Paris, France with an Engineering Degree with emphasis on computer science. He received his Master's Degree from Stanford University in Electrical Engineering with emphasis in telecommunications. Eric currently works for a major international telecommunications company near Paris in the Global System for Mobile Communications (GSM) Research and Development division. He has published two books, the latest in 1994, on French mobile communications, published by Dunod, and he has also written numerous articles on the subject of telecommunications.

William J. Ringle, MS, is president and founder of Star Communications Group, an information technology firm of consultants, trainers, and developers specializing in business applications on the Internet. Experienced in the design and delivery of technical topics to multi-level audiences, Bill Ringle presents to groups in the Philadelphia/Princeton region as well as nationally. His clients include Apple Computer, DuPont Corporation, Crozer Medical Center, the School District of Philadelphia, and various federal agencies and professional organizations. Affiliated with Drexel University, he teaches graduate courses in the field of information science and technology. Bill Ringle is a professional member of the National Speakers Association, and can be reached at ringle@starcomm.com.

Atis Jurka has been in the telecommunications field for over 20 years. After graduating from West Point, he entered the U.S. Army Signal Corp., where his primary responsibility was design and implementation of secure telecommunications systems to support the U.S. Army in wartime. Upon leaving the military after 12 years, Atis worked as a project manager for Electronic Data Systems, as a systems engineer for AT&T Bell Laboratories, and as a digital services Marketing Manager for AT&T Network Systems. He has also consulted to residential customers, small to large businesses, and educational organizations, from school systems to universities, on the subjects of ISDN, Frame Relay, and interactive multimedia. He is presently working for Qwest in Denver, Colorado.

Richard Brennan is the Technology Manager for AT&T Network Systems IMAGE2000 Showcase and Application Development Laboratory in San Ramon California. He has been with AT&T-NS, formerly the Western Electric Co., for 25 years, with assignments in engineering, software design, network planning, marketing, and applications development. He is a past recipient of the AT&T-NS President's Award. He currently provides consultative support of multimedia network design for ISDN and "communications superhighway" applications. He has designed and helped implement applications for the Pacific Bell "Education First" initiative, and for the Pacific Bell/AT&T-NS consumer broadband project. He presents at general and technical trade shows and conferences, including SuperComm, Internet World, Mactivity, the Western Communications Forum, and the National Conference of State Legislators. He has lectured at the University of San Francisco, California State-Hayward, University of California-Berkeley, and Webster University-Geneva. He can be reached at rbrennan@attmail.com.

Farhan Ahmed is a design engineer working for Texas Instruments on cellular technology. He has a Masters Degree in Electrical Engineering from Arizona State University where he specialized in VLSI design and communications systems, and also worked as a network and systems administrator. He received his Bachelor of Science degree in Electrical Engineering from the University of Texas at Austin.

Richard H. E. Smith II has a Bachelor of Arts degree in Computer Science from the University of Wisconsin. He has been solving data communications problems for the last 20 years, working on a variety of systems ranging from slot machines to the custom test equipment used to check out the B-2 Bomber's internal data . Recently, he was responsible for software embedded in radio packet data modems designed by a major aerospace company. He can be reached via the Internet, at "dick@smith.chi.il.us".

Daniel S. McCrary is currently an independent consultant in the field of interactive technologies and formerly director of desktop video communications at Applied Business Telecommunications, San Ramon, CA. A member of the International Teleconferencing Association (ITCA), the Telecommunications Association (TCA), the Personal Conferencing Working Group (PCWG), and the Interactive Multimedia Teleconferencing Consortium (IMTC), McCrary has been a part of the videoconferencing industry since 1990. As a consultant, he has designed, trained, and implemented video teleconferencing in corporate and educational organizations. McCrary holds a Bachelor of Arts in Telecommunications from the University of North Carolina at Chapel Hill and a Master of Science degree in Information and Communications Sciences from Ball State University in Muncie, Indiana. He can be reached at 317-747-0029, or e-mail 00dsmccrary@bsuvc.bsu.edu.

Jeanne Bayless as President of Answersoft, Inc. brings 16 years of management experience, with several years at start-up operations. Prior to founding AnswerSoft, Jeanne Bayless was chief financial officer/chief operations officer for Blyth Holdings, Inc., a client/server software development tool company at which she was instrumental in making the company profitable in a highly competitive market. Additionally, she created a business partner program to promote Blyth developers and their products, fostered investor relations, and raised capital. Previously she worked for Texas Instruments Information Technology Division in the EDI business unit. She holds a Bachelor of Science degree in Computer Science from Texas A&M University and a Master of Science degree in Management from the University of Texas at Dallas.

Greg Schumacher has been with Priority Call Management for two years as Director of Systems Engineering and Advanced Research, defining PCM's long term product directions and researching advanced technologies for use in Priority's One Number System (MSX). Prior to PCM, he worked as a SQA manager with Sequoia Systems, Boston Technology, and Data General. Greg Schumacher has a Bachelor of Arts degree from the University of Maine in History and Computer Science, is a member of the IEEE and ACM, and has a patent pending relating to PCS applications.

Contents at a Glance

Part I The Telephony Environment

Part II Telemanagement: The Focal Point of Telephony

Part III Telephony Connectivity

Part IV Peripheral Equipment and Services of Telephony

Part V Emerging Technologies

Contents

Part II

Telemanagement: The Focal Point of Telephony

Acknowledgments

In order to compile this store of knowledge, a small army of people were brought together to make this possible. The contributing writers really added a feeling of professionalism and expertise to these pages, thank you. The teamwork made it all possible, and the product of everyone's hard work is evident in this book.

I would first like to start with the people at Osborne/McGraw-Hill, who made this project possible. Thank you, Jeff Pepper for giving me the opportunity to write the words, and Ann Wilson for helping to get the project off the ground. To Linda Comer for making so many things better throughout the writing and to John Navarra for making sure we told no lies. To Kathryn Hashimoto for dotting all the "i's" and crossing all the "t's". To Bob Myren and the production staff, especially Rhys Elliott, Jani Beckwith, and Roberta Steele, for recreating the art and making the final product. Lastly, I want to thank Cindy Brown for keeping the glue together on this project from the publisher's side. This book would not have made it to press without her.

There is another group of people that made the final product of this book possible. To my parents for giving me life, I hope I have made you proud. To Richard, Mike, Lee, and Joseph who helped me gain the wisdom to write this book. To Johnny

Huffman for all his words of switching wisdom and fine photography, thanks. To Craig Dunton and Mary Beth Lesko for their additional edits and comments. To Harry Newton and his Telecom Dictionary—it never hurts to have a second opinion. To Creative Labs, Pinnacle Software, Vivo, Voice Information Systems, Teltone, and ACS for providing insider information about the industry. And finally to anyone that I might have forgotten.

PART ONE

The Telephony Environment

Chapter One

The Macro Approach to Telephony

Telephony is a term coined in the late 1980s to cover a wide variety of telecommunications-related equipment, peripherals, and services that have one thing in common: they enable us to communicate. This book takes a "macro" approach to telephony: it shows you the big picture so you can start to understand how the individual components fit together. For example, if you were trying to put together a puzzle and thousands of pieces were scattered before you, the first thing you would probably do is look at the picture on the box to see what the puzzle should look like. If you see the big picture first, the pieces of the puzzle will fall into place more quickly and easily.

Consistent with the macro approach, not every topic that falls under the heading of telephony will be covered in depth; many topics are touched on only briefly, others not at all. If you plan to implement one of the telephony technologies discussed, use this book as a starting point. In Appendix A, you will find a list of qualified vendors to help you take the next step in creating your telephony environment.

Getting the Big Picture of Telephony

Studying a typical corporation's telephony environment is the best way to get the big picture about telephony. To do that, we'll look at the hypothetical XYZ Corporation (see Figure 1-1). Its' telephony network is already in place, but we'll step through the decision-making process by which the network was created. We'll use each of XYZ's locations to focus on a different aspect of the telephony environment covered in this book. To get a slightly more detailed view of the decision-making processes that go on behind telephony, we'll also take a closer look at how decisions for buying and implementing products were resolved at the sales division in Los Angeles. As each component of telephony is discussed, you will be referred to the chapter that addresses that component in detail. By the end of this chapter, you will have been introduced to all the telephony topics discussed in this book.

Corporate Headquarters

XYZ was formed in the mid-1980s, grew rapidly to four major locations in the United States and one in Europe, and now does business all over the world. Its telephony network has been built over a number of years using the prevailing technologies of the time. Each division uses different, specialized components of telephony to meet its unique needs, with the corporate headquarters acting as the focal point of XYZ's operations. The functions described at each location are accessible to everyone in the organization through the network.

XYZ has grown to more than 5,000 employees. In the interest of saving money and increasing productivity, the company restructured its telecommunications and data processing system with an eye toward expansion and making Houston the focal point of its *virtual private network* (VPN). A VPN is a telecommunications network that is wholly owned and operated or leased by an organization for private use. This particular VPN was designed to carry both voice and data. For the following discussion, refer to Figure 1-2.

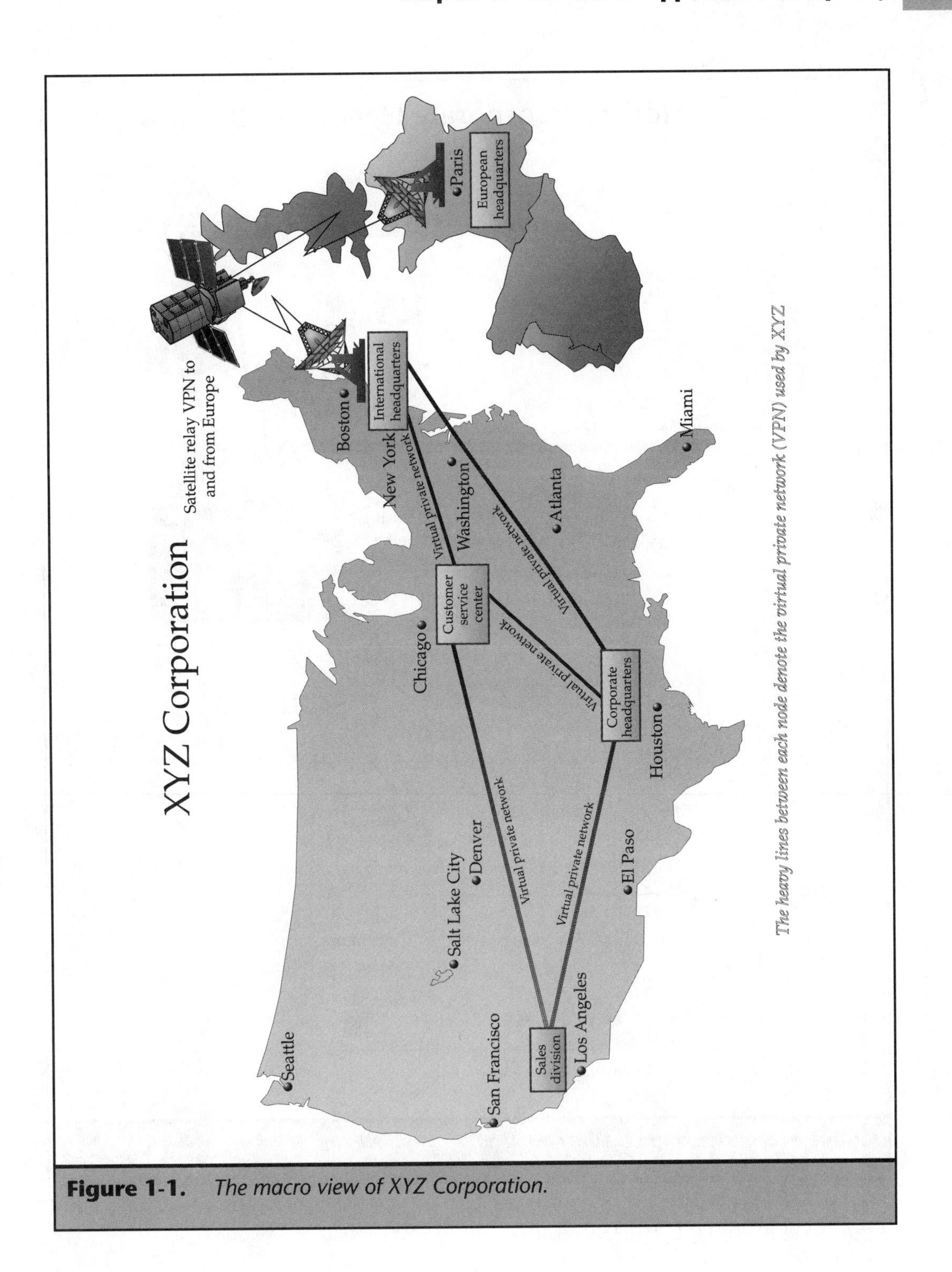

Figure 1-1. *The macro view of XYZ Corporation.*

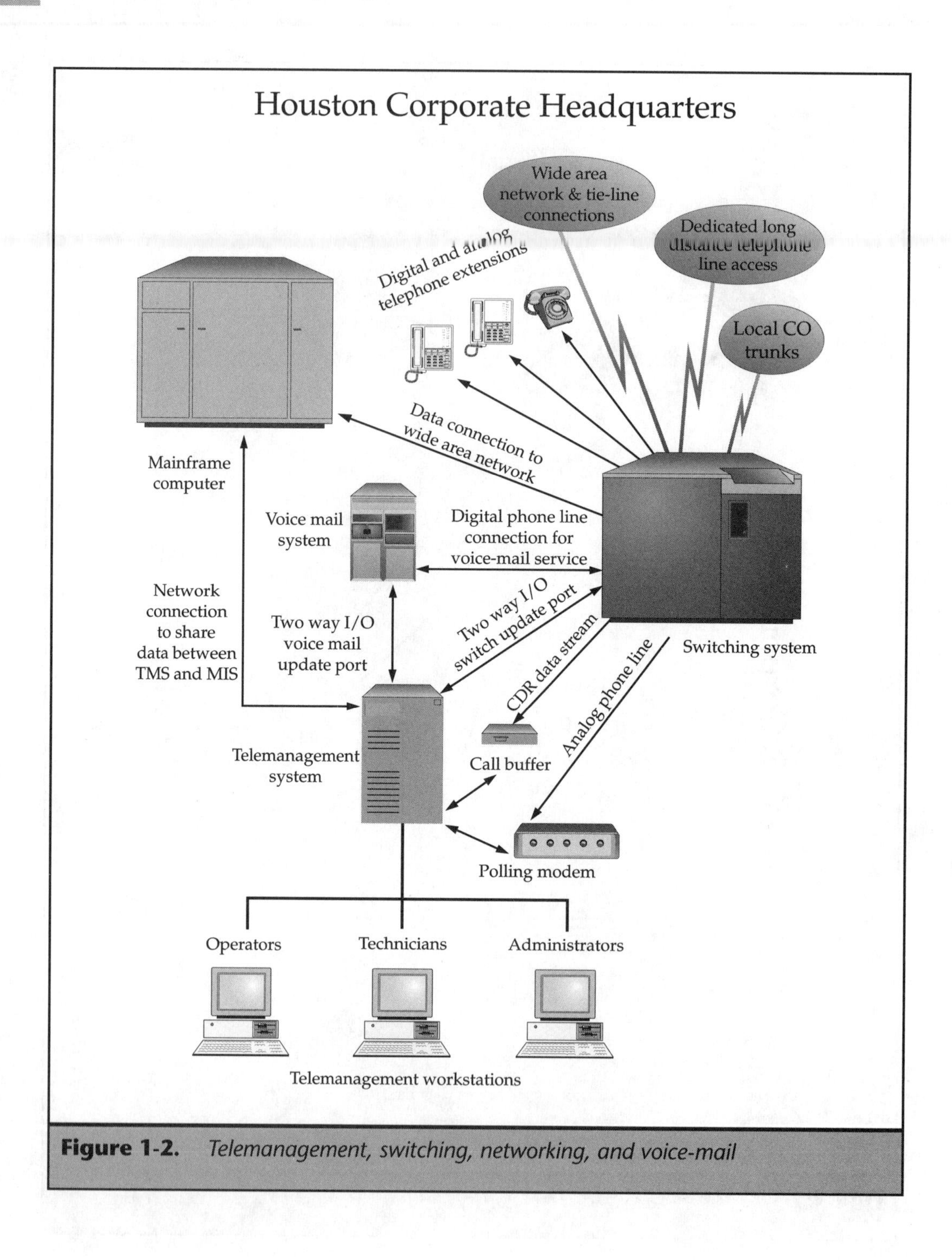

Figure 1-2. *Telemanagement, switching, networking, and voice-mail*

The Telemanagement System

XYZ's management started off on the right foot: They realized that if they were going to control their network, rather than having their network control them, they should first implement a computer-based telemanagement system (TMS). The TMS runs specialized software designed specifically to make possible the management of all telephony-related components, functions, and tasks from one focal point.

Developing a strategic plan was the first step in procuring the TMS. XYZ decided to procure a switching system, also called a private branch exchange (PBX), and a TMS at the same time through a request for proposal (RFP). (Strategic plan development and the RFP process are described in Chapter 5; TMS is discussed in detail in Chapter 6.)

With the TMS, the company can see the macro picture of its telephony environment, because every piece of the network is tracked in minute detail. This is where information is gathered to help make decisions about the telephony network. Without the TMS, all decisions pertaining to the network would have to be made by gathering data on a case-by-case basis.

The primary function of the TMS is call accounting. The PBX produces a call detail record for all calls placed through the system, whether over the tie-lines, long-distance telephone lines, local central office lines (described later in this chapter), or other external connections. The call record is stored in a call buffer before being transferred to the TMS, then it is processed and allocated to the proper department. Network traffic from XYZ's remote locations is also compiled, processed, and stored in the TMS. This information, along with other database information stored on the TMS, is used to compile statistics and reports for managing and optimizing the telephony environment.

Another important function of the TMS is to track and manage all of the telephony equipment, cabling, and infrastructure used throughout the company. Two physical connections help control this function: changes to telephony services are programmed into the PBX through a set of programs and a physical connection to the I/O switch update port; and changes to voice-mail services are automated by a second I/O connection between the TMS and the voice-mail system. (For a thorough explanation of how the TMS centralizes these functions, refer to Chapters 6 through 9.)

One final connection, between the TMS and the management information systems (MIS) department's mainframe, allows accounting information generated by the TMS to be integrated with the data compiled by the MIS accounting package. That information is made available to all XYZ's divisions through the wide area network, or WAN (discussed later in this chapter). Because XYZ knew how important it was to make the TMS information accessible—and realized that the top computer people worked in the MIS department, not telecommunications—purchasing a TMS that simplified integration and database access was critical to the success of XYZ's telephony environment.

The Switch

Behind every telephony environment lies a *switch*, also called a PBX or, more simply, a telephone system. A switch is a telecommunications device that is used to create circuits between two or more points used by people for voice telephone conversations or other telephony-related functions. Switch procurement began with a strategic plan and an RFP. XYZ's requirements—including technical specifications for network, video, and voice-mail connectivity—were incorporated into the RFP. After thorough review, the contracts were awarded to the most qualified vendors.

THE SWITCH VENDOR Obviously, the largest portion of the RFP award was made to the vendor of the switch itself. Because XYZ planned to operate and manage the switch once installation was complete, the company hired a technician who was already certified on the selected switching platform and sent three technical employees with some telecommunications background to the vendor's switch training facility. The plans to meet the switch's site requirements were put into motion at the same time. See Chapter 3 for more information on switches.

The vendor configured the switching system for a specific number of ports and related peripherals (such as telephones, modems, and multiplexors), provided the cabling, switching, and programming for the voice-mail system, and—because the system was designed to have a 100 percent growth path over the next five years—included a quote for expansion components, labor, and maintenance. The installation process was managed through the TMS, with the switch vendor using the TMS work order processing module to document and manage its work force. For more information about work order processing, refer to Chapter 9.

THE CABLING VENDOR XYZ's corporate headquarters is distributed in a campus-like environment similar to the one depicted in Chapter 7, Figure 7-3. Some of the buildings were older, requiring upgrades to their cabling infrastructure; others were newer and needed no further cabling. A second vendor was contracted to upgrade the cabling and inventory the telephony circuits in the TMS. XYZ's telecommunications staff assisted in the documentation process. (See Chapter 7 for more information on cabling and cable/plant management.)

THE LOCAL AND LONG-DISTANCE VENDORS Many of the local and long-distance telephone lines, or *trunks*, were already in place; the switch vendor only needed to *bridge* the trunks from the old system to the new one (Chapter 3 describes switching operations and terms).

However, several local central office trunks had to be added to the existing local trunks. XYZ also determined its long-distance requirements with the help of the TMS and went shopping for a separate vendor that could supply long-distance service at the lowest possible rate. The configuration, ordering, and procurement coordination were detailed in the RFP and the implementation plan that followed, and a long-distance vendor was chosen to supply dedicated T-1 access to the PBX. The T-1 has the capacity for carrying up to 24 voice and data communications simultaneously with a total bandwidth of 1.544 Mbps of data transmission. The long-distance vendor coordinated with the regional bell operating company to complete the circuits. The switch vendor then configured components of the switch, and any additional peripheral equipment, to incorporate the T-1 circuits into the switch platform (see Chapter 10 for more information on T-1s).

The Virtual Private Network

XYZ Corporation wanted to integrate as many of its computer systems as possible. The data, although distributed over several platforms, had to be accessible from anywhere in the network. The solution was a wide area network (WAN).

Realizing that interoffice phone calls were costing the company tens of thousands of dollars, XYZ analyzed its *calling patterns* (see Chapter 6) and determined that creating a VPN was justifiable (see Figure 1-1): the costs to set up and maintain the VPN were far less than the long-distance charges being incurred, and the VPN provided greater functionality, particularly when it came to the computer network. Because XYZ was willing to commit to a long-term lease for the VPN circuits (over one year), the recurring costs of the VPN were reduced even further.

The network would be comprised of a series of T-1 circuits, partitioned into voice, video, and data segments. Due to the amount of voice and data traffic anticipated from each location, the VPN was designed with redundant communications paths between each of the American locations. That way, if one of the VPN links went down, the connection could still be made by rerouting the communications request through a second or even third location and then on to the original destination. Frame relay is the predominant network architecture of the WAN, but XYZ has plans to convert to asynchronous transfer mode (ATM) once the technology becomes more mainstream. For more information on WANs, frame relay, and ATM, refer to Chapter 12.

Long-distance rates quoted by the phone companies (AT&T, Sprint, MCI, etc.) are typically mileage-based: the longer the distance from the origin of the call to the destination, the higher the per-minute charge. But the switches in the VPN can be programmed to shorten the distance a long-distance call travels. Say a call originates in Los Angeles and terminates in Boston. It could be routed over the VPN through Houston or Chicago, over to New York, then off the VPN on the long-distance lines connected to the New York office. The long-distance carrier then completes the call from New York to its destination in Boston, for a savings of more than 50 percent.

The Voice-Mail System

XYZ realized that voice-mail was needed to increase productivity, cut costs, and keep up good appearances with the customer base. A voice-mail system was procured by the RFP process and configured for a desired number of voice-mailboxes and a specified number of hours of storage capacity. The switch vendor and the voice-mail system vendor worked together to integrate the two systems. The voice-mail system has intelligent switching capabilities and, in conjunction with the switch, can forward or route calls to any voice-mailbox throughout XYZ without leaving the VPN. (For more information on voice-mail systems, refer to Chapter 14.)

European Headquarters

The European headquarters is a self-sufficient operating entity. Control of the division remains in the United States, but the vast differences between the U.S. and European marketplaces mean that this office is run differently. It stays in contact with the United States through a satellite link between the two continents. (Chapter 2 provides more information on European communications.)

International Headquarters

The international headquarters in New York is not the largest division of XYZ, but it does provide a closer look at specialized telephony connections and an interactive voice response application being used by XYZ as depicted in Figure 1-3. Obviously, the primary function of the New York office is to act as a liaison between Europe and the United States. XYZ plans to expand to other continents in the near future, with the New York office acting as the focal point for all international communications.

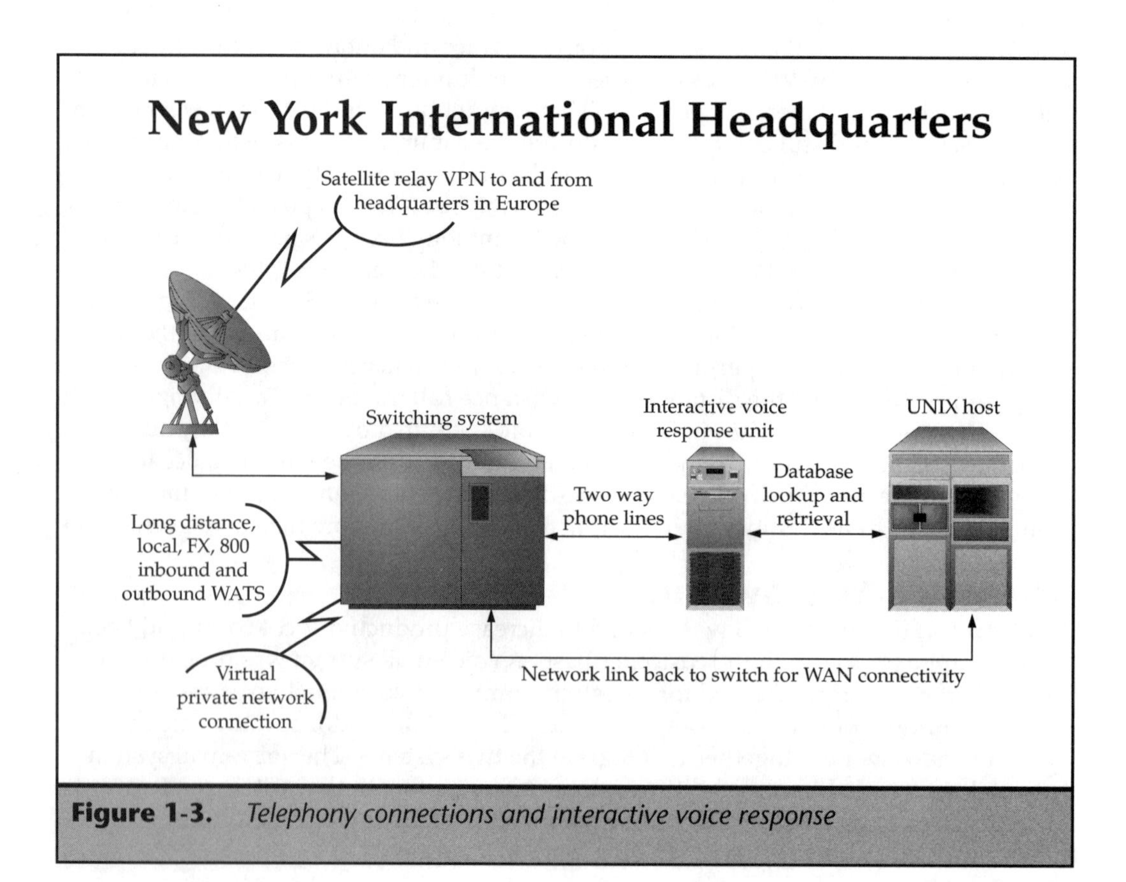

Figure 1-3. *Telephony connections and interactive voice response*

The New York office uses a wide variety of telephone lines provided by a number of telephone companies. The following sections describe each in detail.

Long-Distance Lines

These lines are provided by the same long-distance carrier at each of XYZ's U.S. offices. XYZ negotiated a favorable discount on long-distance rates based on the calling volume that the long-distance vendor could expect from the four sites. The vendor also agreed to bill in tenth-of-a-minute (six-second) increments, rather than by the minute, and to supply T-1 dedicated access to each of XYZ's locations free of charge. Some long-distance vendors actually bill for the entire minute even if a fraction of the minute is used. Those fractions of minutes add up quick. The service rates are mileage-based, and the discount rate was predicated on the vendor's getting all of XYZ's long-distance business (a common practice in the long-distance market today).

Local Exchange Carrier Lines

Every PBX must have local exchange carrier lines, whether or not they're wanted. A central office is where the regional bell operating company supplies telephone service to all users of the national telephone system throughout your area. When a local or local-measured call is made, central office lines are being used. A *local call* is any call made within the central office serving area and is usually included in the monthly service fee for the line; *local-measured* calls are those placed just outside the local calling area and are considered toll calls but are typically billed at a very low rate.

Local central office lines used to be needed to carry *intralata* traffic—those calls that originate and terminate in the same lata. A *lata* is one of 161 geographical areas within the U.S. where a telephone company can have a mini-monopoly on phone service. Intralata calls used to be the most expensive calls one could make—even more expensive than calling across the country—because the phone company had no competition. In January 1995, when an FCC ruling gave long-distance carriers the right to carry intralata traffic, XYZ changed its calling pattern so that all intralata calls were placed over the long-distance carrier's trunks. At home you must dial 10XXX (the long distance carrier code) to avoid the LEC.

Foreign Exchange Lines

Although they are not used as often as they should be, foreign exchange (FX) lines can save businesses a lot of money. These are lines delivered from a central office other than the one that would normally service that location. XYZ decided that the volume of calls from its New York City office to Syracuse warranted having an FX line delivered from Syracuse to the New York office. The resulting reduction in toll charges more than offset the charge to get FX service from Syracuse. The calling pattern was then modified so that the FX lines were chosen whenever a call was placed to the Syracuse area.

800 WATS Lines

XYZ uses 800 inbound Wide Area Telecommunications Service (WATS) to supply free inbound calling to its customers. The nice thing about this service is that one is usually charged a fixed, per-minute rate. XYZ negotiated a favorable rate from its regular long-distance vendor, who was happy to monopolize XYZ's long-distance business.

For outbound WATS service, rates are either mileage-based or *postalized*. With postalized pricing, the long-distance charges are based on the duration of the call, not on the distance the call travels. As it turns out, the long-distance vendor offered the same low rate for outbound WATS traffic as it did for regular long-distance traffic. The only time the WATS line would be used for outbound long-distance traffic would be when no additional long-distance trunks were available, so everybody wins. The only negative side to having two-way outbound WATS lines is that the outbound call could block an inbound call on the WATS line; in XYZ's case, however, the network experiences very little overflow traffic, and the decision was made to have the WATS line provide both inbound and outbound service.

Tie-lines

Tie-lines are part of the VPN described earlier and are leased either through the local exchange carrier (LEC), also known as the local telephone company, or any one of a number of long-distance carriers. They are used to link any two of XYZ's locations so that calls between those offices can be made without dialing an access code and the entire 11-digit phone number. Instead, the employee can simply dial an access code such as "7" followed by a four-digit extension. Tie-lines can also be used to route calls through the VPN and eventually off the network to the local or long-distance carrier from another XYZ location.

The European office uses the tie-lines on the VPN to communicate with XYZ in the U.S. at no additional charge other than the monthly VPN lease price. Calls made off net (call that are made to parties outside of XYZ corporation) are much cheaper because fewer external carriers are used. Although tie-lines are expensive and are usually mileage-based, they are a purchased for a fixed fee every month; if the traffic is expected to exceed a certain level, the savings can be immense.

Interactive Voice Response

The New York office uses one other specialized component of telephony: an *interactive voice response* (IVR) unit. An IVR is a form of automated voice processing that has many applications. XYZ uses its IVR system to automate both order entry/tracking and customer service (through the 800 inbound WATS line). For automated order entry and tracking, the IVR uses a series of automated questions and touchtone user responses. It looks up information on the UNIX host and can provide a wide variety of information about XYZ's products and services; it can also perform order entry if the customer desires. When an order is placed, that information is passed to the UNIX host and transferred over the WAN to the corporate offices in Houston for further processing. The IVR is available 24 hours a day, seven days a week.

The IVR can also handle basic customer-service questions. Callers can check on the status of pending orders, get answers to commonly asked questions, even have information faxed directly to them. If the nature of the call requires a human operator, it can be transferred automatically to a customer service representative in Chicago over the VPN. The IVR more than pays for itself in reduced labor costs and enhanced customer service. For more information about IVR and voice processing, see Chapter 14; for more information about faxing and fax-on-demand, see Chapter 16.

Customer Service Center

The customer service center in Chicago is the newest of XYZ's offices. Figure 1-4 shows the basic configuration of its computer telephony equipment. The Chicago office was built using the latest telephony server application programming interface (TSAPI), created by AT&T and Novell. XYZ already had several Novell networks, so the choice seemed fairly simple. Another reason XYZ chose TSAPI was because the switching system they preferred only supported TSAPI protocol at the time. XYZ is questioning this decision and wondering if telephony application programming interface (TAPI), from Microsoft and Intel, might have been a better choice. The industry as a whole seems to be swaying in the direction of the TAPI standard. Both provide advanced computer telephony integration (CTI).

XYZ's customer service representatives, or CSRs, wonder how they ever did their jobs before they had CTI. Here's how it works: A call comes into the customer service center. The telephone system passes automatic number identification (ANI) or caller ID information to the TSAPI interface, which in turn passes the call along with the ANI to an available CSR. (ANI is the actual telephone number of the calling party; if you want to know what your ANI is, just call 1-800-MY-ANI-IS.)

The CSR answers by clicking the mouse button on an icon on the computer screen. The computer screen displays information about the customer—obtained by cross-referencing the ANI telephone number with the contact management software running on the LAN—so quickly that the CSR can answer the phone with "Hello, Mr. Smith, what can I help you with today?" The CSR is also given the customer's history with XYZ through a second link, this one to a Lotus Notes application designed specifically for XYZ's customer service department. If for some reason the CSR needs the assistance of another employee, such as a technician, a conference call can be established through the CTI link to the PBX switch. For more information on TAPI and TSAPI, see Chapter 18.

Sales Division

Computers and telephony are heavily linked at XYZ Corporation. With all its stand-alone systems, LANs, and WANs, XYZ needed to look at the big picture before building its computer telephony environment (CTE). To get a slightly more detailed view of the decision-making processes that go on behind telephony, let's take a closer look at what is going on at the Los Angeles sales division.

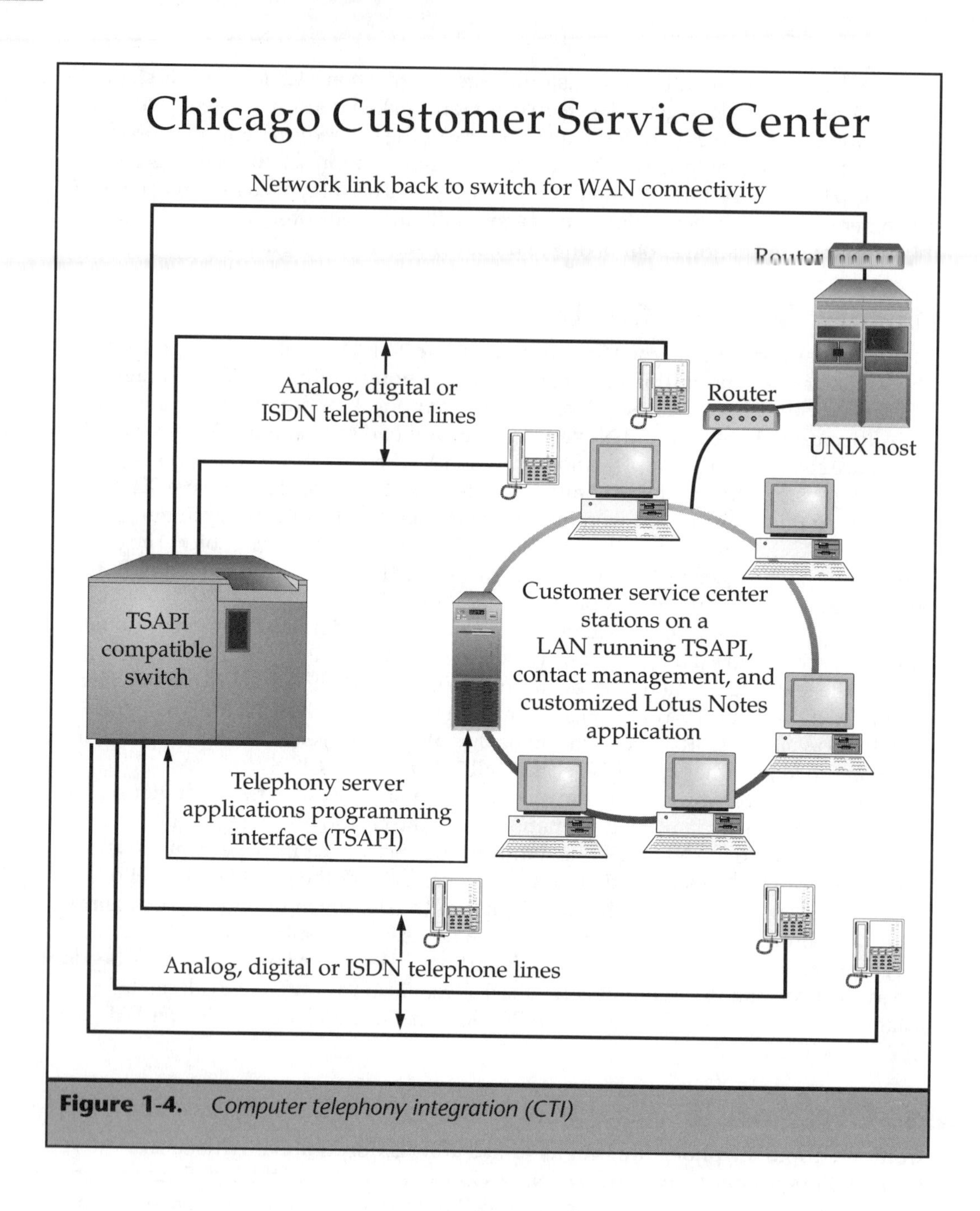

Figure 1-4. *Computer telephony integration (CTI)*

Let's say you are hired as a consultant for the XYZ Corporation and you are responsible for putting together all the components shown in Figure 1-5 for the Los Angeles sales division. First, you'd like to know about the equipment the CTE will be connected to. You begin by gathering information about the telecommunications network. In addition, a good portion of the LAN environment already exists; it is vital that you understand all the existing telephony components before you take action. (Detailed information on sizing up an existing telephony environment is given in Chapter 5.)

The Personal Computer

The personal computer shown in Figure 1-5 (item 1) is an island just waiting to be connected to the rest of the world. To begin with, try to determine which connection will be the most demanding of the PC hardware you purchase. In this case, the telecommuting function is the most demanding because of the data communications speed required for the function, so the Bulletin Board Service (BBS) is the secondary connection and will not dictate the hardware used at the PC. The PCs will be connected to the CTE using external 28.8 Kbps modems or, wherever possible, external ISDN modems. Refer to Chapter 15 for information on installing an external modem and Chapter 11 for information on ISDN.

When choosing a modem, you need to ask yourself what the work force will be doing with this telephony connection. Due to the nature of their business, you can assume most users will be telecommuting—using video, computer, or fax machine rather than physically commuting to the office—therefore you want to create the fastest link possible. If you were adding on to an existing CTE, the maximum speed of the modem, modem pool (a group of modems), or modem gateway (modems used to access multiple computer systems) connected to the LAN might be the determining factor. However, since you're just now creating a CTE, the telephony connections to the LAN don't exist yet.

The 28.8 modems will increase productivity and reduce the connect times needed compared to the slower 9,600 or 14,400 bps modems. They're more expensive, but don't forget about the recurring costs of your CTE—you'll quickly recoup your investment in the form of lower telephone toll charges.

ISDN is a dial-up data connection that requires special lines and equipment. ISDN provides even faster and more reliable dial-up access connectivity than 28.8 modems and has some additional advantages over regular voice-line communications:

- Voice and data transmissions are not analog; all information is sent digitally through the network.
- Two separate channels for transmission are provided for ISDN circuits. These channels can be used separately or combined for higher transmission speeds.
- Faster data transmission—up to 128 Kbps for uncompressed data.
- ISDN can support voice and data over the same line simultaneously.

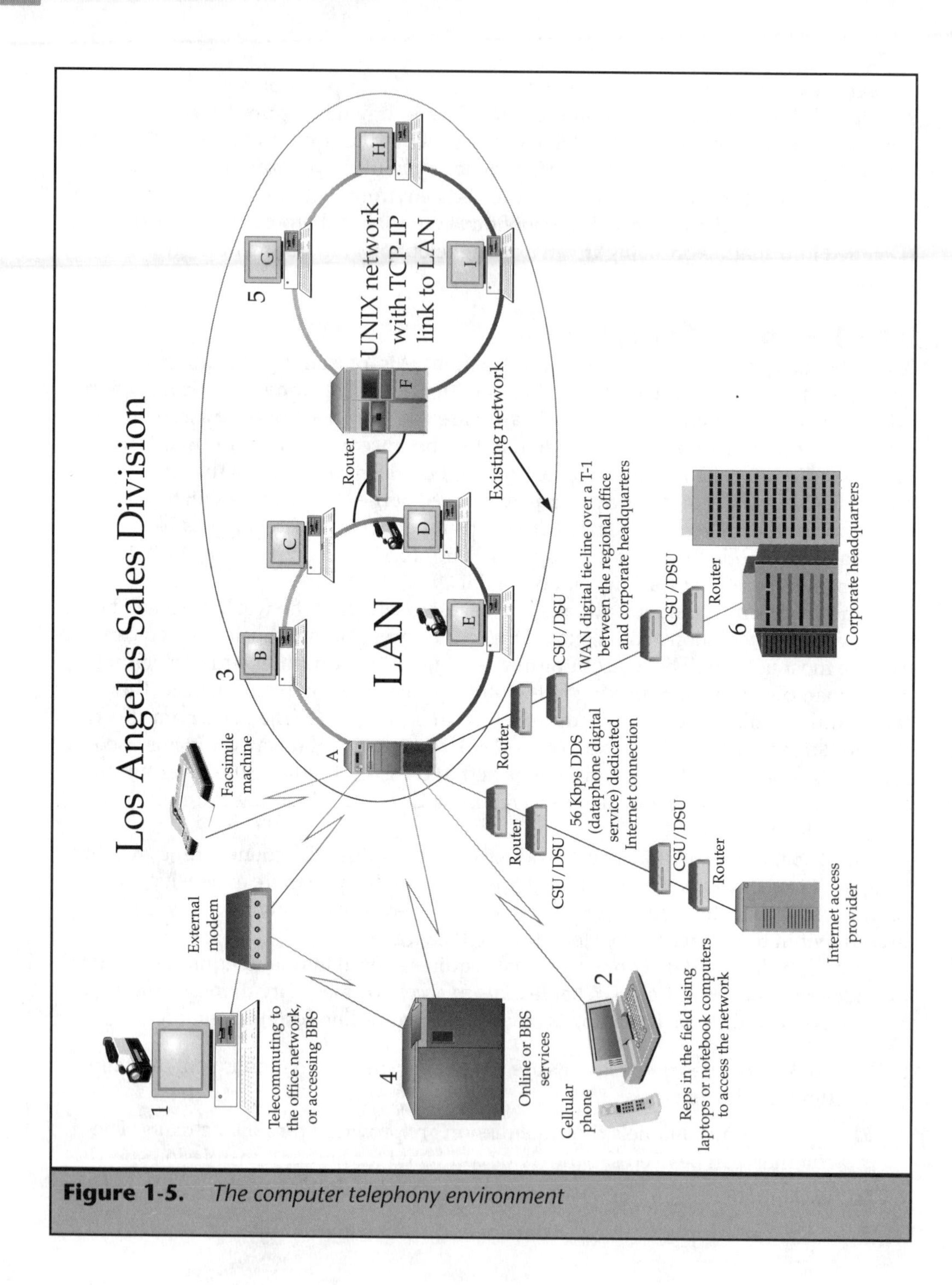

Figure 1-5. *The computer telephony environment*

- Desktop video conferencing quality is far superior to that of plain old telephone service (POTS) lines (the telephone service almost all of us have in our homes), and compatibility with larger video conferencing systems is now possible.

See Chapter 17 for more information on videoconferencing. For more information on ISDN, refer to Chapter 11.

In the present configuration of the CTE, the PC user connects with an online service such as Compuserve, Prodigy, or America Online. The user might also be accessing a specific company's BBS directly. Many people use just one BBS as a gateway to a myriad of other computer networks and services. The Internet is another gateway a PC user may access on a regular basis. In this case, the Internet will be accessed through the main LAN server, as discussed later in this chapter. (See Chapter 4 for more information on the Internet.)

The Laptop or Notebook Computer

When connecting the laptop/notebook computer shown in Figure 1-5 (item 2) to the CTE, the same reasoning applies as for the PC. The new notebook computers use cards about the size of a credit card and a bit thicker called *PCMCIA* (Personal Computer Memory Card International Association) cards which provide all sorts of functionality from additional memory to ethernet to printing—even for modem connections. Cellular PCMCIA modems are also available, so connectivity is easier than ever before. If you're responsible for buying notebook computers, make sure the peripheral connecting components are available in PCMCIA cards compatible with your computer.

Say some of the reps in the field are using older portable PCs that came preconfigured with 9,600 bps modems. PCMCIA cards are not an option for these users. OK, that decision seems to be made for us. But that doesn't mean you should neglect the sales force in the field; you must plan the CTE to work for everyone.

Some of the sales staff, on the other hand, have newer notebook computers that *do* support PCMCIA. You decide that they too will use 28.8 modems, but you're going to make sure those modems are cellular-compatible so users can communicate wherever they might be.

The Local Area Network

The LAN is the focal point of the CTE, and you must not overlook any aspect of its connectivity. Look at the LAN in Figure 1-5 (item 3). Each incoming and outgoing connection in the diagram is a responsibility; evaluate each connection separately and act on the results of your evaluation.

The LAN server represented by item 3A in Figure 1-5 will serve six very valuable functions:

- Connectivity for the telecommuters
- Connectivity for the reps in the field

- The WAN to corporate headquarters
- A modem pool for the on-site personnel
- Faxing
- Connectivity to the Internet

THE TELECOMMUTER AND THE LAN First, look at item 1 in Figure 1-5. One PC connects to the LAN to allow telecommuting. In reality, let's say six people regularly telecommute to the office and another four telecommute occasionally. Three of the regular telecommuters will use ISDN, while the other three are restricted to POTS (Plain Old Telephone Service) lines because ISDN is not available in their area. The occasional telecommuters will all use 28.8 modems. You need to plan for at least four ports on the LAN: three for the regular telecommuters and one for the part-time telecommuters. Three additional ports will be needed for the users connected to ISDN modems and ISDN lines.

The idea behind having a LAN server with a centralized point of entry is to gain control and cut costs. If you configure a one-to-one ratio of computer users to modems on the network, you aren't saving anything; the modems, ports, and telephone lines all cost money, and it is your job to make the CTE perform optimally at the lowest possible cost. Therefore, you configure four 28.8Kbps modems and three ISDN modems for use by the PC telecommuters.

THE REPS IN THE FIELD AND THE LAN Now look at item 2 in Figure 1-5, a laptop or notebook PC connecting to the LAN to access the company's resources while on the road. Now the reps are using laptop computers that can only connect to the LAN at 9,600 bps or slower, while the notebook users have 28.8 Kbps capability. If you purchase 9,600 bps or 14.4 Kbps modems for the laptop users, you can save money, but you are committing money and resources to a technology that will be making an exit fairly soon, and you're not supplying the best possible modem communications for the notebook users. All computer and telephony prices come down as newer technologies fill the top of the marketplace, so putting off the purchase of faster modems until they are actually needed could possibly save money. Ultimately, you decide to spend the money and purchase all 28.8 modems for the sales force to avoid having to split the modem pool. (Modem pools are discussed in Chapter 15.)

Your company has 22 representatives in the field; 12 use the laptops, while 10 use notebooks spread over four time zones. Since you have decided to go with the more modern 28.8 modems, the ratio of new and old portable computers doesn't matter. What *does* matter is how much time the reps need to spend connected to the LAN on a day-to-day basis. If the answer is one hour a day, two modem ports might do the job (but be prepared for complaints from the reps about not being able to access the system when they want to). Three modems are probably sufficient, so in the end you configure three 28.8 Kbps modems for the field reps' use.

You also need to consider cellular phones (see Chapter 13 for more information on wireless forms of telephony). XYZ likes to keep people connected, even if they're out in the field. The power of wireless technology makes this possible. Each field rep has a cellular phone. XYZ's needs did not justify their own private wireless network, so they simply have a contract with one cellular vendor in each area in which the reps are located. Fortunately, the cellular phones are compatible with a 28.8 PCMCIA modem that works with the reps' notebook computers, assuring that the reps can communicate while in the field via both voice and data.

THE WAN LINK TO CORPORATE HEADQUARTERS When preparing to make the jump from a stand-alone LAN to a WAN so that users may share the company's data processing resources, you need to know what you're doing (or know someone who does). A WAN is a data network that extends over dedicated common-carrier lines to link other LANs or computer systems. WANs are usually connected via a 56K circuit or one or more channels on a dedicated T-1 circuit (see Chapter 10).

The jump from a LAN to a WAN is made through a device called a *router* or a *bridge*. This device not only links the two locations but also translates the different protocols that might be used from one LAN to another. In addition to the router or bridge, a *channel service unit*, also called a *data service unit* (see Appendix B), may be required to handle the signaling on the T-1 or 56K data link. At a minimum, you will need a communications port to connect the router or bridge to the LAN. Some systems also require specialized cards to make a WAN connection, and you may need specialized software as well.

THE MODEM POOL Figure 1-5 shows a LAN (item 3) with a few computer users connected. Simple enough, but say you have 47 users and the network is still growing. Now that you've taken care of the telecommuters, turn your attention to those users who come all the way into the office to do their work. The best way to determine their needs is to *ask*. Make up a form, or go cubicle to cubicle and ask users how much time they will spend in the modem pool on average, what types of connections they will be making, who they will be connecting to, what speeds they need, whether they will be transferring large amounts of data, and any other questions that you feel might be pertinent. Ask open-ended questions and let your users fill in the blanks.

Now that you have gathered your data, you realize that half of your work force has been waiting for communications access so they can transfer quotes to their customers, fax letters from their desk, or get to the Internet. A few couldn't care less about modems, while the rest would like occasional access to a modem but don't really need much data communications ability. You determine that the current computer users will need about 45 hours of modem use each day. Given a nine-hour work day, you need to provide at least six modems for your on-site employees, but seven would be even better.

FAXING Two fax modems using network faxing software will probably be sufficient for your users' faxing needs (see Chapter 16 for more information on faxing). Growth is something you'll want to factor in at the end of your analysis, but for now you're only interested in meeting the users' current needs.

Your study has revealed that almost all of your users will be satisfied with 14.4 Kbps modem access, but you decide to spend the money now to provide 28.8 Kbps connectivity and avoid having to upgrade later. So you determine that your modem pool will consist of an additional seven 28.8 Kbps modems and two 14.4 Kbps fax modems (most modems today include faxing capability).

THE INTERNET The LAN users have come to realize that they want dedicated access to the Internet. Because so many people want an account on the Internet, you determine that it is cheaper and more efficient to lease a 56 Kbps link from the LAN server to a local *Internet access provider* (IAP), also called an *Internet service provider* (ISP). Now all of the people on the LAN can get their e-mail and browse the World Wide Web and all the other Internet functions they want without having to make a phone call. Later on, the sales department is thinking about making a Web site for XYZ to exhibit its products and services. For more information on the Internet, please refer to Chapter 4.

Communications Software

Software must be selected to accommodate the different kinds of connections that you will be making within the CTE. Odds are that any software that supports ANSI BBS terminal emulation will be able to handle the job. In addition, most of the major online services have their own, usually free, custom communications software that enhances the user interface.

Your main concern when selecting software is the LAN connection; you need to know what types of connections and emulations are compatible with your LAN environment, which might be running dozens of applications. Since the LAN is set up as a gateway, PC users may be able to simply match the emulation designated for the gateway and be up and running. In other cases, a gateway may not be able to translate a certain application's terminal emulation. Those crazy applications that run on the UNIX computer (item 5 in Figure 1-5) are especially troublesome. Unless your terminal emulation and gateway are 100 percent compatible with the applications you're running, the results will be unwanted characters on the screen, bad data received, modem hang-ups, or even worse. The more complex the application you're trying to run from a remote access point, the more important your communications software and its terminal emulation capabilities.

As you probably know, dozens of communications software packages are available. When choosing a package, make sure all the terminal emulations and transfer protocols that you desire are supported. You may want to purchase software—such as Norton's PcAnywhere or Carbon Copy—that also performs a remote emulation function; that is, a user at one PC or LAN can function as if sitting at another PC or on another LAN via a modem connection. This is a powerful tool for users as well as for diagnostic purposes. In addition, you may want to think about purchasing communications software that supports user verification and/or dial-back, where the computer hangs up and calls back the original caller at a predetermined or user-determined telephone number for security purposes.

Once you have decided on a communications software package for PC users, don't just go out and buy 50 copies of it at your local computer store; contact the manufacturer. You can probably buy a multilicense package at a drastically reduced price. When installing the hardware and the software, try to set it up identically on each PC to cut down on the installation time and make maintenance easier.

You may get lucky and find out that the same software you have chosen for your PCs works with your LAN as well (the better packages work in both). This doesn't mean you should find LAN communications software first and then put the same package on all your PCs; those packages are more expensive, so make your choices separately and just make sure your LAN and PC communications software will work together.

Final Analysis

Now that you've gathered all the information you need to create the CTE, you need to reevaluate the data and make your final decisions. You begin with ISDN users. You will need three ISDN modems for the telecommuters and three for the LAN. You have decided to purchase 31 28.8 Kbps modems—seven for the telecommuting PCs, 10 PCMCIA 28.8 modems for the notebook computers, and 14 for the network. (The laptop users already have modems.) After further review, you decide to combine the inbound traffic of the telecommuters and field representatives with the outbound traffic of the LAN users to obtain some of the benefits of *line consolidation* (see Chapter 3 for more on line consolidation). Now, instead of four dial-in ports for the seven telecommuters, three dial-in ports for the 22 field reps, and seven dial-out ports for the 47 LAN users, you have 14 two-way ports available to all three groups. The modem pool is, however, first-come first-served. You have also decided to get two 14.4 fax modems that you will operate with fax communications software to give users fax capability.

How are you going to hook 14 modems 2 fax modems and 3 routers up to the back of the LAN server? The answer is a *multi-I/O card* (a single card that resides in the LAN server and provides multiple communications port access). This card should have at least 19 RS232 ports and a per-line throughput speed of no less than 115,200 bps. (See the section on multi-I/O cards in Chapter 15 for further information on configuring your communications ports.)

***TIP:** Remember to take into account both one-time and recurring costs when configuring your CTE. A few extra dollars spent today on hardware can translate into thousands of dollars saved in long-distance telephone charges.*

Congratulations! You have just designed, created, and implemented the primary computer telephony environment that the Los Angeles sales division works in. Detailed analysis, such as the one reviewed at the sales division of XYZ, must be carried out every time major decisions about telephony are being made. Chapter 5 should help you appreciate the amout of attention telephony truly deserves.

WARNING: You have just created a CTE that allows people to access a LAN simply by dialing a telephone number. Protect phone numbers and even change them from time to time if necessary. Make sure proper security measures are in place to prevent unlawful access. Also, remember that information is flowing both in and out of your organization through your CTE. Guard against computer viruses by immunizing files and backing up the system on a regular basis.

Cable Television and Telephony

The one technology that was not discussed in this case study but is covered in this book is cable television and telephony. Cable television is obviously not a telephone, but the signal is brought to your house on cable, which is a medium used for communication. Cable television and telephony will have a tremendous impact on the home user in the near future. Both the computer and telephone industries are working hard to integrate their technology with cable television, which provides the ability to send extremely large amounts of information in all sorts of forms, such as voice, video, data, and interactive television, right to your home television, computer, or telephone. Take a look at Chapter 19 to see what is around the corner for cable television and telephony.

Conclusion

Telephony is an ever-expanding and complex environment. This chapter has introduced you to most of the telephony topics covered in this book; use it as a starting point to see what areas of telephony interest you the most. Try to find new ways to use the technologies described here in your own home or business.

If you plan to implement one or more of the telephony technologies discussed in this book, try to keep in mind the level of detail shown in the case study for the Los Angeles sales division. Obviously, even more detail is gathered when preparing to implement new ways to communicate. Chapter 5 is also an excellent source of information and instruction when you are implementing new forms of telephony. Telephony is and will be a growing part of our lives in the time to come. You will do well to understand the modern communications world that you live in.

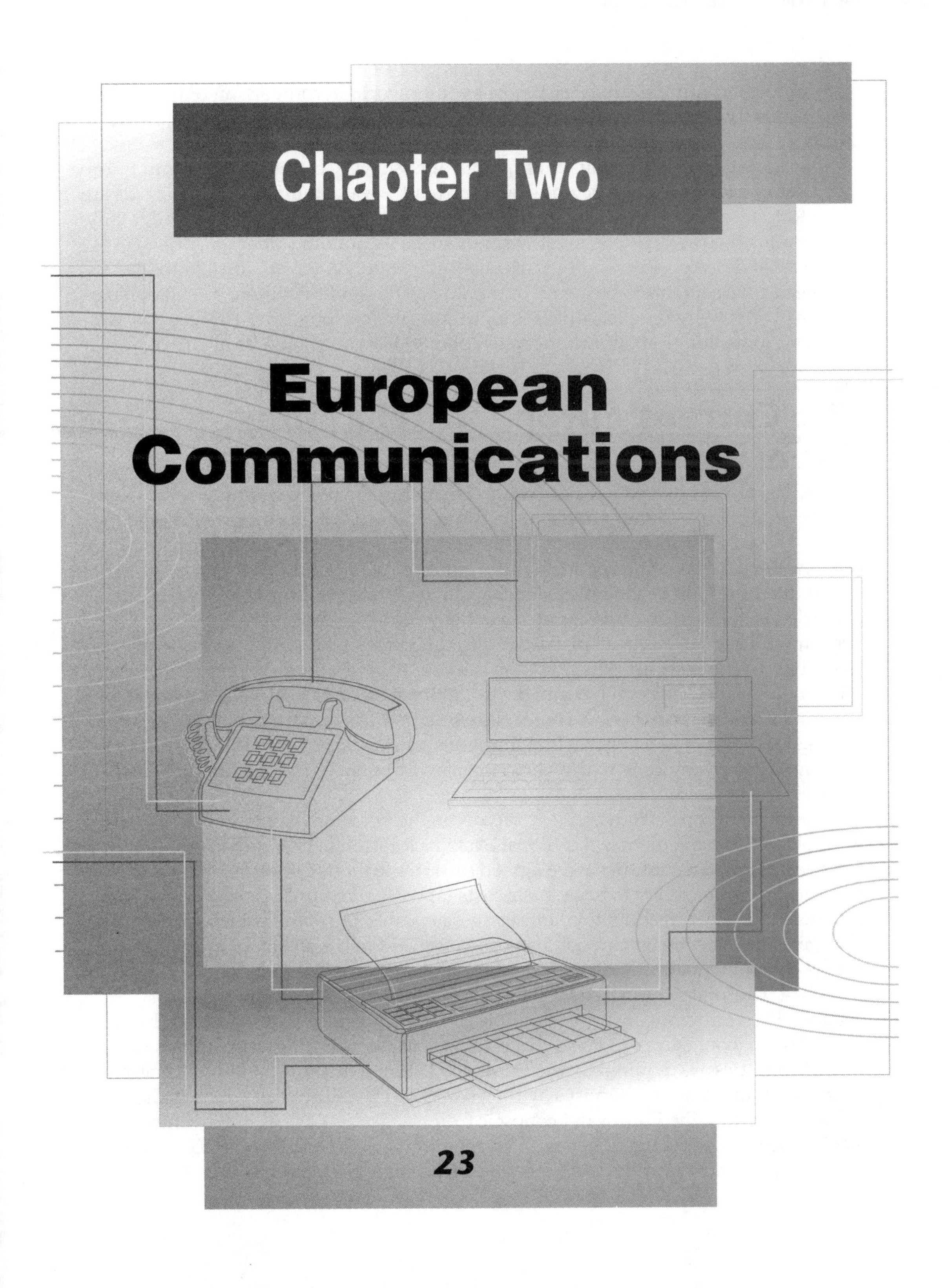

Chapter Two

European Communications

European communications is almost another world in comparison to the United States. Understanding the current European telephony situation is definitely a key issue if you plan to extend your business to this continent or at least have extensive relations with it. You will find that almost every country throughout Europe is its own world. As Europe begins down the road towards unification, standards are being formed, alliances are being made, and telephony is beginning to spread. Many of the more advanced nations must wean themselves from the older, proprietary platforms and protocols to embrace greater unification. As you go forth onto the European continent, try to be aware of the similarities and the differences that make European telephony unique. A little knowledge can go a long way. This chapter will point out some things about European telephony that you should be aware of.

The Current European Telephony Situation

Europe has always played a major role in the international telecommunications market. Indeed, most international telecommunications organizations were (and still are) based in Europe, in cities such as Geneva, Bern, and Brussels. But these organizations focus on international interconnectivity rather than on European issues.

Until recently, there could be as many differences between any two European countries as between any two countries in the world. The concept of European telecommunications has been in existence for only a few years, and even now the way telephony works is highly dependent on the country, with huge differences between the most industrialized countries and the less industrialized countries—not to mention Eastern European countries. Of course, this situation has important consequences on businesses that spread across several European countries: not only is it difficult to get the same kind of service, it is also very expensive, both in terms of equipment and cost of service.

However, Europe is moving very rapidly in the telephony arena. A fine example of this movement is the Global Systems for Mobile Communications, or GSM. All European countries have agreed on a single mobile phone standard, giving manufacturers the ability to drive their costs (and therefore their prices) down. This also gives users the freedom to roam seamlessly across the entire continent. All major European operators (telephone companies), as well as foreign operators, are measuring up the competition to provide high-quality services across the continent and around the globe. The joint ventures of MCI-British Telecom and Deutsche Bundespost Telekom-France T-Sprint are prime examples of multinational alliances that are being formed throughout Europe today to create a new generation of global telecommunications.

All countries are required by the European Union to liberalize (de-monopolize) their telecommunications services by the year 1998. (The European Union is the new name adopted for the European Economic Community, or EEC, as of November 1, 1993.) A few countries have been granted the right to postpone this deadline until the year 2003, but most countries will meet this deadline, and some will even be ahead of the deadline. Telephony infrastructures will also be open to competition in the near future.

Let's look at what Europe has been able to achieve thus far. The French company Alcatel is a world leader in the manufacturing of communications equipment. Deutsche Bundespost Telekom is the second largest telco (telephone company) in the world, just after the Japanese public telco Nippon Telegraph and Telephone (NTT). Minitel is a very popular form of Videotex terminal used in France (see the "Videotex and Minitel" section later in this chapter). GSM is by far the most popular mobile phone technology and has been accepted as the standard throughout Europe today (see the "Mobile Telecommunications" section later in this chapter.) Though European telephony development was and is clearly hindered by Europe's diverse political, racial, scultural and technological history, the general process of liberalization that is taking place around the world should bring forth a very competitive, united telephony environment that will benefit us all.

Because European telecommunications is so irregular and is evolving so rapidly, describing each and every aspect of it would require a book of its own, which would have to be updated monthly. This chapter will therefore cover the mainstream telephony aspects with emphasis on the key differences between the European and American markets. Also, because Eastern European countries are still very different from Western European countries in terms of telecommunications, they deserve a specific section at the very end of this chapter and will therefore not be considered in the various sections that follow. Remember, if you do not see something in this chapter, don't assume that it does not exist in Europe—it may be that there is little difference between the European and American versions.

Plain Old Telephone Service (POTS)

Fundamentally speaking, there are not that many differences between the various European POTS and the American POTS. The telephone you use in your home every single day provides POTS. POTS allows users to pick up the telephone, get a dial tone, dial a number, ring the other party's phone, get connected to the other party, talk to the other party, then hang up and get billed for it. However, when you look more closely, each POTS throughout Europe differs a great deal from the others. Clearly, as POTS are the oldest part of the telecom setup, they are also the part that features the biggest, most numerous, and oddest differences from one country to another. Again, without mentioning each and every detail, we will try to give you a rough idea of how POTS operate in the various European countries, stressing common points as well as major differences.

Telephone Companies

In most countries, there is only one, state-owned, POTS operator service. Even in the U.K., where POTS competition was introduced a long time ago, the secondary operator, Mercury, has only been able to gain a few percentage points of the market share. British Telecom still controls approximately 90 percent of the market. In other words, POTS still remains a monopoly throughout most of Europe.

Quality of Service

In most Western European countries, the telcos provide a good quality of service that is comparable to U.S. telephone calls in that they take the same amount of time to set up and have the same level of clarity. Customer service and technical support are also comparable to the U.S. In countries where the quality of service rates average to poor, we have some indication that it is rapidly getting better, thanks to European uniformization (the alliance between the European nations), which requires governments to invest large amounts of money in telecommunications. Even the telephone companies that still have monopolies in their markets do their best to provide good service and customer support, because they know that the competition is knocking at the door. Getting a phone line usually takes just a few days, but it may take up to a year or more, depending on the country and how remotely you are located. This is especially true in countries such as Greece and Portugal.

Telephone Networks

European networks differ on many points. Some of them are very modern, using fiber-optic links and digital switching equipment. For instance, France Télécom replaced its last mechanical switch with digital equipment at the end of 1994, while others in Europe are making extensive use of older technologies.

As seen from the customer's side, touch-tone dialing is not available everywhere, pulse dialing is not the same in each country, phone jacks differ, and dial and busy tones differ, as well as the ringing tone. There are many stories of people in France calling the U.K. and thinking their party is always on the line and engaged in conversation. This is because the British ringing tone sounds almost like the French busy tone! The French have a special progress tone that starts right after you have finished dialing and ends when the ringing or busy tone starts. Guess what? This progress tone sounds almost like the American fast busy tone. How confusing!

In some countries, such as Germany, phone numbers have a variable length, while in others, like France, all regular phone numbers have the same number of digits. While the British have to dial 010 as the international prefix, Germans dial 00, and the French dial 19. Of course, the same goes for interlata (calls that spread across several local zones): the French dial 16, while Germans and the British dial 0. You might think that emergency numbers, at least, would be common to most countries—no way! Emergency phone numbers are specific to each country. As you can see, even things as simple as placing a regular telephone call can be pretty confusing if you are not aware of the specifics of each country's telephone network.

Fortunately, many of these oddities will soon vanish, as the European Union has requested that dialing be uniformized. For instance, the international prefix should eventually be 00 for all European countries, as it is already in some of them. The emergency phone number should become 112 everywhere. The various tones will also be homogenized to give the feel of unity.

Features

Many operators offer only basic telephone service. In some countries, call forwarding is available, as well as call waiting and three-way calling. Most operators also offer various call-barring features. But advanced features, such as voice-mail, distinctive ringing, and remote call-forwarding, have yet to be introduced. Look for these features to creep into the European market over the next few years.

Itemized billing and caller ID are only available in a few countries, but in many cases it is not only a matter of technology. Legislation in a number of countries stands in the way of providing these features. The information provided in itemized billing and caller ID is often viewed as a violation of the laws protecting individual privacy. For instance, itemized billing has long been available in France, but until 1994, the four digits of every phone number listed were systematically omitted. Caller ID suffers the same problems as it does in some parts of the United States. The jury might be out on this one for a while.

Pricing

Because competition does not exist, prices are very simple. There is a price for obtaining new line service, ranging from $30 to $200. That is quite a range, and the price varies from country to country and depends on whether the physical line exists already or not. Monthly service fees range from $8 to $18. Per-minute rates also fluctuate, as they do in the U.S., according to the location of the party called and the time of day. Table 2-1 shows a sample list of calling charges from the big three European Countries: United Kingdom, France, and Germany.

There are almost no special calling plans available in Europe, which is why international callback is so popular (see the "International Calling" section later in this chapter, and also see Chapter 14 for more information). The American nightmare of choosing the cheapest long-distance carrier for a given call does not exist in Europe, and neither do the bickering, bragging, and slamming that go on among the American long-distance carriers. Well, for once, Europe has made something simpler in the telecommunications arena. Unfortunately, this is the one area in which complexity can really save you money. Look for pricing structures to change in the future as competition for market share begins to heat up.

There Is No Such Thing as a Free Call!

Recently, the move towards an open market and the introduction of international competition has led to significant changes in pricing. Telco operators have been forced to revise their international charges. Most of them have reduced their rates several times a year for the past few years. This has forced some operators to raise the prices of local calls to keep their budgets balanced. As you can imagine, this practice of

	U.K.	France	Germany
Operator	British Telecom	France Télécom	Deutsche Bundespost Telekom
Local calls	$0.02 to $0.06*	$0.02 to $0.05	$0.01 to $0.03
Intralata calls	$0.03 to $0.10	$0.06 to $0.18	$0.08 to $0.16
Interlata calls	$0.05 to $0.16	$0.15 to $0.44	$0.22 to $0.44
Calls to the U.S.	$0.74 to $0.80	$0.86 to $1.10	$2.20 at all times

* All figures are in U.S. dollars. Rates are only approximations, and are subject to change.

Table 2-1. *Typical Calling Prices in the United Kingdom, France, and Germany in U.S Dollars*

passing the buck has been generating many complaints from the residential users who were most affected by these reforms, while companies ended up being the big winners of the pricing wars.

Videotex and Minitel

Videotex is a network of services that you can access throughout the public telephone system. Figure 2-1 shows a picture of a *Minitel,* the French-made version of the terminal used to access Videotex services. Minitel is to Videotex terminals what Kleenex is to facial tissue. Alcatel, Philips, and a number of other companies manufacture Minitel Videotex terminals and market them mostly in France. The principles of Videotex are rather simple. Thanks to a single low-cost terminal, the user is able to access online services of all types. This is not very different from accessing an online service in America using a computer equipped with a modem.

Since the creation of the service, France Télécom has been handing out one Minitel per residential phone line for free. Bildschirmtext is the commercial name of Videotex services in Germany (the connection provider is Deutsche Bundespost Telekom), while British Telecom is the connection provider in the United Kingdom. Terminals must be purchased or leased in Germany and the United Kingdom, which is probably why they are not as popular as in France.

The precise deal in France is as follows: Since Minitel was first introduced as a means to access the electronic phone directory, you have the choice to receive a Minitel or the white pages. Minitels remain the property of France Télécom. This was a risky bet, but France Télécom came out a winner. Approximately six million Minitels are used by 22 million people in France today.

Another key point in the amazing success of Videotex is the way billing is handled: France Télécom simply bills users on their regular phone bill for the use they make of the various toll services, and then France Télécom pays back the service providers.

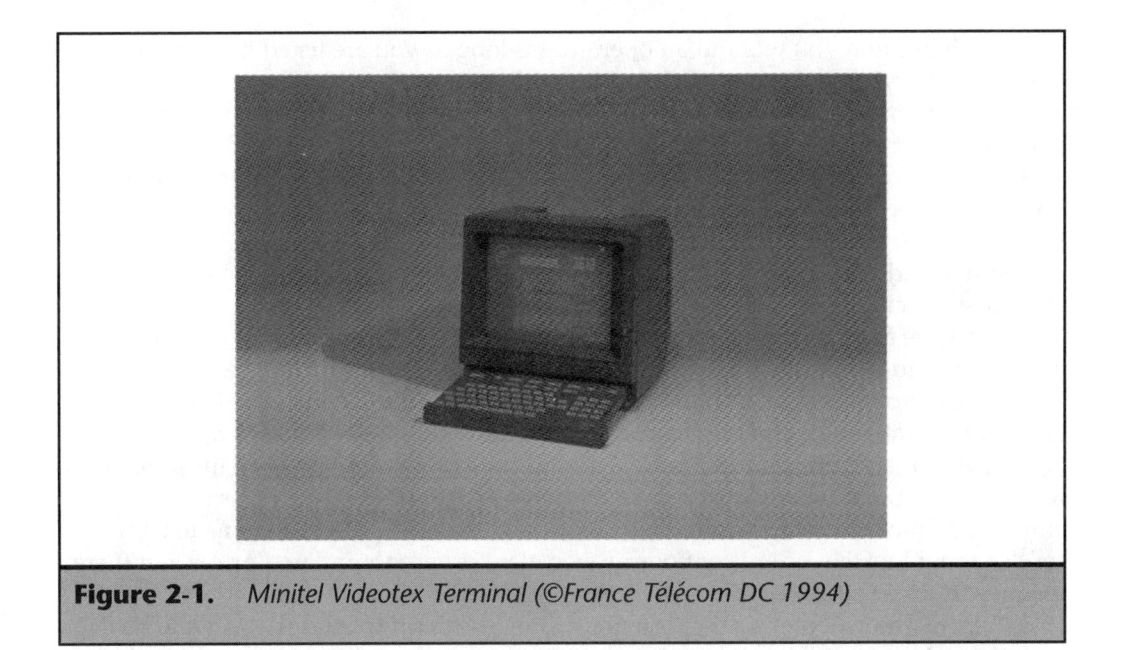

Figure 2-1. *Minitel Videotex Terminal (©France Télécom DC 1994)*

This is, as you may guess, very convenient, for both users and service providers. The former do not need to subscribe to each and every service they intend to use. The latter do not have to manage their customer base, nor do they have to handle billing. France Télécom is the terminal and connection service provider, while thousands of other companies provide their specialty services over the connections France Télécom makes possible. Some services are toll-free (the service provider actually pays for the connection), while others can cost up to $1.70 per minute. A typical session on a Videotex service will cost around $0.25 per minute. In all cases, the operator keeps approximately $0.08 per minute to pay itself and passes the rest to the service provider.

Services Available

What kind of service is available? Almost anything. Think of any service that could reasonably be provided on an ASCII terminal with some graphic capabilities and a 1,200 bps modem, and it surely already exists. Around 15,000 different services can be accessed with a simple Minitel. Minitel is used as both a professional and residential tool. The most important service (in terms of connection time) available in France is the electronic phone directory, which is free (the first three minutes of each connection, at least) and very convenient. It takes no time to find somebody's phone number if you have some idea of their last name and where they live. Moreover, you may also look for products and services using your native language to express your request. For instance, you may look for "a good restaurant" or "a company that can fix your refrigerator." This is definitely a very powerful tool. But there's more. Another service gives access to several foreign phone directories, including the U.S. Can you believe the French can find you without an operator (as long as you are listed in your local phone directory) in seconds with their Minitels?

Banking services are also a must on Minitel. These services will give your account balance and let you schedule transfers and payments to other accounts. Some Videotex services will even let you buy and sell shares on the stock market, right from your home. Other professional services include travel services, from which you may book train and plane tickets, insurance services, information services, etc. The Internet itself can be accessed with a Minitel, though this might not be the most convenient or the most cost-effective way to do it.

Games are another reason for its popularity—there were numerous cases of children running up huge bills playing games on the Minitel. Problems with these services are probably similar to problems experienced with American 800/900 telephone numbers. Companies have also been hit with Minitel abuse: employees have spent time playing games, some with the hope of winning prizes. This problem has come to a point where most companies have programmed their PBXs to prevent employees from accessing Minitel-enhanced services. Much has been done to ban service providers that offer licentious or worthless services at inflated prices, but there remains much to be done.

All in all, the Videotex service is a great thing. Most companies that need to deal with the public or with locations spread across the country will enter into the realm of the Minitel business. If your company has business relations with the French public, you should consider becoming a Minitel service provider. Several Videotex companies offer good support, from consulting to full turnkey Minitel services (please refer to Appendix A). You will gain a better image, and you might even make some money.

Videotex in other countries is definitely not as popular as in France, because other operators have not followed France Télécom's steps and have usually set a rather high price for the Videotex terminal. Videotex services do exist but are usually reserved for professionals, who access them with a computer and a modem, thanks to available Videotex terminal software emulators, which come free with most modems purchased in Europe today or can be bought separately. Most Videotex networks either use the French standard (Teletel) or the British standard (Prestel). Teletel charges for the time spent, while Prestel charges by the number of pages viewed. Despite many differences between the systems, gateways have been established between the various European Videotex networks.

What remains to be seen is how Videotex and the Internet will evolve in Europe. Videotex uses older technology, but service providers have gained a lot of knowledge and a huge market share. Internet providers may find it very difficult to pry the public away from the Videotex services currently offered. Will the two worlds compete or meld? It depends very much on whether Videotex and especially the Minitel (because of the number of units currently installed) are able to evolve to include Internet access. Services offered on Videotex will expand onto the Internet as well, or the competition between the two information platforms may heat up. You will definitely see things change fast in this area of telephony in Europe.

Fax Machines and Modems

Fax machines are very popular in Europe in the business world. They operate exactly the same way as, and are fully compatible with, American facsimiles. Fax manufacturers are now pulling the business market towards the high end, with expensive, plain-paper, laser fax machines, while offering new low-cost individual fax machines to personal users.

Up until 1994, modems were not very popular in Europe for a few reasons. To begin with, personal computers themselves were not very popular until recently. Furthermore, regulatory issues, such as the need for modem manufacturers to get their modems approved in each and every country, resulted in a very expensive and time-consuming experience. Which in turn led to very high modem prices and few choices. Finally, the existence of Videotex, which at first took advantage of this phenomenon, eventually turned out to discourage BBSs and other services that might have boosted the modem market.

It seems that the European Union has finally realized that imposing heavy regulations on telephony is not the right way to go. Modem approval, like all approvals of this type, has been sped up (from around 18 months to a couple of months), simplified, and are now (most of the time) uniform for all countries (approvals can cost anywhere from $10,000 to $25,000; you can easily imagine what consequences on prices it can have to require one European approval instead of dozens of national approvals). Modems are therefore becoming more affordable and popular. A typical 14.4 modem can now be purchased for around $200; it used to be over $500. Prices are still coming down, but at least they are now affordable, and they include software (various terminal emulations, including Videotex emulations, fax software, etc.) and documentation in the language of the user. This is just the beginning. In any case, European modems are compatible with American modems. They both now use the same ITU-T (formerly CCITT) standards.

Integrated Services Digital Network (ISDN)

ISDN in Europe, because it is much more recent than POTS, does not differ much from the American ISDN. It serves the same purpose with roughly the same terminology (see Chapter 11 for more information on ISDN). Actually, the main difference between the U.S. and Europe is that ISDN was already available almost everywhere in Europe in 1993. At least, this is true for Germany, France, and the United Kingdom.

There is one thing you should know about ISDN, though. Until recently, there were as many ISDN definitions as countries. Indeed, the lack of precise normalization of ISDN, compounded with the desire of many European telcos to offer ISDN as early as 1987, had led most manufacturers to make choices that were not always compatible. The consequences of this are quite important. It is not a matter of international connectivity. All operators have set up the appropriate gateways to make sure that ISDN calls from their networks can be routed to most foreign countries, including the U.S. Rather, it is a matter of equipment compatibility. If you are buying ISDN equipment, make sure it is compatible with the PBX or public switch that you will be hooking it up to. There are many subtle differences that you should not overlook.

This relative incompatibility among ISDN services (mainly concerning advanced features) will not remain for ever. Europe has now agreed on a common ISDN standard, called *Euro-ISDN*. Euro-ISDN does not differ from ISDN in terms of services; it only clearly states how all services must be provided by the equipment involved to ensure perfect interoperability. All European telcos are required to switch to the new standard. Some of them are already fully compatible, especially those that have just opened ISDN access recently. But others, like France Télécom, which opened its ISDN access some time ago, face the problem of converting a large number of subscribers who have invested large amounts of money into ISDN equipment that is not compatible with the new Euro-ISDN standard. The telcos must therefore provide old-standard lines while opening new Euro-ISDN access so that users may switch to the new standard in a reasonable time frame.

ISDN is already widely available throughout Europe. Getting a new ISDN line is usually rather expensive (from $100 to $400, depending on the country and whether the line exists already or not). However, the monthly fee (around $40 in all countries) is not that high, and the communications costs are in general quite close to the POTS equivalent (see the section on Plain Old Telephone Services earlier in this chapter), which means that you can save a lot of money for data transmissions. Primary access is also available for competitive prices.

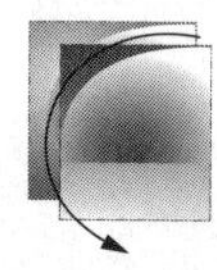

TIP: *If you find the need for ISDN services in Europe, be sure to ask for Euro-ISDN lines and invest in Euro-ISDN technology and equipment. If you plan to interconnect with European sites thanks to ISDN, make sure your American ISDN provider is interconnected with the corresponding European ISDN providers.*

Americans tend to believe that Europe has made some strong advances in the field of ISDN. This is true, but, at first, European telcos were surprised to see a relative lack of interest in ISDN. People tended to think that ISDN was not worth its price for voice communications. Concerning data, decision makers were not ready to move to ISDN when they heard buzzwords such as ATM (Asynchronous Transfer Mode), SDH (SONET), or the Internet, which all promised them throughputs in orders of magnitude much greater than ISDN in the near future. However, ISDN has not had its last word yet. With costs being driven down, thanks to the uniformization and globalization of the European market, with analog lines not providing the nice features ISDN has to offer, and with the information superhighway still on paper in parts of Europe, ISDN might well boom in the future. Many of the larger companies in Europe are connected using ISDN, and the telcos are now targeting small companies, offering them dual (analog-digital) access, so that they do not have to throw away all of their existing equipment at once.

There is not much to be said about other data networks. All countries have X.25 networks, most of them being of high quality, rather widespread, and interconnected with each other. Frame Relay is also available in a number of countries. A few operators have started offering ATM (see Chapter 12 for more information). And of course, leased lines are also available in most countries.

E-1

E-1 is the European equivalent to the American T-1 (see Chapter 10 for more information). While the fundamentals of E-1 and T-1 are the same (they both convey 64 Kbps channels), they differ in many aspects, including the following main differences. E-1 uses 32 timeslots instead of 24, which leads to a total throughput of 2.048 Mbps. Refer to Figure 2-2 for a simplistic view of E-1 versus T-1. One entire timeslot is devoted to synchronization instead of a single bit as in a T-1, which leaves us with 31 usable timeslots. When in-band signaling is used, another timeslot is used for signaling (but in-band signaling is rarely used, and separate CCS7 signaling is far more common).

We could speak about differences for hours, mentioning for instance that the encoding laws used for voice transmission are different, as are the impedances. But if you need to mix E-1s and T-1s in your network, there are devices that will do the conversion for you. Please refer to Appendix A for a list of vendors.

The Internet

Until recently, the Internet was not very popular in Europe, except perhaps in a few countries like Finland, and later in the U.K. In Germany, the Internet had gained some popularity also, but nothing close to the success it has had in the United States. French

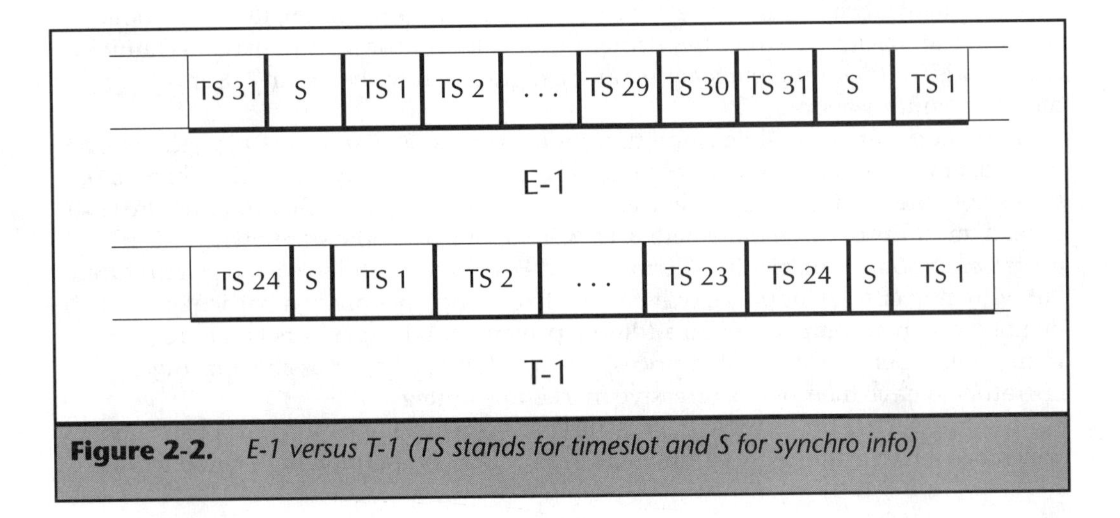

Figure 2-2. *E-1 versus T-1 (TS stands for timeslot and S for synchro info)*

people have ignored the Internet for quite some time as well, mainly because getting an Internet connection was very expensive, so that only very big companies could afford one. Videotex was also slowing down the growth of the Internet, because many people were already getting the connectivity they needed.

Videotex and the Internet do not serve the same purpose at all, the former being thoroughly organized, controlled, and fee-based, while the latter is less organized, almost out of control, and virtually free. However, Videotex and the Minitel have one big advantage for French people: The vast majority of services are provided in their native language. This is true in every country that offers Videotex services. Actually, until recently, in most European countries, only universities and large corporations could have some form of access to the Internet. But all that has changed. In 1993 and 1994, several small Internet service providers have entered the market and are providing inexpensive access for both small companies and private users. Internet access can now be purchased by the minute or by the month, around $0.35 a minute or $35.00 a month per connection. The cost of modems going down has also helped a great deal. With access now available and the entry price lowered, the public has started discovering the Internet and the web of information that is circling the globe.

Accessing the Internet is now easier than ever, and the costs have really been driven down. In 1995, major companies such as Microsoft and IBM are expected to announce the opening of many access points to the Internet throughout Europe. This should lead to another price war from which all users will benefit. But, even if the European people do talk about and begin to embrace the Internet, the European web has a long way to spread before it covers the continent as the Internet has done in the U.S. For more information on the Internet, see Chapter 4.

Mobile Telecommunications

Until the beginning of the 1990s, only a handful of European countries were serious users of mobile technologies. The U.K., because of real competition, utilized mobile communications much earlier than other European countries. Scandinavian countries too, because of their inherent weather and geography, have been early users of mobile communications.

In other countries, mobile communications were considered a luxury, because of the quasi-monopolies running mobile networks that artificially kept prices high and the use of different technologies in each country, which led to high infrastructure costs. Indeed, most European mobile networks were operated by the state-owned POTS operators. In some countries like Germany and France, competition had been introduced, but with little effect. Duopolies (where only two companies are in a particular market) do not create real competition. In addition, private mobile operators had to rent lines from public operators at inflated prices, making it virtually impossible for private operators to have their own aggressive marketing strategy.

The first barrier of the regulatory issues has been tackled by the European Union, which has forced national governments to introduce competition in the mobile marketplace. Countries that do not introduce competition are sued by the European Union. The second barrier of standardization has been tackled by the European Union as well, which has created a common digital standard for mobile networks. The Global System for Mobile Communications, GSM, is a good example of a comprehensive and thoroughly devised standard, now available in practically all European countries with differing coverage from one or more mobile operators in the region. Its success has crossed the boundaries of Europe, with many countries adopting it as their new mobile network standard. But before getting into more details about GSM, let's have a quick look at analog services.

Analog Mobile Networks

Many analog mobile networks are used throughout Europe. They are based on several standards such as TACS, NMT, and Radiocom 2000. These networks will eventually disappear, but they are still very valuable in countries whose GSM network(s) are not fully operational, as is the case in Spain and Greece. Cellular phones for these networks used to be very expensive, because they were sold on a small scale, but the advent of GSM has forced operators and distributors to reduce their prices drastically.

GSM, DCS 1800, and PCS 1900 Networks

GSM is being rolled out in several phases, each new phase adding features to the previous phase. GSM was extensively thought out, and has an incredible number of features to implement. This section will stress the key issues of GSM.

GSM is fully digital. Voice as well as data and signaling are handled digitally on the air as well as by the network itself. Digital signaling has many positive ramifications. GSM achieves a more efficient use of the frequency spectrum, allowing for many more users per square mile than analog systems can achieve (see Chapter 13 for more information on cellular technology). There is a better voice quality produced, as you would expect. The ability to implement sophisticated features, including inviolable authentication and communication encryption, fits very well in the GSM digital network model.

GSM allows roaming between various GSM operators, enabling users from different European countries to use their phones in any European country. Actually, GSM totally isolates the handset from the subscriber's identity, thanks to the subscriber identity module (SIM) card. This credit card-sized card holds the information about the user. The user can therefore use any GSM phone anywhere as if it were his or her own mobile phone. More and more American cellular operators are offering GSM roaming in this way too. Ask your mobile operator; they may be able to deliver a SIM to you that you will be able to use anywhere in Europe with a GSM handset that you may rent or find in your rental car.

GSM features are too numerous to mention. They include voice-mail, call forwarding (unconditional, on busy, on no answer, on unavailability), call barring (outgoing international calls, incoming international calls, etc.), and so on. One major feature is called the short message service. It allows users to send and receive short messages. Applications vary from acknowledged paging to local advertising (depending on the geographic location of the mobile phone), to a voice-mail notification system. These messages have one drawback compared to standard pagers: they can only reach you within the GSM coverage area, which is usually smaller than conventional paging network coverage, but they are buffered and acknowledged. If for some reason you are unreachable while a message is being sent to you, the system will hold it for you until you return in range of the system. For instance, let's imagine that you turn your phone off while having lunch. Somebody tries to reach you. They end up reaching your voice-mail instead and leave a message. This automatically triggers a short message telling you that you have a voice-mail message waiting. As soon as you turn your phone on again, this short message reaches you and lets you know that you should review your voice-mail.

DCS (Digital Communications System) 1800 is just a minor adaptation of GSM to another frequency band. This allows for even more subscribers, lighter handsets, and better indoor penetration, but it is only cost-effective in high-density areas. Dual-mode (GSM and DCS 1800) handsets and SIM cards are expected out soon, allowing users to benefit from both networks.

PCS (Personal Communication System) 1900 is another minor adaptation of GSM, just like DCS 1800, but targeting the North American market where the frequency band used by DCS 1800 was not made available. Some American operators have already adopted PCS 1900, but others have turned to CDMA (Code-Division Multiple Access), which is another promising technology for mobile communications. Which technology—PCS 1900 or CDMA—will eventually take the lead in the U.S. still remains to be seen. CDMA definitely has some nice advantages over PCS 1900, because it is much younger and incorporates some more modern concepts in wireless. But its youth also has a drawback. While many manufacturers are able to offer second- generation PCS 1900 systems at relatively low prices, there aren't many CDMA manufacturers yet. Refer to Chapter 13 for more information on wireless communications systems in the U.S.

Almost all European countries have at least one GSM network. The most advanced ones, such as the U.K., already have two GSM networks and two DCS 1800 networks. The usual monthly fee ranges from around $20 to $60, depending on the plan chosen. Communications prices are usually around $0.40/min, but this depends a lot on the marketing strategy of each operator. One British operator has been offering free local calls during nighttime and weekends for quite some time. With the appearance of growing competition, you will see new pricing schemes develop in the cellular market just as they have developed in America.

Paging Networks

The history of pagers is similar to that of mobile communications until the 1990s. Pagers weren't popular, and the use of different standards in the various countries was primarily responsible for the lack of popularity. Once more, only England and the

Scandinavian countries had a number of pager users that was not ridiculously low. But while cellular phones boomed at the beginning of the 1990s (thanks to GSM), pagers have not experienced the same prosperity, and not because of the lack of a common standard. European Radio Messaging System (ERMES) is the equivalent of GSM for paging systems, but the European market just doesn't seem to be ready to communicate all of the time, yet.

History has shown that a growth of the mobile phone market leads to a growth of the paging market. While this might not be the case with GSM, because of its inherent paging capabilities, some paging operators have started offering ERMES services. The success of these services might come from their capabilities: ERMES allows the transmission of messages that may be hundreds of characters long, but what will certainly boost the paging industry is the new marketing approach that certain service providers are choosing. They waive all subscription fees forever. In other words, once you have paid for your pager, you will never get a bill again. Only people who send messages to you are billed. This approach was adopted by a paging service provider in Scandinavia some time ago and has led to a significant increase of their subscriber base. There are already many paging networks in all European countries, all with numeric and alphanumeric capabilities. Some of them may be usable across borders, though roaming should really become seamless with ERMES only. Pagers cost around $250, monthly subscription fees vary from nothing to $50, and sending messages costs from nothing to one dollar.

Other Mobile Networks

Several European countries have tried to introduce CT2 (cordless telephony 2) based services. These services offer a low-cost alternative to regular mobile phone services, but with several big drawbacks. The coverage of the network is usually very limited, and the user has to stay within the service cell (which is no bigger than a few hundred feet in diameter) during the entire communication. Most systems only allow outgoing calls, that is, they are not able to route incoming calls.

In practice, the U.K. had started such a service in London, but the very small number of cells made it almost useless. Germany was thinking of starting such a service also but has given up the idea. The Netherlands has a CT2 service, which has had some success, as does Belgium. But the biggest success in Europe definitely comes from France Télécom Bi-Bop (the France Télécom CT2 system) has almost 100,000 subscribers and a rather large coverage area. All of Paris and its suburbs are well covered (that is, as much as they can be with such a technology). Strasbourg and Lille also have good coverage. The basic subscription costs approximately the same as a POTS line but is only an outgoing line. For an additional $5 per month, the user is given a personal phone number and can receive incoming calls, but only under special conditions, for instance, if he or she stays within the same cell while waiting for the call, which, you must admit, is not very convenient. Since a very good voice-mailbox comes with it too, this is not as much of a problem. Communications costs are computed from the price of the POTS call, to which a $0.15 per minute fee is added.

CT2 has one nice advantage over the more traditional mobile phones. Because its technology is rather simple (but still digital), companies as well as private users may buy their own base stations for a reasonable price (a private base station costs around $180), which they hook up to their PBX or a regular phone line in order to create their own little mobile network. This way, the same terminal can be used as a high-quality cordless phone at home, in the street, and at work.

DECT (Digital European Cordless Telecommunications) is another standard that seems more promising than CT2. It basically does not have the various negative points CT2 has, such as the lack of handover between cells, as well as the inability to receive incoming calls efficiently. But DECT is more sophisticated and therefore more expensive. Which standard will eventually survive, squeezed as they are between real mobile solutions such as GSM and DCS 1800 and the plain old telephone service? The answer remains to be seen.

There are also a number of private mobile radio services available from several local operators, as well as some digital packet radio networks, though the latter are just at their rudimentary stages and are still rather expensive.

Finally, let's mention that most high-speed European passenger trains are equipped with phones. This is particularly interesting, since passenger trains are far more common in Europe than in the U.S., mostly because distances are shorter, giving a competitive edge to trains over planes for quite a few destinations. Europe has also agreed on a common aircraft phone standard, but seeing a phone in a plane is still very uncommon in Europe.

International Calling

From all Western European countries, international calls are reliable (at least, as much as they are from the U.S.). The real problem with international calls is their price. Luckily, this is one of the very few domains of telephony that was deregulated early on. Of course, the U.K., once more, has made some advances in this field. Competition between the various operators led to a reduction in costs. More generally, in most European countries, there are several ways to reach a foreign phone number.

Calling Cards

Calling cards are one way to make your away-from-home calls. By contacting any large American long distance provider, such as AT&T or Sprint, you can obtain a calling card that can be used to call almost any phone number from any country. Beware, though, that using such cards might only be advantageous when calling the United States. Calls to other countries will probably be routed through the U.S., and the total charge paid from your location to the U.S. and then from the U.S. to the place you are calling usually exceeds the price you would have paid if you had used a European POTS operator. Of course, you might consider the inherent advantage of being able to call anywhere. And there is the added advantage of having an English-speaking operator available to assist you. These advantages may be worth the extra price you pay for European calling via an American long-distance provider.

International Callback Operators

The other way to reach foreign phone numbers at a cheap rate usually involves a callback operator. The principle of international callback is rather simple. It relies on the price difference that might exist for a given communication, depending on which side the communication originates from. In the case of Europe, communications from Europe to the U.S. are a lot more expensive than communications from the U.S. to Europe. Callback operator services offer Europeans wanting to call the U.S. the ability to be charged as if they were calling from the U.S.

How does it work? Figure 2-3 describes an international callback scenario. First, you simply register. You are then assigned a unique American phone number that you will have to dial to signal your intention to place a call to the U.S. This signaling call does not cost you anything, since it is not answered. A few seconds after you have hung up, your phone rings as you are being called by the callback operator's switching machine. When you pick up the phone, you are provided with an American dial tone, and you can then proceed to dial the number you want to reach.

Your callback operator will bill you for both the call that was made to you and the call you placed (all in one item), but the sum usually happens to be smaller than the regular price you would have paid using your European national operator. Savings range from 30 percent to 70 percent, depending on your calling plan. For example, calling the U.S. from France with France Télécom costs from $0.86 to $1.10 per minute, depending on the time of day. If you have subscribed to a low-traffic callback service (which usually involves no installation fee, no monthly fee, and no minimum charge), your callback operator might bill you $0.60 instead, which already corresponds to a 30 percent to 45 percent savings. If you have subscribed to a high-traffic callback service (with usually involves an important monthly fee), your callback operator might bill you $0.40 instead, which corresponds to a 50 percent to 65 percent savings (but you would have to take the monthly fee into account to compute the real savings).

There are several points worth pondering before signing up for a callback operator. Fraud is usually not much of a problem with callback operators, because placing a call using your account requires that you be physically present when the callback operator calls you back. Nevertheless, you should avoid registering a phone number that corresponds to an extension that many people have access to. Also, beware of call forwarding and call remote forwarding. People might then be able to steal your American dial tone by forwarding your line to theirs.

Another problem with callback systems is that all you have to do to ask for a dial tone is call a phone number in America and let it ring a couple times. If you are using callback at home, you might end up being called at weird hours of the night, simply because some automated telemarketing computer called your signaling line in the U.S. thinking it was a normal line. Some callback operators offer the option to use a personal identification number (PIN) when signaling. This is to make sure the signaling

Figure 2-3. *International callback scenario*

call is indeed originating from you and is not a bogus call, but then the signaling call is not free anymore. Callback operators that offer this capability usually make sure that you can enter your PIN very quickly. Actually, this signaling call should not cost you more than a couple of cents and may save you some hassles. It all depends on whether you are concerned about receiving bogus callback calls.

You should read carefully the section about billing in the callback agreement that you will enter into. Is an unanswered call billed anyway, or does your callback operator only charge you for answered calls? When does billing begin? What is the timing increment (seconds, tenths of seconds, or minutes)? Most of the time, billing begins when you first get the American dial tone, so do not spend too much time dialing the number!

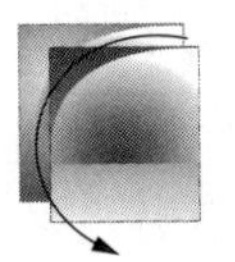

TIP: Don't just rush into a callback solution. Shop around and find a company that does it right, and provides the features you desire.

Finally, note that some callback operators now offer cheap rates not only to the U.S. but to many countries, including other European countries. This can be very interesting for companies with locations throughout Europe. However, always keep in mind that callback is not very convenient. Are you and your employees ready to go through the hassle of signaling? How about faxes? Sending a fax with a callback service is not straightforward. There are some devices that will take care of automatically routing your international calls to the cheapest operator. These machines may even take care of signaling a callback operator and waiting for the dial tone, but they are still awfully expensive (we are talking tens of thousands of dollars).

So what is the bottom line? If your international telecommunications needs are important, you have to think about callback. But if most of your calls stay within your organization, you might be better off signing up for an international virtual private network (see the "Virtual Private Network" section in this chapter). Altogether, callback usually turns out to be interesting, a bit cumbersome, but also a real money-saver.

Cable TV

Cable TV is discussed in this chapter only as a front door to the merging of the cable and telecommunications industries (refer to Chapter 19 for more on the cable TV industry). It is hard to talk about European cable TV in a general way, because each country has had a very different approach to cable. If we look at the three major European countries (France, Germany, and the United Kingdom), the cable TV setup is very different in each.

In Germany, cable TV is available virtually everywhere and is very common: 14 million Germans subscribe to it. This success is mainly due to a constant government agenda, which has always set cable TV as a major goal. No new TV channel could be created if it was not available on cable. The monthly fee is about $10. But up to now, despite this success, there has been very little interaction between German telephone and cable TV, mainly because of regulatory barriers.

In the U.K., as in Germany, cable TV is also available virtually everywhere and is very common. But in the U.K., cable TV operators have been allowed to sell phone service too. This has resulted in even more local competition, driving costs down, especially for local calls. The British phone companies are still denied the right to provide video channels, but they do provide video-on-demand and related services.

In France, cable TV is a little bit behind. It is available to only 30 percent of the population, and only 1.4 million French people actually subscribe to it, for about $25 per month. The main reason for this situation is the lack of interesting TV programs on cable. Until recently, cable TV was simply just not worth it. Several new channels have appeared on cable, and it seems that French people are finally moving towards cable TV. The interaction between French cable operators and the telcos is still rather poor. Some experiments are being carried out to merge the two industries, but nothing serious has happened yet. But in France more than in other countries, cable operators are looking for alternate ways to use their networks. We'll see what happens.

Virtual Private Network (VPN)

Virtual private networks (VPNs) are a reality in most Western European countries—POTS operators offer this kind of service. A VPN is when an organization uses public connections between several locations to form its own telephone network. The savings can be quite significant.

International VPNs, on the other hand, are not a reality yet. Many countries do not yet have fully digital networks, which hinders the development of feature-rich virtual networks. Many public operators are actually afraid that these virtual networks will steal most of their business.

Big alliances, such as Unisource (an alliance of various European operators) and Concert (from MCI and BT) are already marketing international VPNs, but the number of countries that are concerned is still rather small, though growing very steadily. Basically, you have the choice between a feature-rich virtual network available in a small number of countries and a network that will cover more countries but with fewer features. The multinational VPN, in reality, ends up in a complex mixture of public and virtual networks, which may be a nightmare to administer. With the rapid roll-out of Euro-ISDN and the various alliances between European operators, real international virtual private networks should become a reality in the near future.

Eastern European Countries

Most Eastern European countries have rather old POTS networks, which are rarely digital and most of the time do not work very well, especially when it comes to international calling. It usually takes several months, if not years, to get a telephone line. Many telephones in Eastern Europe are still rotary dial. Some governments (notably Polish, Czech, Slovak, and Hungarian), sometimes with the help of foreign private investors but more generally thanks to European funding, have acknowledged the importance of modernization of these networks and have started to massively introduce modern technology into their networks. The German public operator, Deutsche Bundespost Telekom, thanks to the experience it gained upgrading the former East Germany's network and because of some other political considerations, is often present in one way or another in the various initiatives that are taking place in Eastern European countries these days, helping to speed the advancements of telephony.

The only reasonable alternative for a company that wants to have reliable voice and data links between Eastern Europe and the rest of the world is to use satellite telecommunications, notably VSAT (very small aperture terminal). These terminals provide bidirectional communications with other locations through a satellite communications link.

What has attracted most foreign investors to Eastern Europe are the mobile networks. In almost all Eastern European countries, there are such networks, with the service and quality comparable to what you may expect from a mobile network in the U.S. In contrast to the POTS situation, it only takes a few hours or days to open up new service on these networks, but these networks are usually too expensive for the average citizen. Costs will continue to go down, and POTS will become more readily available. However, it will be some time before the general populace will be connected to telephony.

Conclusion

European communications has many similarities and differences with its North American counterpart. When doing business in the European marketplace, you have to be prepared to encounter diverse barriers in language and technology. Even with standardization, there are still many differences among European countries with regard to telephony issues. Be prepared to make adjustments and modifications when dealing with European telecommunications, and don't be afraid to take your business abroad. The vision of a global economy is quickly becoming a reality. Your decisions about telephony can help bring about change and prosperity for your organization, Europe, and the world.

Chapter Three

Switching Platforms

To understand telephony, you need to start at the beginning: the telephone and a switched or dedicated network. In this chapter, you will learn about the various types of switching platforms, the form and function of the switch components, and telephony connectivity and how it is accomplished. You will learn about the cabling that is used to make connectivity and what the right cable for the job is.

After reading this chapter, you should have a fairly good understanding of how telephony connections are made and of the telephony components that make mass communications possible. The information contained in this chapter is the basis for all the telephony technologies discussed in this book.

Line Consolidation and Routing

Before we talk about the actual switch, you should be familiar with the concepts of *line consolidation* and *routing*. Whether you have your own key system, PBX (private branch exchange) or you use Centrex for telephone service, you're using line consolidation: the servicing of many phones by fewer telephone lines or circuits (see Appendix B for definitions of these terms). An example of line consolidation on a micro level is the telephone you have in your home. Most homes in America have only one or maybe two physical phone lines, with four or more telephones connected to it. You can't pick up all four telephones and make four separate phone calls; only one connection can be made per telephone line.

Just as it would be unreasonable to have a telephone line for every telephone in your house, it would be unreasonable for a Centrex switching system to have one outgoing line for every telephone number it serves—the cost would be too great. PBX and key systems also take advantage of line consolidation. What this means to you is that just having a telephone line doesn't mean you will be able to make your call. The telephone network is only designed to handle so much traffic at any given time. The phone company figures on about one two-way (can accept inbound telephone calls or make outbound telephone calls) telephone line, called a *trunk*, for every ten to fifteen telephones it serves. For PBXs, the ratio is usually closer to 1:10, and for key systems the consolidation typically runs between 1:1 and 1:10. Behind the key system, PBX, or Centrex lies the phone company's switching network. The concept of line consolidation is used for the entire switching network. Figure 3-1 provides a simplified example of how line consolidation is used throughout a telephone network.

Routing is how the switching network can connect one telephony device (telephone, fax, modem, etc.) to any other. Did you know that your telephone number is like an address? Your area code (also called the NPA, or number plan area) identifies a certain area in the country where a phone call must be sent to get to your specific telephone. The second three digits of your number, the exchange (also called the NNX or NXX), identify the specific community that is being served, while the last four digits identify your specific telephone line. For example, NPA 213 serves the greater Los Angeles area, NNX 850 serves the community of Hollywood and the last four digits of

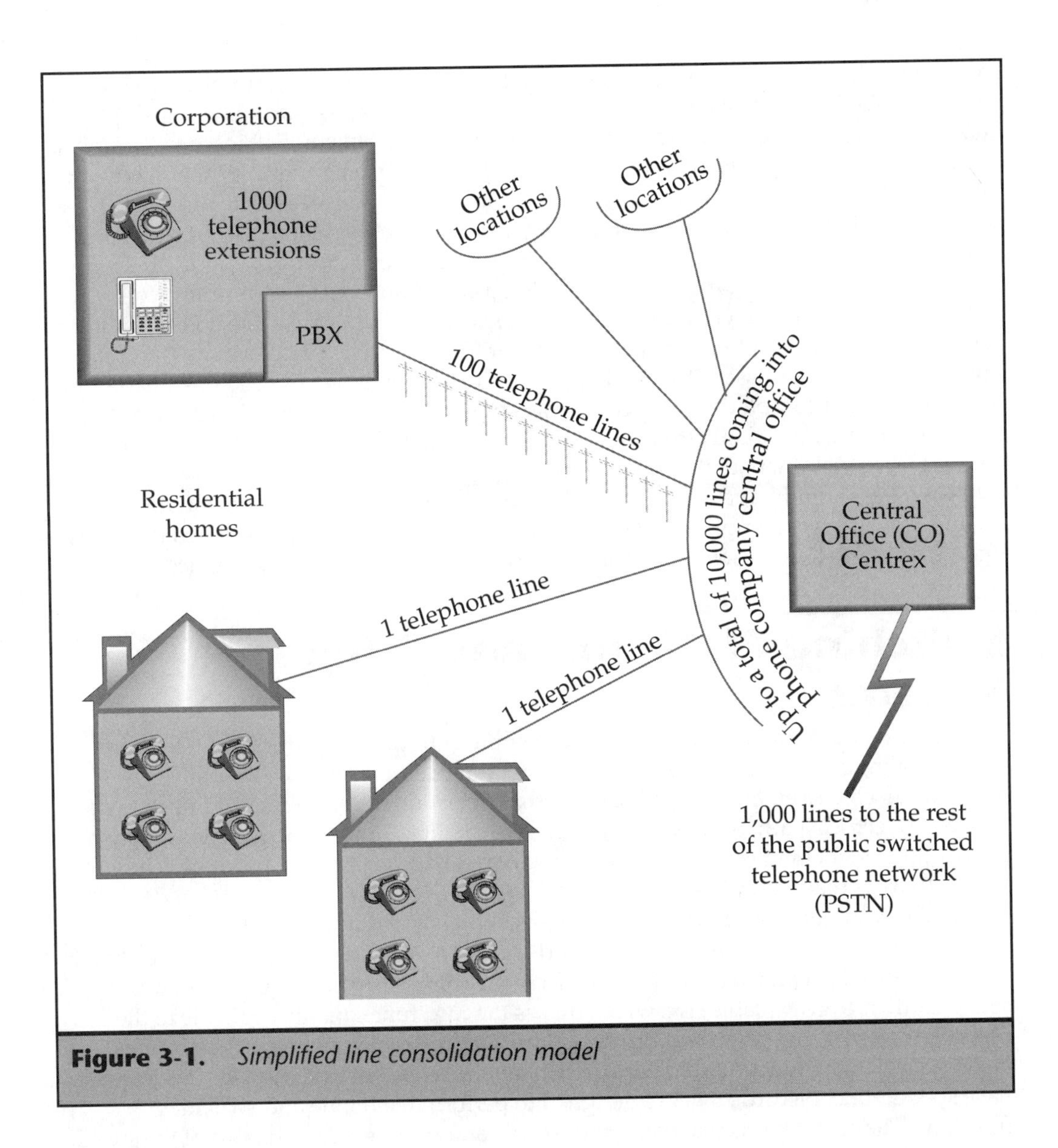

Figure 3-1. *Simplified line consolidation model*

the phone number specify the actual telephone. International dialing plans work in a similar fashion to complete your call.

Now, let's take these two concepts of line consolidation and routing and apply them to a community or to a company. The telephone number provides a map to the switching network of how to route the call to get to the right destination. From one location to another, a telephone call can take hundreds, even thousands of paths to reach its destination. If for some reason one path is full and the circuit cannot be completed, then the call is rerouted and another path is chosen automatically. Have you ever called an area that just experienced a natural disaster, such as an earthquake, and gotten the message "We are sorry, but all circuits are busy now. Please try your call again later"? This happens when the number of people in a community that have telephone lines and are trying to use them, simultaneously, exceeds the number of connections that specific area or community can make to the rest of the world at one time. If you try your call again in a few seconds and it goes through, that means a circuit was available to complete the call. But, aside from extreme examples, you almost never get the "all circuits are busy" message; this is because line consolidation uses traffic management (explained in Chapter 6) to statistically determine the right number of trunks necessary to provide a specific *grade of service* (a ratio of call attempts to successful call completions). In companies where this consolidation is done on a more micro level, your call attempt might be blocked more often, but line consolidation techniques are usually very accurate.

Now you should have a general understanding of how a telephone network is set up and how calls are completed. The next phase is to understand how a switch performs these functions—line consolidations and routing—to provide telephony services.

Switching Platforms and Telephone Systems

Have you ever asked yourself how you can connect to precisely the right telephone just by dialing a number? The answer is switching technology. A *switch* is a mechanical, electrical, or electronic device that opens or closes circuits, completes or breaks an electrical path, or selects paths to make circuits. Along with line consolidation and routing, switching is how the telephone network is supported and how telephony connectivity is made possible.

To help you get a grasp of what a modern switch is, think of a switch as a large computer with hundreds or even thousands of bus slots. Bus slots give you the ability to add special functions to computing devices by adding a card, such as a modem or sound card, into an available slot, which makes the card functionality available to the central processing unit of the computing device. Figure 3-2 shows a switching cabinet (left) and switching function cards in a switch bus. All switches and most key systems have special function cards that are designed to perform telephone and switching functions. The circuit paths that connect the cards and the overall operations of the switch work on the same basic principles as the computers you are probably familiar with.

Switches provide a great number of additional functions and features that make up a telephone system. Centrex and PBX are the two major types of switching telephone systems. A key system provides many of the features of a traditional telephone system, but it does not have a switching function. Before you learn about the different types of telephone systems, however, let's discuss the various types of features that you can expect from modern telephone systems.

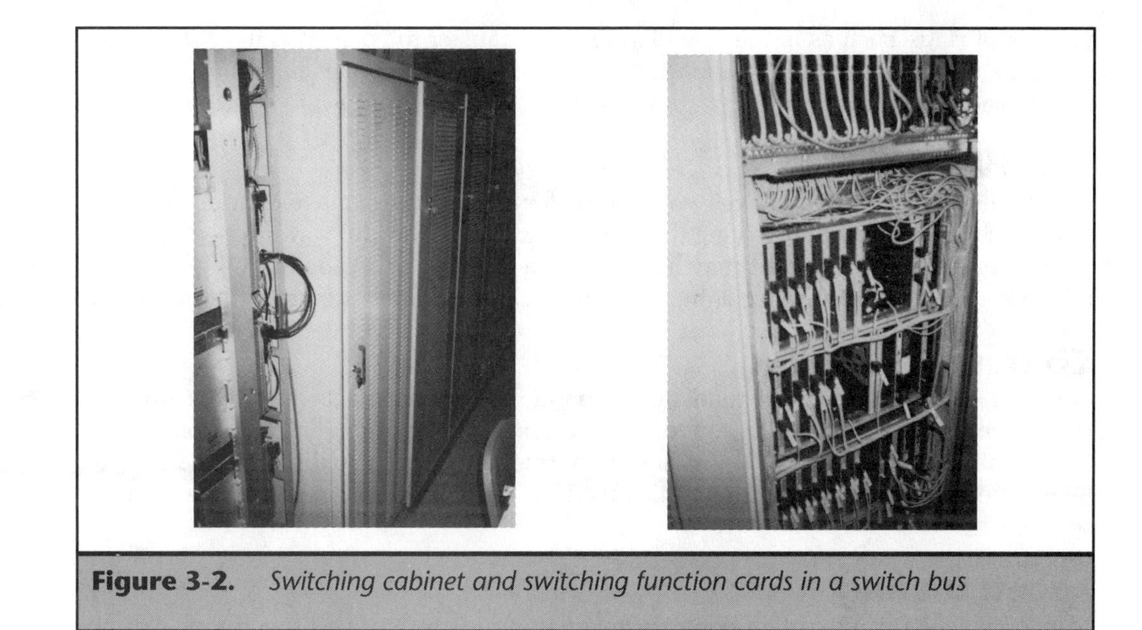

Figure 3-2. *Switching cabinet and switching function cards in a switch bus*

Telephone System Features and Functionality

Each telephone system has its own set of features and functionality. Try to be aware of the overall features available, so when you are shopping around for a telephone system, you will be assured of getting the functionality you need. It should be mentioned here that all the features discussed are available on Centrex and PBX systems. Some features listed are not available in key systems, but the dividing line between the different platforms has become difficult to see; in fact, many Centrex functions can be added to key systems in the telephone line configurations provided by the CO (central office), the telephone company's office of operation. The following is a limited list of popular features and functions found on telephone systems today.

Intercom

The *intercom* function is only available on systems that support telephones with built-in speakers or on systems that connect to building intercom systems. An intercom gives you the ability to page someone throughout the entire operating environment of the telephone system. This is a very common feature supplied on all three types of telephone systems.

Call Forwarding and Call Transfer

Call forwarding instructs the telephone system to send any calls that are received to a specific telephone number or extension and to redirect that call to another telephone number or extension. Sometimes call forwarding is programmed to occur after a certain number of rings, such as in the case of automatic transfer to voice-mail if a person doesn't answer. *Call transfer* will manually transfer a call from one telephone number or extension to another. These functions are typically not found in key systems.

Call Waiting

Call waiting is the ability to have two requests for a call to the same telephone number or extension. The way of notifying the user about the second call varies based upon the equipment used. The user may hear a tone in their ear or a beep from the telephone unit, or see flashing light. Again this feature is rarely found on key systems.

Conference Calling

Conference calling places more than one telephone call and connects the two or more calls to one another so that the conversations can be shared among all telephone lines in use. In the Centrex and PBX environments, this function is usually provided by the telephone system. In a key system, this function is usually performed by the telephone set itself.

Speed Dialing

Speed dialing programs certain phone numbers for one- or two-digit touch-tone dialing. All the digits of the telephone number are dialed by the system; the user simply dials the location of the programmed function. Some telephones have special buttons that can be programmed with telephone numbers, while others require the user to push a function key and then one or two digits on the touch-tone keypad. In most Centrex and PBX systems, the feature is supplied by the telephone system; in key systems, the feature resides in the telephone.

Automatic Number Identification/Caller ID

Automatic number identification (ANI), also referred to as *caller ID,* allows the called party to be informed of the telephone number of the person making the call. Many phone systems also allow the user to assign a description, such as a name, to the telephone number. This feature is not just a function of the telephone system but also a function of the public switched telephone network (PSTN) to which the telephone system is connected. If the caller ID function is not supported in your area, it does not matter if the telephone system you are using supports the function or not. You will not be able to get the service, except for maybe in the case of station-to-station calling with the telephone system. In some municipalities, this function has been deemed an invasion of privacy and has been prohibited by law.

Call Block

Call block allows the user to define telephone numbers that are not permitted to be connected to a particular telephone. This feature is extremely useful to block harassing phone calls or unwanted telemarketers. You will find that this feature is primarily used on home telephone lines supplied in a Centrex environment and is rarely applied to business applications using a PBX. Most key systems do not offer this feature.

Digital Telephones

A *digital telephone* is a phone that converts the analog sound picked up by the receiver in the telephone to digital binary data for transmission through the network instead of just sending the sound in an analog format. The conversion is performed by sampling the analog signal 4000 to 8000 times a second. Eight to sixteen bits of linear data are needed to represent the sample. The majority of business telephone systems are digital. Digital phones are usually proprietary and come in a wide range of styles and prices. Make sure you are aware of the individual handset costs and the features they offer. You may be purchasing or leasing many more over the years.

Voice over Data

Some telephone systems provide a data circuit directly through the digital telephone unit or a secondary device that is coupled to the telephone system. Incredible as it may seem, many digital telephone systems allow multiple functions over one or two pairs of wire. They do this using *time division multiplexing*, which allocates part of the 8000 time slices available per second to voice and part to the data. Neither the voice nor the data is affected by the multiplexing. Speed of the data might be slightly reduced, but this is usually not the case.

ISDN Circuits

Many modern telephone systems make use of ISDN (Integrated Services Digital Network) technology (see Chapter 11) to connect their digital telephones and peripheral devices, such as modems and fax machines. The telephone system might only offer one 56 or 64 Kbps B voice/data channel and one 16 Kbps D control channel or the standard two B and one D channel basic rate interface (BRI). The latest in digital telephone systems is bandwidth on demand, usually used for videoconferencing over a LAN that is connected through the telephone system or possibly a dedicated or dial-up circuit. Bandwidth up to one-half of a T-1 (see Chapter 10) can be requested on the fly. ISDN technology is definitely one of the directions in which modern telephone networks and telephony in general is heading.

TAPI or TSAPI Interface

Probably the biggest advance telephone systems and computers have made is the ability to provide telephone access and control to computer systems. Telephony application programming interface (TAPI), an offering from Microsoft, and telephony server application interface (TSAPI), offered primarily by Novell, both provide a number of features and functionality to the desktop computer and LAN that were never before possible. (For more information on TAPI and TSAPI, see Chapter 18.)

Now let's look at the similarities and differences between these three types of telephone systems and why you might choose one system over another.

Centrex

Centrex is business or public telephone service that is provided through a local CO or that resides on the customer's premises. Both configurations are supplied by the local or regional bell operating company (see Chapter 6). Complex codes must sometimes be memorized in order to use the Centrex features. However, there are many reasons why a company would rely on Centrex service: it offers powerful features; if you are not in the position to purchase your own equipment, Centrex is cheaper to get started with and is much easier to manage; and repair costs are very low, because many times you do not own the equipment.

Centrex systems also have the advantage of size. It's almost impossible to outgrow a Centrex, and if you can, you should probably own a PBX or even a network of PBXs. COs house multiple Centrex switching systems, which are connected to many other COs providing not only backup but access to multiple locations for distributed organizations. Centrex offers a large number of lines (telephone numbers, stations, extensions—whatever you want to call them), and it takes up very little space at your location. Your site requirements to house your switching solution are reduced to almost nothing, because 99.9 percent of the time the switch is located and maintained by the CO.

Centrexes are extremely reliable. All sites feature power generator backup and 100 percent DC battery backup. Someone is there 24 hours a day. Centrex has double redundancy switching capability, and is designed to be down less than three hours over 40 years. Virtually all of the lines supplied to you are provided through a Centrex system anyway, whether you are PBX or key system-based. Technology is another good reason to choose Centrex: Centrex systems are constantly being upgraded to the latest technology. A few older Centrexes are still out there being run by various telephone companies, so check to see which features are currently available in your area.

One of the more recent advances is the ability to manage the Centrex system yourself. You can now change phone features with programs like MACStar and CDC along with a modem connection and terminal emulation software (see Chapter 15). Another advance is the delivery to the customer of CDR (call detail record) data, also called SMDR (station message detail record) data, which is a telephone call record similar to the one you see on your phone bill (see Chapter 6). This gives the customer the ability to perform telemanagement (see Part 2) for internal reporting of telephone traffic. Both of these features are standard functions of PBX systems.

These are some of the bigger reasons to look at Centrex telephony solutions, and all are valid under certain circumstances. One thing is certain: we all depend on Centrex service to provide us phone service virtually every day of our lives. Every call you make outside of your organization definitely goes through a central office that provides Centrex service. From a business standpoint, you need to decide what combination of telephone system and service is best for you.

Private Branch Exchange (PBX)

PBX systems have grown steadily in popularity with businesses. When you invest in a PBX, you are put in the driver's seat—you now own your telephone system. PBX systems can be purchased from companies such as AT&T, Fujitsu, Intecom, NEC, Nortel, Rolm, and Siemans, just to name a few. All of the functions listed above and more must be carried out by your telephone system, and *you* are now responsible for the telephone service to be provided throughout your organization. More importantly, you are in control of your switching needs. PBXs can be configured for as few as 24 ports (telephone extensions), with even fewer in use. Running your own PBX is a formidable task but the rewards are usually worth it.

As always, there are significant cost considerations when purchasing a PBX over a Centrex. In the long run, PBX-based solutions are less expensive in almost all situations. Once you've paid for it, it's yours, which has additional financial merit. The major money-saver when using a PBX is the reduction of telephone line costs. Remember, the PBX can also take advantage of line consolidation, as described earlier. Just think of all the services that Centrex and PBX systems must provide. The equipment, the labor, the electricity, the environment, the features, etc. all become yours, and thus are subject to more cost-effective management. Your goal is to provide better and more responsive service and more features, while at the same time saving money.

PBX systems are very reliable. When properly implemented, it can be argued that PBX systems can be almost as reliable as Centrex systems. Special electrical, environment, and space requirements must be supplied as in the case of Centrex systems in order to ensure reliability.

Connectivity is an important benefit of PBX systems. If you aren't at the stage where your computers are hooking up to your phone, you probably wouldn't be reading this book. Having the equipment that you are trying to interface with in an adjacent building instead of 5, 10, or 20 miles away has distinct advantages. If you want to add a voice-mail system, automated call distributor, or other voice-processing application, you have the power to configure the PBX to integrate with the application optimally.

PBX telephone systems provide very advanced functions to the user through the telephone unit. Digital readout, speed-dialer buttons, conference buttons, hold, speaker phone, etc. are all common functions available on PBX telephones. The ease of use and high functionality of the PBX telephone is a distinct advantage over the standard Centrex telephones typically offered.

Expansion needs can usually be met more easily with PBX. If your PBX was purchased correctly, the system should have been configured to allow for future growth (see Chapter 5 for PBX acquisition tips). Technological obsolescence can also be avoided by making sure the PBX you purchase is up to date and the manufacturer is planning to make further upgrades available. You have the ability to put the wheels of operation into motion faster than the phone company. If you are well prepared, service can quite often be delivered in less than one day. The phone company can take days, weeks, or sometimes longer to meet certain needs.

Control—it is all yours, if you want it. Functionally, this is what a PBX offers you and it is the second best reason to own your own PBX telephone system (behind cost.) You have the control to be flexible and to do things when you want, the way you want; and you have the control to perform upgrades at your discretion, do MACs (moves, adds, and changes) when you want, and provide the features that you want. You can now rely on yourself instead of a third party, or you may subcontract with a PBX vendor to provide the maintenance services, but on your terms instead of the phone company's.

Key Telephone Systems

Key telephone systems handle the lower end of the telephone systems market today. Systems start with as little as two ports, and some may grow to over 100 telephone extensions. The main difference between key systems and Centrex/PBX systems is the switching capability: key systems do not have the ability to transfer a call to another extension or telephone number unless the function is supplied by the CO. In smaller organizations, this is not a big deal. Many key systems can allow 20 or more actual telephone lines to appear and be accessible on each telephone connected to the key system. Just pick a line and dial—no access code such as "9" is required. Different telephones can be configured with different sets of lines to reflect divisions within the phone system and throughout the organization. Through this function, a key system can take on even more of the characteristics of a PBX system.

The majority of key telephone systems utilize analog telephone lines rather than digital. Hybrid key systems add many functions that were traditionally supplied only by PBX systems; many can support analog or digital telephones and analog lines. Some can even be configured to handle digital telephone service such as ISDN from the phone company.

Key systems usually do not require any special environmental conditions because of their small size and power requirements. Just plug them into a reliable electrical outlet and they run. However, they are more susceptible to down time than PBX and Centrex systems, primarily because of their size and the unjustifiable costs involved with making them more reliable with peripheral equipment. However, enhancing the environment you install your key system in is not a bad idea.

Size, performance, and price are the three best reasons to own or use a key telephone system. If your organization is small and/or has multiple locations, key systems may be the way to go. Voice-processing applications can be fairly well integrated with key systems, but the switching function is still lacking. Key systems do, however, provide some very advanced functions these days. Some even offer voice over data and network connectivity. An advanced key system can supply better functionality than a Centrex system for less cost, providing you don't outgrow the key system you invest in. The prices for key telephone systems start at around $700 and around $100 to $400 per phone. The larger digital key systems can end up costing over

$35,000 fully configured. Undoubtedly, key telephone systems do have a place in the business market both now and in the future.

Components of a Telephone System

The components of a telephone system give the system its tremendous capabilities. The components described in this section refer primarily to Centrex and PBX systems, but many of the components listed can be found in key systems as well.

This discussion about the components of telephone systems begins in the switch room, also commonly referred to as the main distribution frame (MDF). Let's assume we are talking about a PBX, so the division between telephone systems and phone company components, and the corresponding responsible parties can be more clearly defined. Not venturing past the telephone room itself, let's look at the major components of a telephone system in basically the order that they are used to connect a circuit and provide telephony service (see Figure 3-3).

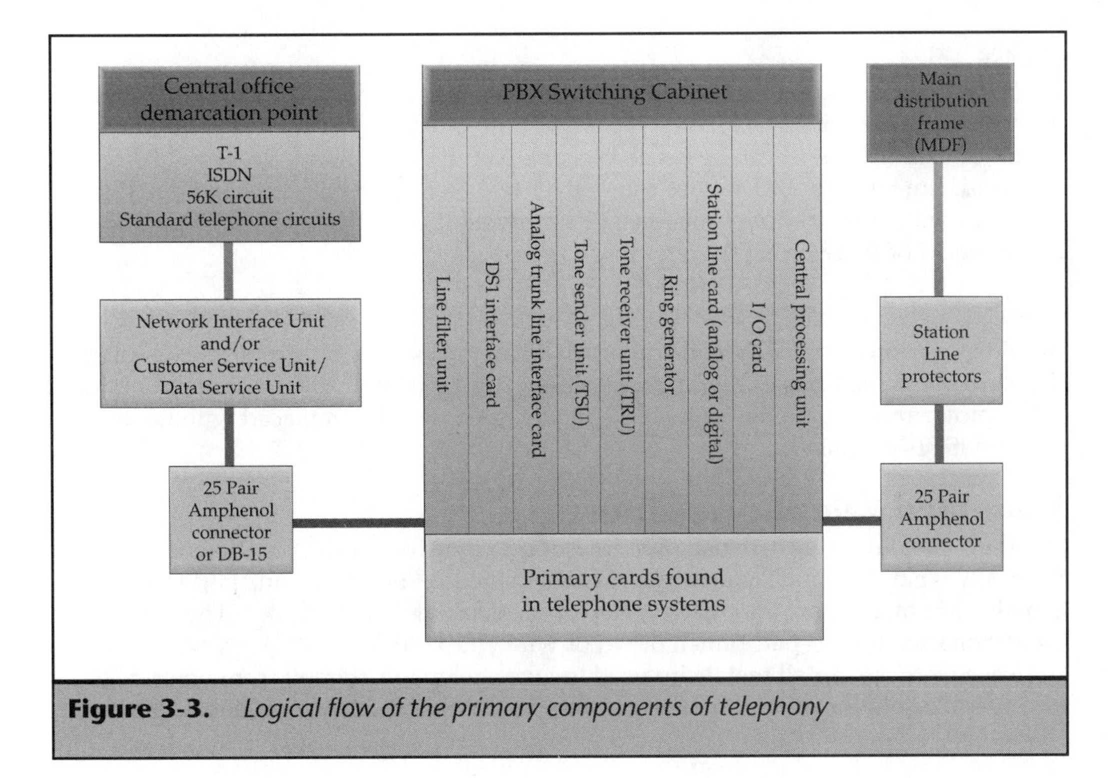

Figure 3-3. *Logical flow of the primary components of telephony*

The Switch Room

Our journey begins at the central office demarcation point (the place where the phone company delivers your telephone lines and telephony services) delivered to the MDF in the switch room.

As you go through this section, try to keep in mind that some lesser links and components of the telephone room environment might be left out for simplicity; use this information only as a base to give you a general understanding of the telephony environment.

The CO can provide four basic types of circuits to a customer's premises.

T-1

A T-1 provides 1.544 Mbps of data transmission capability. This equates to 24 simultaneous digital telephone conversations at 64 Kbps each. This circuit is supplied either by fiber optics or over two pairs using, typically, an RJ-48 or DB-15 connector. At the end of the circuit provided by the telephone company you will either find a network interface unit (NIU) or a customer service unit (CSU) or possibly both. Basically, the NIU and the CSU both perform the function of error correction, signal boost, and loop back testing. An NIU is supplied by the phone company, and a CSU must be supplied by the site. (For more on T-1s, refer to Chapter 10.)

ISDN

ISDN circuits can also be provided by the phone company. ISDN service can range from BRI (basic rate interface) to PRI (primary rate interface) service (ISDN service over a 24-channel T-1), it can be carried on one pair of wire, and an RJ-48 or DB-15 connector can be used. BRI service can deliver two 64 Kbps B channels and one D channel used for controlling functions. PRI provides 23 64 Kbps B channels and one D channel. (For more on ISDN, refer to Chapter 11.)

56K

An older version of an ISDN-type circuit, 56K can provide a single multiplexed pipe that can contain 56 Kbps of digital transmission, and it is typically used for data rather than voice communications. Many 56K circuits are now being replaced with the more modern ISDN circuits.

Standard Telephone Lines

Standard telephone lines are usually brought into a site on an RJ21 connector, which is usually what the phone company provides to the PBX switch technician. The RJ21 contains 25 pairs of wire and up to 25 separate telephone lines (trunks). The RJ21 is then connected to a 50-pair punch down or wire wrap block.

Bridging clips (a small metal clip used to take a telephony circuit from one side of a 50-pair block to the other) are then used to pass the circuit from one side of the block to the other side of the block. The block can then be patched to an Amphenol connector to be passed to the switch.

25-Pair Amphenol Connectors

A 25-pair Amphenol connector is another name for an RJ21 connector. They are listed separately because the RJ21 supplied by the phone company is usually the end of the circuit that the phone company is responsible for. Sometimes a network interface unit is also included at the end of circuits provided by the phone company. The RJ-48 circuits used for T-1, ISDN, and 56K circuits can be converted to Amphenol connectors if desired. The Amphenol connector is also used to take lines out of the PBX so they can then be connected to blocks on the MDF.

DB-15

In many cases, a DB-15 connector can be used in addition to or instead of the Amphenol connector. The DB-15 connector performs the same basic function, but it can contain a maximum of only 15 wires. The connector used will vary based upon the PBX manufacturer and the card the circuit is being connected to. There are many other connectors used within a switch room environment. However, these are the most common connectors used in telephone switching today.

The PBX Cabinet

A PBX can contain thousands of computer cards designed to perform specific functions. All of these cards are inserted into bus slots connected to the switch's central processing unit (CPU). In some *chassis* (the casing or frame used to contain PBX components), the connections between the various bus card slots and the CPU are built in, or preconfigured. In these systems, the right card must be put in the right slot. In other chassis, the connections between the cards can be made by the technician, allowing for greater flexibility.

In most PBX systems, the CPU can be found on a card, but in some cases the CPU brains are built right into the system. The PBX relies on the CPU to supply switched connectivity between the various cards that reside in the PBX. Here are the cards used by the switch (in the order they're used) and the functions they perform.

LINE FILTER UNIT The *line filter unit* is used to take as much background noise off of a circuit as possible. It is used to ensure that the highest quality of signal is supplied to the CPU and sent from the CPU. Line filter units are used when telco lines are brought in and when station line card signals are sent in and out of the PBX. Basically, anything that goes in or out of a switch goes through a line filter unit. In some systems, this function is built directly into other cards in the system. Key systems do not usually include line filter units, but secondary devices can be added to the key system to provide this function.

DS1 INTERFACE CARDS DS1 interface cards are used to bring T-1 circuits into the PBX, as shown in Figure 3-4. Once the T-1 is connected to the PBX, its resources are

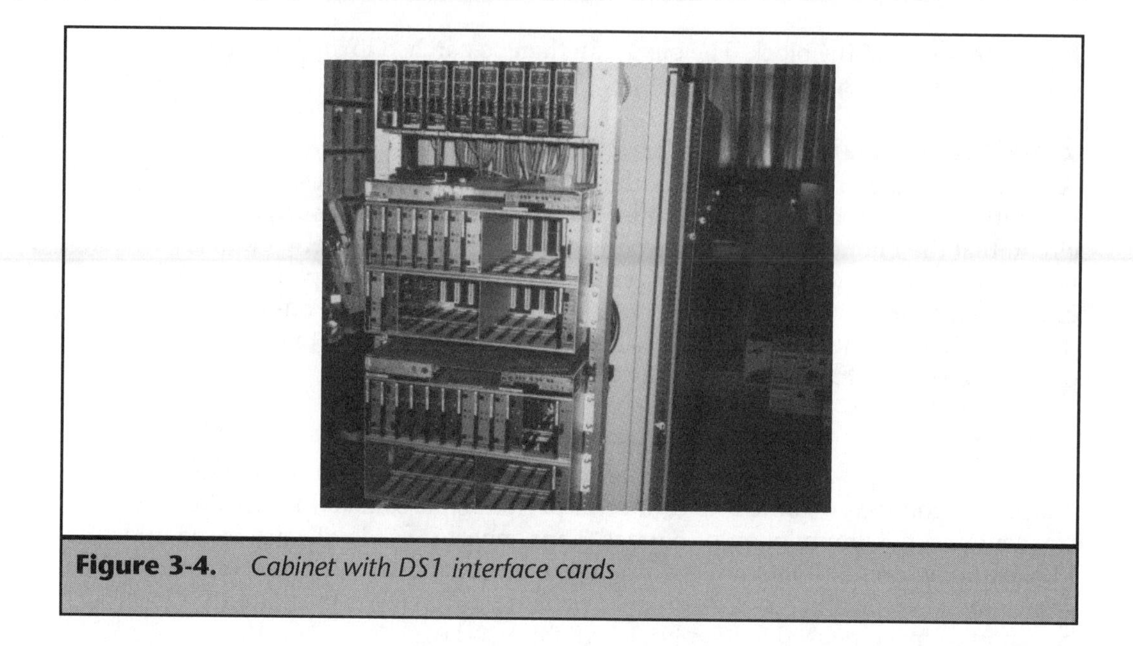

Figure 3-4. *Cabinet with DS1 interface cards*

made available to the PBX CPU. The different channels in the T-1 can be broken out or combined in any fashion that the technician wishes. Obviously, the lines must be configured in kind by the central office to achieve the desired results.

For example, a T-1 has 24 channels. The configuration for the 24 lines might be 12 for Sprint long distance, 8 for local CO access to make local telephone calls, and 4 for a tie-line circuit over to another site that is part of the company. Although one T-1 brings the circuits in, there are three logical trunk groups.

ANALOG TRUNK LINE INTERFACE CARDS *Analog trunk line interface cards* are used to connect the 25 circuits found on an Amphenol connector to the switch CPU. They serve the same basic function as a DS1 interface card does for the different RJ21 line service provided.

TONE SENDER UNIT The *tone sender unit* (TSU) is used to pass along the touch-tone digits to specific circuits on the interface cards or on the station cards. The CPU can use one TSU to perform the tone sending function for multiple station and interface cards. Because of the switching function provided, there is not a 1:1 ratio of TSUs to station or line ports.

TONE RECEIVER UNIT The *tone receiver unit* (TRU) is used to receive and interpret touch-tone data received on a specific circuit. The TRU basically performs the reverse of the function that the TSU performs. The switching function makes it possible for TRUs to be shared by cards in the same way that TSUs are.

RING GENERATOR The *ring generator* is used to send the ring voltage (20 Hz, 90 volts) necessary down a telephone circuit to indicate that a complete circuit is being attempted. The telephony unit at the other end receives this ring voltage and makes the telephone ring (or gives some other indication). If the phone is answered, the circuit is completed.

STATION LINE CARDS Station line cards come in a wide variety of configurations with varying features. Station line cards can be analog or digital. Some even support ISDN functions. Many of the features and functionality described above are provided to the telephone user by the PBX through the station line cards in conjunction with software running on the CPU. These cards typically can support between eight and twelve telephone station ports each. The port on the station card is usually the beginning of the cabling circuit tracked by the facilities management system (see Chapter 7).

I/O CARDS Since a PBX is a computer, it would sure be nice if we had a way to get information in and out of it. Thank goodness for the I/O card. The I/O card serves three main purposes: to allow for programming input to be made, to allow for data or reporting information to be output to a printer, and to generate SMDR data to be collected and processed by a telemanagement system (for more information on SMDR, refer to Chapter 6). One I/O card can be configured to supply all three of these functions and a few others.

STATION LINE PROTECTORS Station line protectors serve a minor, but valuable, function. One or more station line protectors are used to guard against electrical spikes, such as lightning strikes. There are two types of station line protectors: one is for standard circuits, usually black, and one is for special circuits, usually red. An example of a standard circuit is a telephone line. An example of a special circuit is an alarm system.

This section has only covered the major cards and operations of a switching telephone system. Due to the scope of this book, several components and many switching functions have not been described. Use this section only as a starting point to understanding how switching functions are provided. Now it is time to look at how telephony service is provided to the end user.

Cabling and Physical Plant

Cabling is the wiring used to physically connect the various telephony components. Your physical plant is all of your existing cable infrastructure, the various locations such as MDFs, IDFs, manholes, and junction boxes where connections are made, and the equipment used to make those connections. In order to understand cabling, it might be helpful for you to understand what a circuit is and how all of the connections are made to make a circuit.

A *circuit* is a physical connection, or path, between two points, including all equipment necessary to complete the circuit. In telephony, the circuit is designed to carry electrical current in both a send and receive capacity. Circuits can also contain other components, such as resistors, capacitors, filters, etc. Circuits can also be created with a closed path where electrical current is designed to flow through the circuit, as is the case with alarm circuits.

Let's walk through the creation of a fairly long circuit path in an attempt to touch upon most of the possible legs (connectivity) that a telephony circuit might pass through and help you gain a better understanding of how your telephone is connected to a telephone system. If you come across a term you're not familiar with, try looking it up in Appendix B.

Installing a Telephone

OK, put yourself in the shoes of a telephone technician for a moment. First of all, you are going to have to strap on the tools of the trade, some of which are shown in Figure 3-5. These items will help you make a telephony circuit.

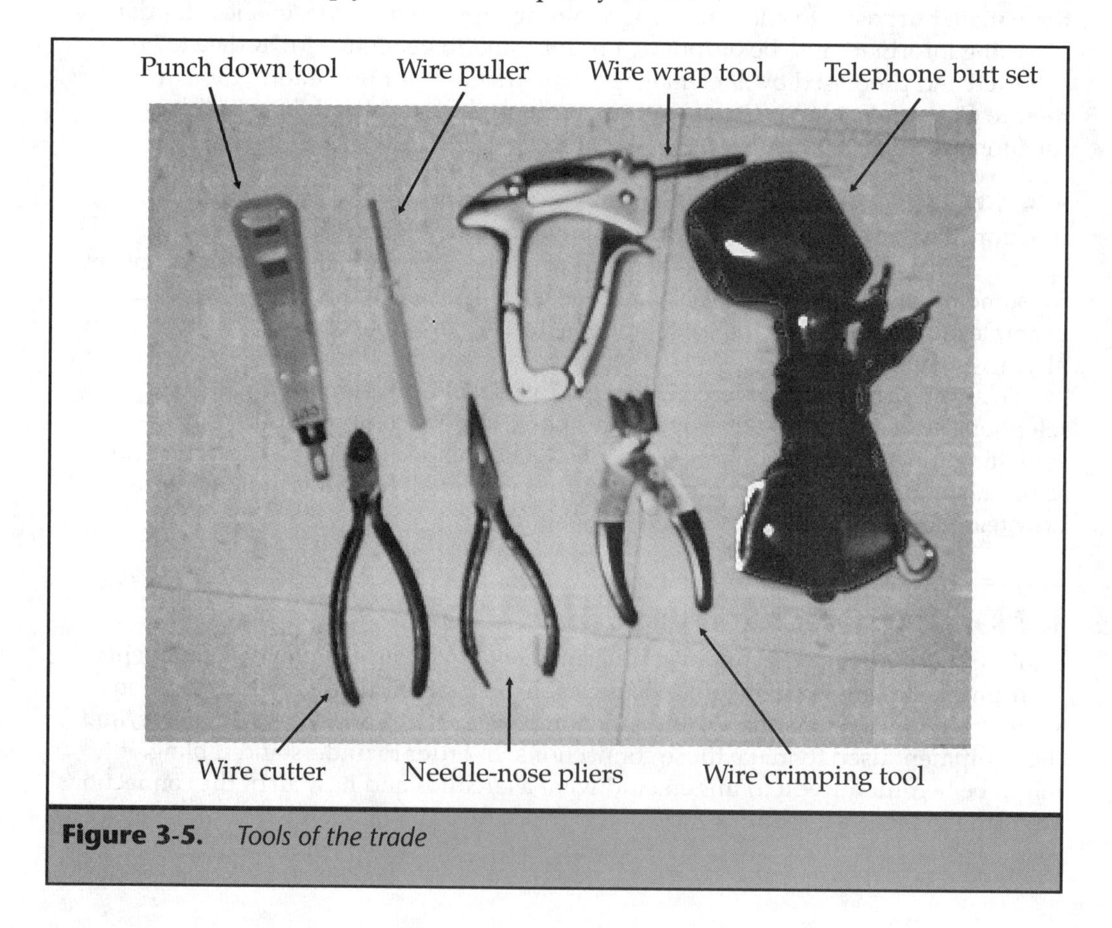

Figure 3-5. *Tools of the trade*

Let's say you need to install a telephone on the third floor of the library on the north end of campus. Your switch room (of course) is on the south side of campus. Your goal: to create and test a telephone circuit. Figure 3-6 gives you a basic map of how the circuit will be layed out. You start your quest in the switch room, where you find the MDF, a large wall of blocks and wire that is divided up into different sections. Each section contains cable destined for a different location. You start in the section that contains cables connected to ports on the switch. You then find a pair of wire on the block that is available for use. You can tell that the port is available because at the

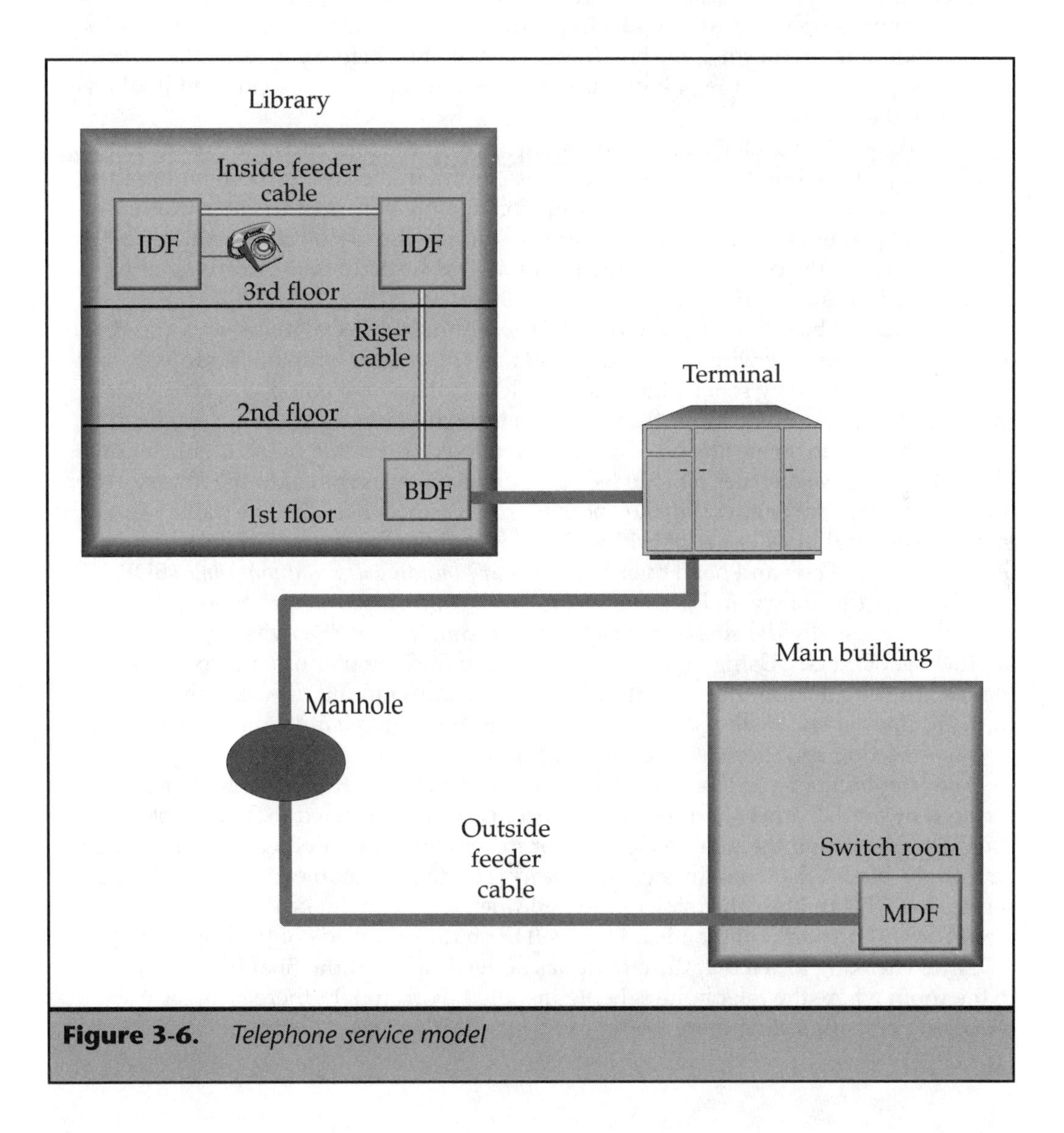

Figure 3-6. *Telephone service model*

moment the wires don't go anywhere but from the port to the block. Using your *wire wrap tool,* you make a *cross-connection* using station wire between the port and... hey, now where does this circuit need to go?

After long deliberation, you determine that the best way to get this circuit from the switch room to the north side of campus is to use the north *outside feeder cable.* You find the section of the MDF that is connected to the north outside feeder cable, and wire wrap the opposite end of the station wire on the next available cable pair. You then put *bridge clips* on both ends to complete the cross-connection. Great! The telephone circuit has just left the building and is now heading north underground in the outside feeder cable. The next stop we need to make to get the phone service working is a *manhole* about a thousand feet north of the building you are in. But, before you leave the switch room, you put a *toner* on the pair of wire that you are using, so you can find it when you get to the manhole.

You take a little stroll over to the manhole, pry off the top, and climb in. In some manholes, you can find blocks similar to those found in the switch room, but in most cases, manholes are used strictly for *splices*. Fortunately for you, this cable is already spliced and presumably the circuit you are creating is already on the north side of campus. You put the cover back on the manhole and walk up to the north *terminal,* also referred to as a *junction box.*

You open the big green metal terminal box with your special set of keys and find, of all things, telephone wire on blocks. What a surprise. You whip out your *probe* and search for the tone being sent by the toner you left on the circuit pair back in the switch room. Luckily for you, you find the tone with little difficulty—the circuit has been made all the way to the north end of campus. You look for a different set of punch down blocks that connect another outside feeder cable from the terminal to the library. After finding an available pair, you make another cross-connect between the pairs you intend to use and install a set of bridge clips to continue the circuit. You close and lock the terminal box and head over to the library *building distribution frame* (BDF).

You enter the library and then enter the small room, also referred to as a *closet* (because it is usually the size of a closet), which contains the BDF where you find yet another set of blocks. Using your probe again, you find the pair of wire you are using to make the circuit. You then find the blocks that connect to the *riser cable* that goes to the third floor *intermediate distribution frame* (IDF) through *conduit*. You again make the cross-connection and then head to the third floor.

The telephone you will be installing is on the other side of the building and is serviced by an IDF other than the one the riser cable is connected to. You must again use the probe to find the pair being used for the circuit, and you then find an available pair on the blocks that contain the *inside feeder cable* (also contained in conduit) that services the IDF on the other side of the building. You again make the necessary cross-connection and you then head to the IDF on the other side of the building.

After checking to see that the circuit has arrived safely at the final IDF, you go to the room where the telephone is being installed. Fortunately, there is already a telephone jack just sitting there waiting to be used. If no jack were installed and available, you would have to cut a hole in the wall, then, using a *wire snake,* install station wire up the wall and over to the IDF and then install the telephone jack.

The station wire would then be punched down on an available block. The IDF might be labeled with the location of the station wire, or might need to use a toner at the telephone jack to find the station wire pair at the IDF. After making, hopefully, a final cross-connect, you would go to the telephone jack and verify that the circuit has been made.

You get lucky—your coworker is in the switch room just finishing the programming on the switch for the telephone you have just installed, saving you a trip all the way back to the switch room and then back to the room where the phone is being installed. You request that the toner be removed from the circuit and the proper *station line protector* be installed so you can test the line. You then plug in a telephone set that you brought along for the journey to see if everything is OK. Everything goes smoothly, and the phone works on the first try. The telephone has been programmed properly in the switch by your counterpart, and the telephone circuit appears to work properly.

Did you get all that? In our example, all the components needed to provide the service were available in the existing physical plant. You weren't sure that you had everything you needed until you actually went out into the field and surveyed the situation, one step at a time. Many times you find that the circuit can be created, but you must send the circuit in less than optimal directions to take advantage of existing cable facilities. If your organization used a facilities management system (discussed in Chapters 7, 8, and 9), you would have known precisely where you needed to go and what equipment was needed. Even in this example, you would have been able to save several steps by using a facilities management system.

This project could have been extremely difficult and time-consuming if any component of the circuit was missing or you did not have the telephone equipment in stock to fill the order. Planning ahead and knowing where you need to go and what you need to have is a very valuable tool when you are building telephony circuits.

Now let's take a closer look at the different types of twisted-pair cable that might be used by your organization.

Twisted-Pair Cable

A number of types of twisted-pair cables can be used in the telephony environment. There are two major types of twisted-pair cable: standard and Plenum. Standard cable can be used in most telephony situations. Plenum cable is special cable designed specifically for use in a plenum (the space between the floor and the ceiling panels used to circulate air back into the heating and air conditioning systems). Plenum cable is specially insulated to produce low flame and little smoke in the event of a fire. The EIA/TIA 568 Commercial Building Wiring Standard also defines five major categories of cable types that are available in both standard and Plenum versions:

- Category 1: This is traditional, unshielded twisted-pair telephone cable that is primarily used for voice and is not suitable for data transmission. This type of cable is not used in modern telephony.

- Category 2: This is unshielded twisted-pair acceptable for use in data applications up to 4 Mbps. This type of cable has four twisted pairs and is similar to the IBM cabling system type 3.
- Category 3: This is unshielded twisted-pair rated for local area networks such as token ring, 10 Mbps ethernet and 10Base-T topologies. This cable contains four pairs and boasts three twists per foot for better transmission integrity.
- Category 4: This is unshielded twisted-pair rated for up to 16 Mbps token ring topologies. This cable has four pairs of wire and is the most common type of twisted-pair cable used in telephony today.
- Category 5: This is unshielded twisted-pair rated for data transmission at 100 Mbps and used in fast ethernet and ATM applications. The twists in the cable must be maintained all the way up to the jack in order prevent signal crossover and interference problems.

Conclusion

You have been exposed to an incredibly broad set of information that spans decades of use, modifications, and technological advances. The actual cabling, telephone system, and features provided at every site will be as unique as a snowflake. There are thousands of providers out there selling hundreds of different versions of telephone systems and related services. Try to use the information provided in this chapter to round out your overall understanding of telephony. As you read through the other chapters, remember that behind all of the technologies discussed in this book lie the telephone system and the switching network that supports it.

Chapter Four

Telephony and the Internet

What is the Internet? Think of the Internet as the world's largest computer network. It consists of over 720,000 host nodes and tens of thousands of miles of cables connecting over 30 million people. The Internet has grown from a private, small-scale experimental project started by the Department of Defense in the late 1960s to the sprawling global network that can be used today by anyone with the right equipment for about a dollar a day.

An Internet connection creates the potential to amplify your communication channels in the broadest sense for personal enjoyment, professional networking, and business opportunities through simple telephony connections. A writer can question multiple background sources for a story simultaneously, and without requesting a travel voucher. A botanist can download a scanned color image of a new hybrid leaf that won't be published in the professional journals for months with far greater detail than a fax could offer. Researchers from different universities can exchange lab results or collaborate on projects. A manager can retrieve a collection of articles from a diverse set of journals pertinent to a current proposal, providing fresh insights and viewpoints. Consultants and technicians can swap experiences in using new techniques or equipment.

Organizations, as well as individuals, are going beyond internal LANs and WANs to discover the potential of the Internet. By providing services internally as well as externally, companies are realizing innovative ways to do business in the '90s. The proof rests in the results being achieved on a day-to-day basis, such as Arthur Andersen consultants using e-mail to manage projects or Microsoft providing a public site for their customers to download the latest software patch or update for free. These are examples you would expect from high-tech organizations, but did you know that organizations of all kinds are finding ways to lower costs, expand their markets, and differentiate their products and services through the Internet? If you haven't, then you probably have not registered for a seminar, searched for a job, read about pending legislation, queried a medical database, or sent a business report over the Internet lately.

Internet Origins

In 1969, the Advanced Research Projects Agency (ARPA) of the Department of Defense funded a project to create a network that could withstand wartime conditions, such as an enemy attack on a particular computer site. While an installation could be protected from a bomb dropped upon it, the simple wires connecting one site to another could not be protected in a cost-effective manner. Can you imagine the taxpayers' response if miles of cable were encased in lead with a concrete housing? Or the diagnostic and repair bills when troubleshooting was needed!

Instead of stringing a network connection from one computer to the next, then to the next (forming a simple serial connection, which is highly vulnerable) or connecting every computer to every other computer on the network (prohibitively expensive on a national scale), a new approach was needed.

ARPAnet, as this first big network project was called, explored new ways of linking military institutions and university research centers into a seamless, sturdy network. Stanford University, UCLA, UC Santa Barbara, and the University of Utah were selected as the four original ARPAnet nodes.

Decentralization is what made this early ARPAnet special. Instead of the network cables dictating the flow of information, each computer was assigned equal responsibility for receiving, interpreting, and transmitting network information. Also, ARPAnet marked the first implementation of the Internet Protocol (IP).

To understand IP, consider the scenario of sending a letter from Boston to Phoenix. You do not hand it to someone in the post office who then gets in a truck and drives it to the proper address in downtown Phoenix. That kind of direct connection would cost a lot more than 32 cents for delivery, because it is terribly inefficient. A more realistic scenario might be that your letter gets brought to the closest metropolitan post office, which sends your letter along with a bunch of other mail headed for that zip code range to the airport to be flown to a regional post office in Arizona, which then sends your letter to the closest city post office, which routes it to a mail carrier, who delivers the letter to its destination. Many intermediaries would be involved in the process, and a letter going from one point to another might take a different route from time to time. On its next trip, the plane leaving Boston might stop in Chicago to consolidate the Arizona mailbags into another plane headed for the Southwest.

IP works in a similar way, in that it establishes connections from one point to another on a network via a route that is determined by the networked computers. The protocol part of IP is a set of rules for determining how a process should be carried out; in this case, the rules pertain to how information is sent efficiently and dynamically over a network. Messages, divided into small segments for easier handling, are put inside an IP packet, which is addressed and transmitted. Like the post office example, the message packets may make several "hops" from source to destination. The fewer the number of hops, the faster the message travels. Of course, other factors also play a part in the speed of transmission, which will be discussed shortly.

During the 1970s, ARPAnet grew as other universities and nonmilitary researchers were permitted to join. So many sites across the country joined that the original standards and communications protocols could no longer support the traffic—they needed to evolve. A new standard, called Transmission Control Protocol (TCP), was proposed in 1979 as a supplement to IP, and it was adopted in 1983 as TCP/IP.

The IP portion of the joint protocol takes care of routing the packets to their destination. The TCP part performs integrity checks on the data and enhances the reconstruction of the packets into the original message or file at the destination end. In this sense, using Internet communications is more reliable than sending mail through the post office, since packets can't get damaged, lost, or misrouted without the TCP/IP finding out and resending the necessary packets of information.

Today, TCP/IP is a foundation for over 100 different protocols for how data and messages should move among computers on a network. Part of the driving force behind this protocol's development was the desire to allow different kinds of computers to communicate with one another over a common network. If you are on a Sun workstation and want to send a file to a friend who is on a DEC Alpha workstation, you could do so over a TCP/IP network connection. Any computer that can use TCP/IP can communicate over the Internet.

Growth Factors

ARPAnet was still a few years of explosive growth and rebirth away from being called the Internet—a turning point in which TCP/IP played a key role. Many other organizations were sponsoring their own networks in the early 1980s and establishing connections to ARPAnet. CSNET established a network serving the engineering and computer science departments of various academic institutions. BITNET (Because It's Time Network) was created in 1981 to link instructors and researchers at colleges and universities using IBM mainframes. BITNET's European counterpart, EARN (European Academic Research Network), was formed shortly thereafter. Even though these networks were not strictly TCP/IP-based, they set up gateways to ARPAnet so that electronic mail messages could be exchanged. Other non-IP networks, such as the popular commercial services America Online and Compuserve, at one time simply had gateways to allow e-mail exchanges. Now that is changing for both the independent and commercial networks as they begin to offer full-service translators between their networks and what is known today as the Internet.

Industry terminology describes the act of connecting networks as *internetworking*, and uses the term *internetwork* for the network of networks that results. The term *Internet* was coined to refer to this global "network of networks" outgrowth of ARPAnet. *Bridges*, *routers*, and *gateways* are devices for linking networks to one another, as shown in Figure 4-1.

Computers that could be connected directly to the Internet were called host computers, or hosts. Typically, hosts were mainframes or minicomputers, which then had numerous users who connected through VT100-type terminals. In the early 1980s UNIX workstations proliferated and could support the same direct-to-the-Internet type of connections as the mainframe hosts, since the workstations had built-in IP. During this period people were connecting ethernet LANs, which could be comprised of many individual workstations, directly to the ARPAnet. Using *routers*, different kinds of networks could be connected together, so a token ring network could be linked to an ethernet backbone, for instance. This capability led to an even greater acceleration in Internet growth.

In 1983, the Internet was formed when the military operations portion of ARPAnet split into its own network, MILnet, unburdening ARPAnet of some of its traffic load, yet allowing it to continue to serve as a research communications tool and information highway. Computers were assigned IP addresses, and gateways were set up to handle TCP/IP packet forwarding between ARPAnet and MILnet networks. As new networks and individual nodes were added to the Internet, they adhered to TCP/IP network conventions.

In 1987, the National Science Foundation (NSF) began a program to encourage the use of supercomputer resources. NSF funded a national high-speed backbone to connect its five supercomputer centers, thus forming NSFnet. Realizing that their

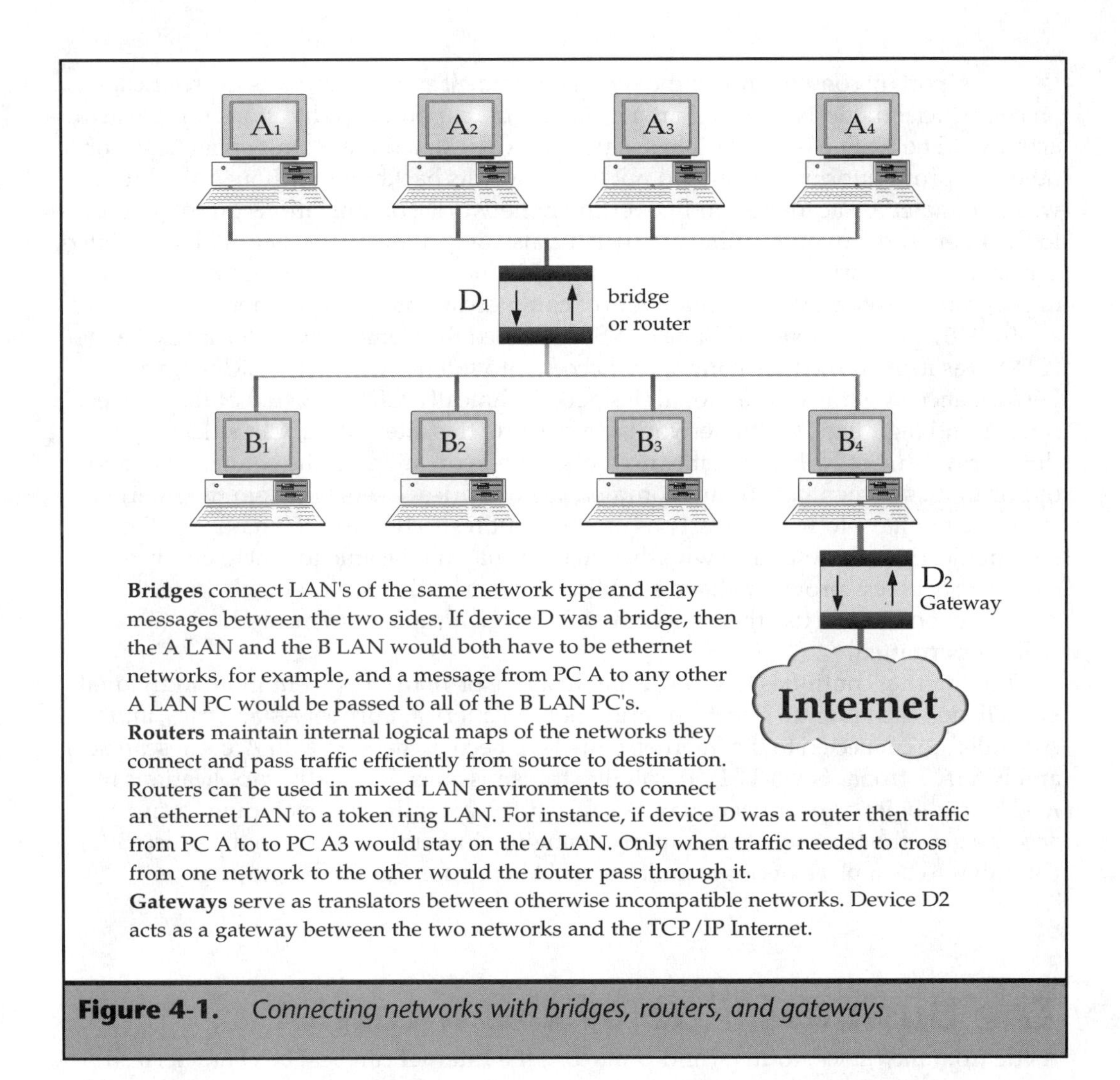

Figure 4-1. *Connecting networks with bridges, routers, and gateways*

purpose and structure were similar, NSFnet and ARPAnet began not only to interconnect but to merge their resources. ARPAnet's mission was declared complete and it ceased to exist as an active project in 1991. In November 1994, due to rapid commercial network expansion over the past three years, NSFnet started directing universities and government sites to find other network carriers, such as MCI, SprintLink, and ANS, to provide Internet services. By April 1995, NSFnet had withdrawn from its Internet role. The sponsorship resources that supported NSFnet are earmarked for future gigabit bandwidth projects, which will be crucial for providing access over the Internet to services such as video conferencing, home shopping, entertainment, and multimedia databases.

An important consequence of the early government sponsorship was the restriction on commercial traffic. NSFnet's charter called for the support of "education and research" activities. The NSFnet's acceptable use policy (AUP) stated that commercial traffic or other for-profit information should not travel over its backbone. Although the Internet was dominated by academic and government networks, organizations offering access to the Internet to anyone willing to pay a fee also began to appear. The AUP prohibited commercial network traffic to travel over the NSFnet backbone, so these users could only communicate with other members of commercial Internet providers.

In 1991, an alternative backbone to NSFnet, called the Commercial Internet Exchange (CIX), was formed by three commercial access providers, AlterNet, CERFnet, and Performance Systems International (PSInet). Technically, CIX consisted of high-speed routers and high-bandwidth networks. In creating this alternative, CIX sidestepped the thorny issue of NSF acceptable use policy while opening up the connectivity and opportunities to members of this commercial enterprise. Users who join the Internet today do not have to worry nearly as much about restrictions on the nature of the communications they send or who they may or may not be able to reach, because commercial access providers have established avenues that have a much wider latitude of acceptable use than the United States government-sponsored policies permitted.

The Internet continues to grow today, at a rate of 10 to 15 percent more individual participants each month. North America, South America, Europe, Asia, Africa, and Australia are connected to the Internet; e-mail has been exchanged with Arctic researchers and NASA astronauts via TCP/IP satellite transmissions. Innovative applications of new technologies appear every day. Because of new standards developing to ensure access across the Internet, it is a rapidly changing domain—but far from perfect. But it continues to be a place of opportunity and challenge for those who participate.

The Business Side of the Internet

Aside from the cables, routers, and protocols, the Internet can best be characterized as a community. Estimates mark worldwide membership in early 1995 at between 20 and 35 million users—more than the population of many cities, and for that matter, countries.

In order to tap into the Internet for your own purposes, you should understand certain practical fundamentals. Since the Internet is composed of thousands of networks, there is no one central authority to contact and ask questions.

To establish connectivity, you will need to find an Internet Service Provider (ISP), hook up your computer to the network, acquire the right set of tools, and develop your understanding of how you can travel safely and confidently on the Information Superhighway (ISH). See Figure 4-2 for an example of a typical individual connection.

Before connecting, consider the different levels of access appropriate to your needs.

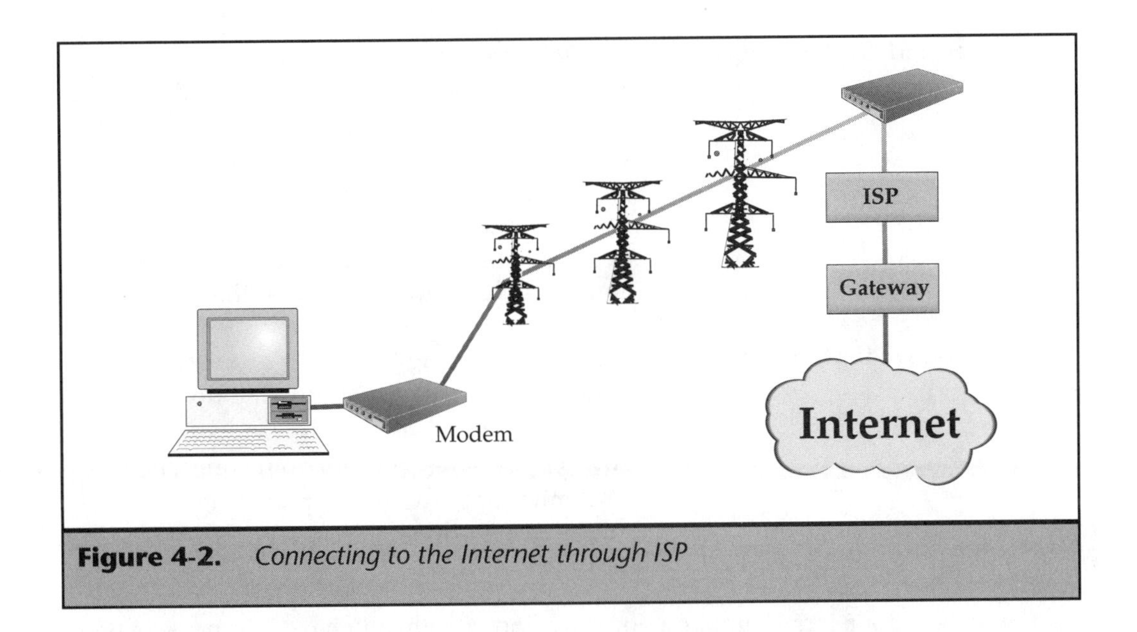

Figure 4-2. *Connecting to the Internet through ISP*

Internet Access Levels

Although anyone can use the Internet, there are different levels of access; the greater the functionality of your connection, the more you will pay for it. Here is an overview of the four levels.

Level 1: Internet host account	Perhaps shell access No graphical clients accessible Limited support, usually Modem speed access: 9,600–28,800 bps Network connection also possible May be restricted by available dial-in ports Can use command line interface for e-mail, Netnews, FTP, WWW
Level 2: Commercial account	Graphical access through a commercial vendor Commercial providers have support available Modem speed access: 9,600–28,800 bps May be restricted by available dial-in ports Can use an implementation of many Internet client services, but has limited abilities for businesses serving their customers

Level 3: SLIP/PPP connection	All TCP-based graphical clients Internet service providers have limited support available Modem speed access: 9,600–28,800 bps Access may be restricted by available dial-in ports Learning curve ahead, but opportunities and control multiply dramatically at this level
Level 4: Dedicated node	Welcome to the big leagues High-speed access: 56,000 bps and up Dedicated line Always a spot reserved for your connection Maximum opportunity for providing interactive services

If you work for a large company, you may find you already have Internet access of some sort. Many times, engineering or research departments connect their mainframe to the Internet and begin using it to explore information resources and to communicate with colleagues. Unfortunately, they do not always do the best job of informing the rest of their organization that this fantastic new resource has been made available. Check with the CIO's office, MIS department, or technical support group for details. Not having any kind of network access is informally referred to as having "Level 0 access."

Level 1 Access

Level 1 access is characterized by a command line interface through either a PC-based telecommunications package or a host terminal. Sometimes users of these systems have access to a shell, which means that you can manipulate the account environment and store files. Internet services such as e-mail, telnet, gopher, FTP, and World Wide Web can be used, but in limited ways and without the GUI interfaces which most PC users expect. These uses of the Internet will be covered in greater detail in the section "Internet Client Software." Sometimes when accessing this type of account through a dial-up connection to a host, you will only have an account that lets you download mail. Level 1 access has very limited use for most business applications of the Internet beyond e-mail.

Level 2 Access

Level 2 access is the most popular type of Internet connectivity known to the general public. Currently the "Big 3" commercial service providers (America Online, Compuserve, and Prodigy) offer comparable Internet services. For $10 to $15 per month, they give you the chance to explore their databases, download shareware files, and "chat" in "rooms" designed for meeting people and exchanging messages in real time through your keyboard. Compuserve has the most extensive information resources online, ranging from detailed guides and reviews of major city restaurants to beta test forums for Windows 95 users—but not all of these forums are available to all users, and when you enter some of these special areas, you begin to incur a higher hourly charge rate. These vendors provide you with proprietary software that has a graphical interface. Each time you enter an area of America Online that has been changed since your last visit, new resources, such as icons, are downloaded automatically to your PC-based program. A nice feature for staying up to date, but while the downloading takes place you are unable to continue on to other areas and you are paying for every second of connection time.

Another advantage of this level of access is that these commercial providers have enormous resources behind them to run customer service and technical support departments, ensure system maintenance and reliability, as well as to research and develop new services. However, these service improvements are designed to appeal to *their* customers, not yours, as illustrated in Figure 4-3. Prodigy's rates are subsidized by commercial advertisers whose slow-loading graphic messages appear on many

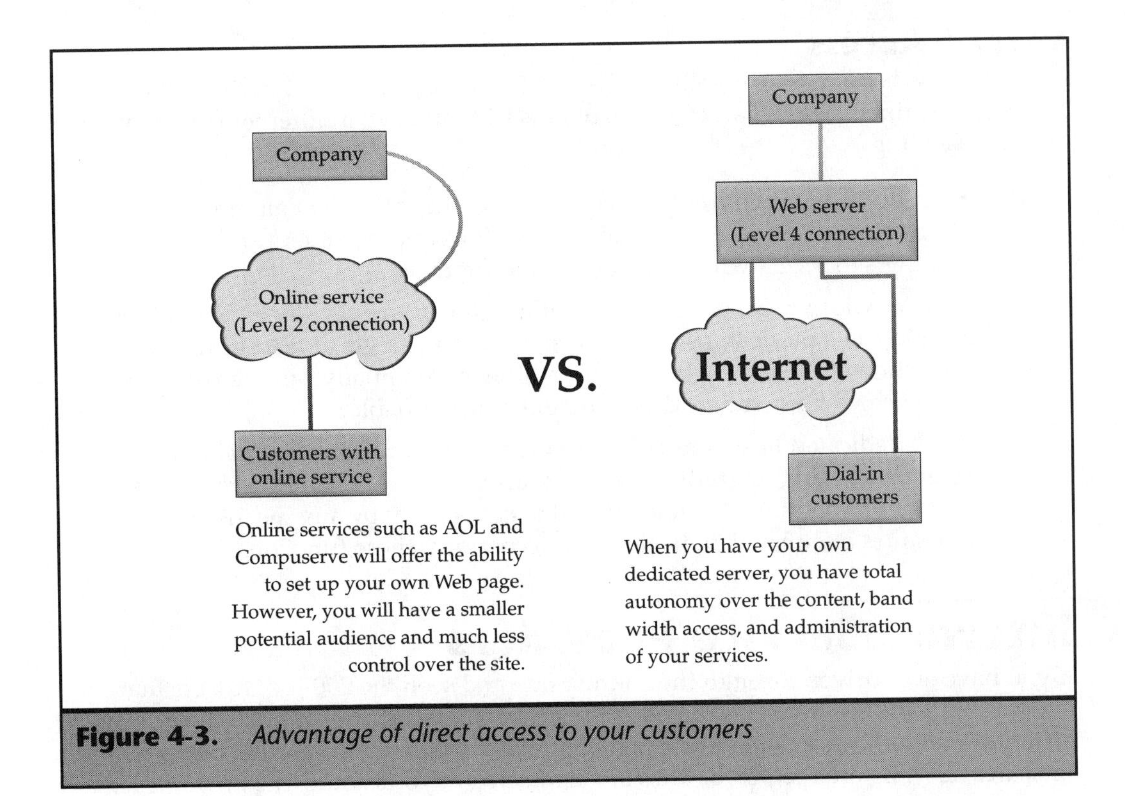

Figure 4-3. *Advantage of direct access to your customers*

screens like miniature billboards. While the added support and friendlier software allow you to learn about online resources in a structured fashion, the desire for more control over the online environment is what drives many Level 2 users to higher levels over time.

Level 3 Access

Level 3 access is where your computer becomes part of the Internet, a peer node with TCP/IP connectivity and an assigned IP address. If you are connecting via a Level 3 dial-up connection from a Mac or Windows PC, you will need to understand the concepts covered in the"PPP and SLIP Connections" section later on. When using commercial Level 2 services, you are limited to using the software provided by the company. On the one hand, at Level 2 you have just one piece of integrated software to learn and use for your online needs; on the other hand, if you want different features for some reason, you are stuck. With a Level 3 TCP/IP connection, you can run whatever software you choose to exchange messages and files, browse and search for information, and make connections with other Internet users. For a small setup fee and then about $50 a month, a home or small business can become a node on the Internet through a dial-up connection to an Internet Service Provider (ISP). See the "Internet Service Providers" section, next.

Level 4 Access

Level 4 access has similar capabilities to Level 3 in that you create a node on the Internet. The three major characteristics of Level 4 that make it different from Level 3 are the following:

- With Level 4, your computer is connected to the Internet at all times. You can set up information resources and services for your customers and potential customers to access when it is convenient for them.
- The bandwidth is generally better. Although you can get a dedicated phone line and run a modem, usually an organization will get at least a 56 Kbps connection with Level 4. It may cost a little more initially, but the performance difference will become readily apparent and justifiable.
- With a dedicated host, you will have more responsibilities for maintaining the system, including upgrading software, applying security measures, archiving, and troubleshooting. It is not a trivial responsibility by any means, and companies often have dedicated staff to perform these functions.

Internet Service Providers

If you have ever driven through the countryside and seen the 200-foot towers that support high voltage transmission used by utility companies to conduct power from station to station, you will readily understand the purpose of an Internet Service Provider (ISP). Before the power lines are connected to your home, the power is fed through a series of step-down transformers, so that when you plug a toaster into the socket, it is connecting to 110 volts. You could not use, you do not need, and you would not want to pay for the higher level of power. In a similar way, an ISP provides the level of bandwidth that your personal or professional needs require.

Bandwidth refers to the speed with which data travels, measured in bits per second (bps). Over a network connection, bandwidth makes an enormous difference in how much data you can send and receive in a given period of time, as Figure 4-4 shows. Bandwidth becomes increasingly significant as you learn to exchange not just text messages, but more space-demanding types of data, such as graphics, audio, and video. While a brief e-mail message takes up only 2.5K, a typical scanned picture occupies 300K, audio takes up about 475K per minute, and one minute of video can require upwards of 1,800K.

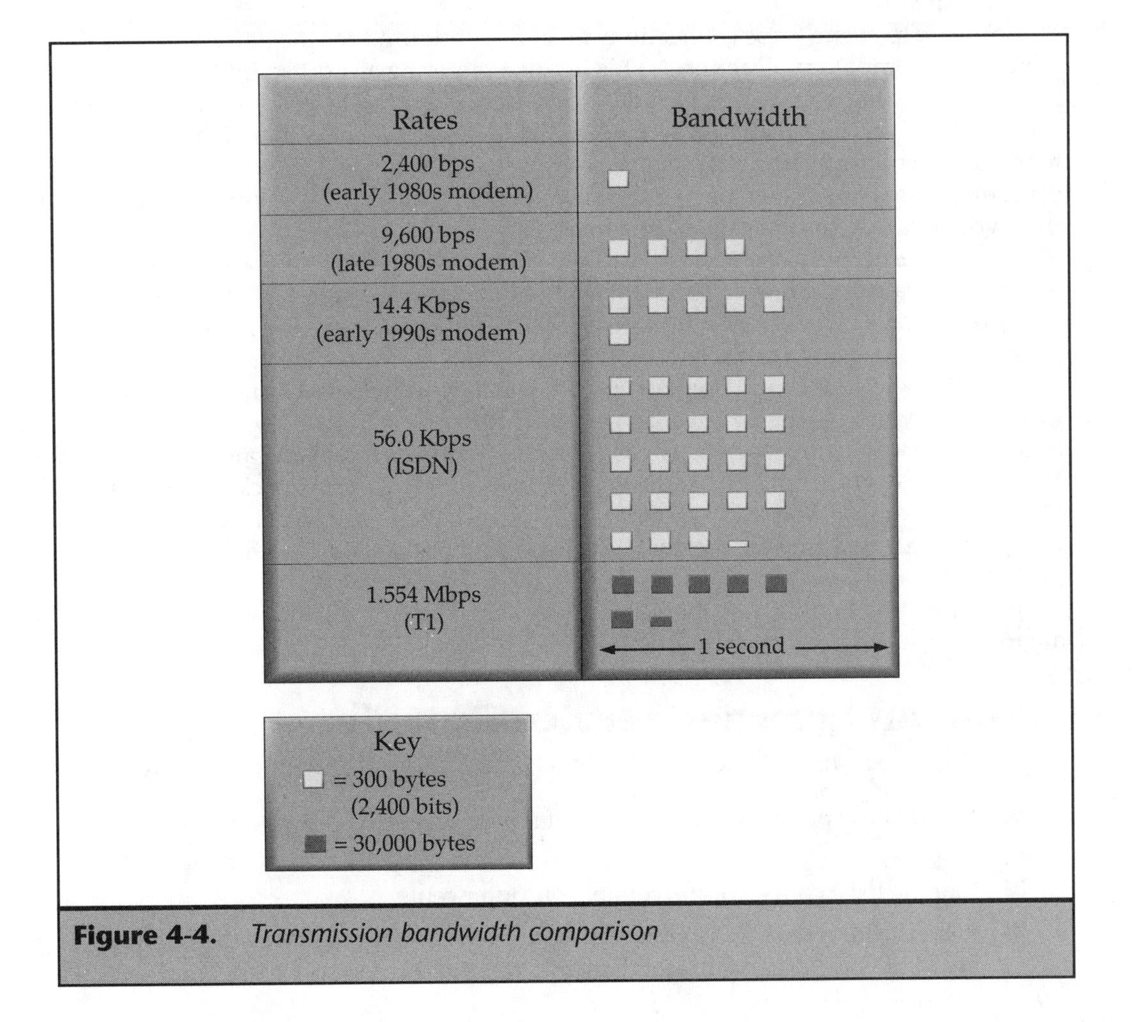

Figure 4-4. *Transmission bandwidth comparison*

The goal of Level 3 access is to establish TCP/IP connectivity on the Internet through either of two methods: SLIP or PPP. It's important to understand how this portion of the connection works.

PPP and SLIP Connections

Point-to-Point Protocol (PPP) and Serial Line Internet Protocol (SLIP) are communications standards that allow a computer connected via a modem to act as a node on the Internet network. In addition to supporting TCP/IP, a modem connection from your PC to the Internet must support the underlying protocol for communicating over a serial line, such as is found with plain old telephone service (POTS). The serial line protocol is responsible for error control, packetizing the data, and other activities that allow TCP/IP programs to work together. PPP is the newer protocol, and it is considered to be more robust than SLIP.

Software to create these connections must be configured for the particular host to which your computer will connect. This involves entering a short script to interface with the host computer, which includes your user ID and password for an account on the host. Once this information is verified by the host, you will have a modem connection to the Internet through the host that appears just as permanent a connection as if your computer were physically a part of the network, even though it is set up on demand from your desktop through the phone lines.

Many Internet Service Providers can provide you with the appropriate public domain software or shareware for establishing a SLIP or PPP connection. Alternatively, many bookstores carry Internet bundle titles, i.e., packages that contain both a reference book and a disk of related software.

Once you get your connection established, you may want to check an FTP archive to make sure you have the latest version of your SLIP/PPP software, as explained in the "Internet Client Software" section. For Macintosh computers, the shareware programs MacSLIP and MacPPP fulfill this purpose. Windows users can install packages such as PC/TCP or Trumpet Winsock to meet this need. Check with your system administrator to find out what packages are recommended and supported before making any purchase decisions.

Now that you know more about the protocol and what it enables you to do, you are ready to start researching Internet Service Providers.

Selecting an Internet Service Provider

To get a list of local Internet Service Providers, check these common sense sources first:

- If you know people who have a similar service, ask who is providing their service.
- Call the local college or university computer center.
- Read the local user group newsletter, the college newspaper, or the bulletin board at your local computer dealer. If you have access to e-mail, Peter

Kaminski maintains a file called the Public Dial-Up Internet Access List, which contains primarily phone dial-up ISPs listed by area code. Send mail to Peter's server at info-deli-server@pdial.com and put the words "send pdial" (minus the quotation marks) in the subject line of the message.

In addition, below are several key questions to ask when comparison shopping for an ISP.

DOES THE ISP SUPPORT THE TYPE OF COMPUTER(S) THAT WILL BE CONNECTED? Whether you are connecting a UNIX, Windows, Mac, or other platform, you should make sure that your provider has experience with your type of computer so you can get good support. It is a good sign if they have a computer platform of your type that they use themselves at their location or that they at least have available for reproducing tricky problems.

DOES THE PROVIDER HAVE MULTIPLE POINTS OF PRESENCE? A *point of presence* is an entry point for users to dial in for service. To avoid paying toll charges to the phone company for your line connection (in addition to the service charge from your ISP) you will want to check whether you are making a local or toll call to your ISP at whatever point of presence you may need. Having multiple points also indicates that the provider has a regional presence and is a larger operation, which has greater resources to draw upon and also greater stability as a business. Find out whether they have been established for more than a couple of years and how many subscribers they now service. If you travel and need to dial in from different cities, you should investigate national ISPs, such as NetCom at (408) 554-8649, PSInet at (800) 827-7482, or AlterNet at (800) 488-6383.

DOES THE PROVIDER PLACE ANY RESTRICTIONS ON BUSINESS USES OF YOUR ACCOUNT? Check their acceptable use policy in advance, and make sure the ISP is a member of the Commercial Internet Exchange (CIX). You can request a list of CIX members via e-mail at info@cix.org, and a standard information sheet will be sent automatically. Providers offering service to the public for a fee should understand that their users will have different needs for both personal and business matters. It is important to ask, though, because you would not want to invest your time setting up a system and find you have to move due to a policy violation.

WHAT IS THE ISP'S MAINTENANCE SCHEDULE? HOW OFTEN DO THEY BACK UP THEIR SYSTEM? If you are using this system to exchange business correspondence, you should be sure that daily backups are the norm. Recently, a system administrator's error wiped out over 150 pieces of mail to Star Communications Group one morning. Luckily, the messages were restored from a system backup that was only a few hours old. When disaster strikes, your system is only as good as its latest archive. Make sure your ISP is at least as vigilant in backing up data at their end as you are about safeguarding data at your end.

HOW WILL THE ISP RESPOND TO SERVICE PROBLEMS? DO THEY HAVE A MEANS OF ESCALATION FOR TECHNICAL SUPPORT QUESTIONS? Do they have a direct number to call, or must you work your way through a phone tree? Are support technician phones staffed beyond 5:00 P.M.? Many telecommuters who work from home need support during evening and weekend hours. Do they respond to questions posed via e-mail or newsgroups as well? Is service guaranteed in writing, or are their support capabilities less structured and predictable than a written policy would indicate?

When you have a problem to report, ask if there is a help desk tracking number system. The larger and better-organized support departments assign unresolved questions a log number and keep statistics on how quickly problems get fixed. When the frontline staff gets stumped, they have resources they can draw upon to get an answer. Often, an ISP tech support person will use the Internet to research questions. Sometimes, an ISP will maintain good relationships with outside consultants (who are often subscribers) to help answer questions. Find out what kinds of resources your potential ISP has available.

WHAT WILL THE START-UP AND MONTHLY COSTS BE FOR USING YOUR INTERNET CONNECTION? ARE THERE ANY ADDITIONAL COSTS OR DISCOUNT PROGRAMS? You can expect a start-up fee for setting up and activating your account, then a monthly maintenance fee. Businesses should be particularly interested in applying for a custom domain name—your mailing address on the Internet (for more information, see the next section, "Internet Domain Names." While some network consultants scoff at the idea as little more than vanity license plates for a business, you should understand that your domain name creates an impression about your organization. It is becoming common practice to include an e-mail address on business cards, and it reinforces your appearance as accessible and established to have your own domain name. For a nominal fee ($50 to $100), your Internet Service Provider will discuss an appropriate domain name, fill out the forms, and make the application for you.

Other services that may have fees associated with them are setting up mailing lists, making files available for downloading by anyone at any time, and setting up and maintaining World Wide Web pages. These services (and why they matter to you) will be described in greater detail in the "World Wide Web" section. Again, the best idea is to call a few ISPs and comparison shop to determine the going rates. Ask if they offer discounts for paying an annual fee instead of a monthly fee.

CAN YOUR ISP MEET YOUR FUTURE NEEDS? As your familiarity with the Internet grows, so will your use of it. Before you are ready to upgrade from dial-up to a dedicated connection, find out what options the ISP can offer. Can they assist you in setting up services such as FTP and World Wide Web servers? Will they provide wider bandwidth dedicated lines at a competitive cost? Ask around at your local user group or chamber of commerce for others who have upgraded their capabilities through a particular ISP, and find out their experiences.

Internet Domain Names

One of the first service matters you will conduct with your ISP is to get an Internet address so others can contact you. A fully qualified address includes the user ID and a complete domain description. For example, the address president@whitehouse.gov is a fully qualified address. The user ID, the part to the left of the "at" symbol (@), describes the account owner; the part of the address after the "at" symbol is the domain name.

Whether you opt for dial-up or dedicated line access, it is worthwhile to apply for a custom domain name for your business. Having your own domain name will make it easier for customers, contacts, suppliers, and others who want to reach you to remember your address and associate you with your organization. Individuals, on the other hand, may not wish to go through the trouble to apply for their own domain name, in which case they will be able to use the ISP's domain name. Consider the differences between the following two possible addresses for the same person (e.g., Dan Rather). On a custom system, dan_rather@news.cbs.com could be his address; on a system in which he used the ISP's domain, he might be cbs_rather@netcom.com.

A domain name has multiple parts, separated by a period. The rightmost portion of the name is used to indicate the highest level of the domain, known as the top-level. Working backwards from that portion of the name, you get more and more specific information about the host. Take the address "dunx1.OCS.drexel.edu," for instance. Reading from the right, the .edu indicates an educational institution, Drexel University, within the department of OCS, and the specific machine name is dunx1, a Sun 670 UNIX host. The following table lists the most common top-level domains and gives a description and example of each one.

Domain	Description	Example
.com	Commercial organizations (business use)	starcomm.com
.edu	Educational organizations (colleges, universities, some K-12)	dunx1.ocs.drexel.edu
.gov	Government sites	whitehouse.gov
.mil	U.S. military	army.mil
.net	Network resources	rs.internic.net
.org	Organizations (usually nonprofit)	npr.org

Also, country codes are being added to the list of top-level domains. There are over 300 country codes, and over half of those countries have Internet connectivity at present. Recent changes in the way domain names are assigned opens up "geography" domains as opposed to "category" domains. For example, the domain name for a host at the School District of Philadelphia, sdp2.philsch.k12.pa.us, has a country code as its top-level domain. If it had been assigned a couple of years ago, it most likely would have had an .edu suffix instead of the country code.

Associated with each domain name is a numeric address, known as the IP address. *IP addresses* are 32-bit numbers divided into four parts, sometimes called octets. Each octet represented as a decimal number can range from eight binary 0's to eight binary 1's—in other words, from 0 to 256. Sometimes these IP addresses are called "dot speak," since you pronounce the period as "dot." For instance, the IP address 128.182.62.99 would be read "128 dot 182 dot 62 dot 99."

By now, you may be wondering how these numbers and domains are assigned, especially given the unprecedented demand from new users. Though no central authority runs the Internet, various committees and sponsored organizations manage various aspects. Among those organizations is the Internet Network Information Center, or InterNIC for short. The InterNIC, started in 1993, is a five-year project funded by the National Science Foundation to provide network information services to the Internet community. The InterNIC is one of dozens of working groups and research groups guided by the largely volunteer Internet Activities Board.

Previously, the names and IP addresses of machines could be kept in a single file maintained by the InterNIC, and that file was distributed to every host on the network. But as the size of the network grew, so too did the size of the file, not to mention the time to process domain name registrations. Now, the process is more decentralized. Instead of assigning individual names, the InterNIC assigns an Internet Service Provider a block of IP addresses, from which they assign a number to a unique domain name. These blocks of addresses are given out in three sizes. A Class C block is the smallest and most common. It would look similar to 192.3.44.*, and would allow a network administrator to assign up to 254 IP addresses. A small business or school district might be interested in obtaining one or more Class C addresses. A large organization or a regional ISP would be better off with a Class B address, which would give you the 128.3.*.* number space of 64,516 nodes. Very rarely are Class A blocks assigned, since they take the form of 0.*.*.* and hold the potential for over 16 million IP addresses. Class A addresses start with a number between 0 and 127, Class B addresses start with a number between 128 and 191, and Class C addresses start with a number between 192 and 255.

Generally, human beings are more comfortable with the domain names than the IP numbers. Computers, however, translate the name into an IP address so they can process it faster. This translation process takes place automatically when you send mail or request a service of a host computer via the domain name system.

It is important to realize that sometimes you will request an Internet service such as a Web page or FTP directly and not be able to use it. Several reasons may account for this, including the following:

- The host providing the service is offline for maintenance or repair.
- Too many other people are using the service at this time, and you will have to wait for an open port.
- A network connection is having problems.

- Your specific IP number or domain name has been declined access because the administrators of the host no longer wish to offer public service due to resource abuse or excessive traffic loads.

Before you judge this capricious nature of Internet services to be unfair or poor customer service, consider the fact that the service is provided voluntarily in the first place. From a service provider's viewpoint, you might want or need to shut down a machine for system updates or to restrict users for security reasons. Or, if a service has become so popular that it begins to impact system performance, then it becomes necessary to restrict outside access to allow the organization's internal users to have priority access to information and resources.

Connecting to the Internet

The following is what you will need to connect to the Internet:

- Personal computer with a modem connection
- Phone jack
- Communications software

One of the myths surrounding Internet connectivity is that you need the latest and most powerful CPU to take full advantage of these services. Almost any IBM-compatible computer that is capable of running at least DOS 6.0 or Windows 3.1 can be used to connect; similarly, any Macintosh that can run MacOS 7.1 meets the minimum qualification for driving on the Information Superhighway (ISH). In other words, if your computer is running a current operating system and is meeting your other day-to-day needs, you should not have to upgrade your CPU just to get on the ISH. When cruising the ISH, CPU speed is usually not the limiting factor in your travels—modem speed is much more critical. The two areas you should consider first investing in are hard drive space and a fast modem. A modem is a device that translates the digital signals produced by your computer to analog signals, which can be handled by telephone lines. See Chapter 15 for more on modems.

Because of the vast number of files and programs available on the Internet, it makes sense that you have a sufficiently large hard drive to store downloaded information. The process by which a file is transferred from a host computer to your local computer is called *downloading*; *uploading* means sending a file from your local computer to the host. Although 100MB may seem like a large amount of free space, it really is not, once you consider the size of the audio, graphic, and video files that are available for downloading.

A phone jack is necessary to connect to a dial-up or dedicated line network. The jack should be a standard RJ-11, and with standard twisted-pair wiring you can exchange information at reasonable speeds. For speeds higher than 14.4 Kbps, you should investigate whether your local phone company supports Integrated Service Digital Network (ISDN), which still runs on standard copper wiring, but up to five times faster than a 14.4 Kbps modem, and with greater tolerance for line noise. ISDN comes in different packages. The Basic Rate Interface (BRI) is known as 2B+D capability, because it offers two B channels ("bearer" channels), which can handle 56 Kbps data flow, and one D channel for negotiating the connections. Primary Rate Interface (PRI) is the next step up in ISDN packaging, and it offers 23B+D capability. Using a technique known as *bonding,* two B channels can be combined into one larger channel; a BRI can be bonded to offer 128 Kbps capability, for instance.

Though it costs more than a standard phone line and requires special equipment to interface to desktop computers, ISDN is becoming increasingly popular in metropolitan areas as it becomes more cost-effective. ISDN handles voice, data, studio-quality sound, and still and moving images at high speeds as illustrated previously in Figure 4-4. Up to eight devices (any combination of phones, fax machines, and computers) can be connected to a BRI line as well. For more information, check out Dan Kegal's ISDN Web page at http://alumni.caltech.edu:80/~dank/isdn/.

Larger organizations should consider dedicated lines such as a T1 line (1.54 Mbps), or perhaps a T3 line (45 Mbps). If you are looking to upgrade an existing connection, consider how fast the line must be, how frequently it will be used, and the capacity it will be required to handle in terms of the quantity of users and the type of data being transported.

You will need two kinds of software when using the Internet over a dial-in, Level 3 connection. First, you will need communications software to establish a SLIP or PPP connection. Once that connection is established, your machine will be recognized as a node on the Internet, and you will be able to run TCP/IP-based software for the particular Internet services you wish to use. Level 4 users will have a TCP/IP connection established already and will be more concerned with providing client software and setting up Internet servers.

Internet Client Software

Once you have Internet access, you will want to become familiar with the tools that will enable you to make connections with people and resources.

During your explorations, you will encounter the terms "client" and "server" used individually as well as together. *Client-server* describes the way a software tool is used. The client part of the software makes requests of the server side of the software. The server side performs the requested action, then returns the results of that action back to the client part.

When you use client-server tools, think of the two ends of a screwdriver. The handle is the front end you interface with as a user; the tip, which does the actual work, is the back end of the tool. In client-server terminology, the front end is the client, and the back end is the server. For instance, when using e-mail you can command the client software on your computer to connect to the host machine and issue a request to the mail server to check for your mail. The mail server would then send the client any mail stored in the electronic mailbox.

The client software is generally responsible for providing the user with the interface, the connection, and the translation. Client software is responsible for opening and maintaining a connection to the server, translating your commands into instructions suitable for the server, and interpreting the results of the server's work back into a useful format. Many current Internet programs employ a graphical user interface (GUI), complete with icons, pull-down menus, toolbars, scrolling windows, and mouse control. These functions are handled by the client application as they would be on any other applications you run on your Mac or PC.

The following table provides an overview of the seven most common Internet services and a brief description of how they can serve your needs.

Service	Description
Electronic Mail (e-mail)	Exchange messages with individuals/groups
Listserv	Publish participatory text magazines; use any e-mail program to join and participate in listservs
Netnews	Electronic bulletin board; over 8,000 interest groups
Gopher	Access text in a hierarchical structure; can be accessed via WWW clients, such as Netscape or Mosaic
Telnet	Remote login to other host computers; useful for remote computer administration of UNIX systems
FTP	Retrieve and send files across different platforms
World Wide Web (WWW)	Hypertext/hypermedia access to a variety of resources

Though these services all have their place, a contemporary perspective encourages us to explore four of the services in-depth due to their popularity and usefulness: e-mail, Netnews, FTP, and WWW.

Electronic Mail (e-mail)

By far, the application most widely used by Internet users is electronic mail, or e-mail. In the early days of electronic correspondence, e-mail served researchers who kept each other apprised of project development, asked each other technical questions, and logged requests for more equipment and funding.

To understand the versatility of e-mail, consider the following capabilities available to users of many e-mail programs presented by functional groups.

Options available when sending mail include:

- **Send to an individual** One person sending a message to another. The content of the message can be anything at all. The size of the message ranges from a few words to a few pages of typed text. Messages can also have files attached, the same way you can ship a package along with a cover letter. If the size of the attached file exceeds 300K, people will tend to favor FTP as a more efficient transfer method.
- **Send to a group** Here is one of the most efficient ways to make announcements either within an organization or to a relatively small group. Without the time or additional cost of photocopying and addressing envelopes, e-mail can distribute a message to a list of people with approximately the same effort as sending to a single person. You can create a list of individuals and their full e-mail addresses and then create a nickname, such as Service_Team. Whenever you send mail to that nickname, all members of the list will receive your message.
- **Send to a machine** Sometimes you will want to participate in an electronic journal or magazine. This is an ongoing discussion to which people can subscribe and can even contribute articles and opinions, and ask questions of other readers. Rather than distribute the large and ever-changing mailing list to each participant, the address of the list server, or listserv, is distributed. In effect, people are sending mail to a computer program, and that program is making sure that everyone who has requested to be a part of the list is forwarded the appropriate mail.

Here are options generally available to users receiving a piece of mail.

- **Read** View the contents. Sometimes there are limitations on the size of an individual message. In this case, the message is received and divided up into segments of the appropriate size. This limitation is usually a function of a gateway restriction when sending from the Internet to another network or of the actual mail program on your personal computer.
- **Discard** You are finished with this piece of mail and no longer need to have it taking up space on your host account or hard drive.
- **Forward** This mail message contains important information that needs to be shared with others. It could be feedback on a project by an important customer that other members of your team should be made aware of or simply an interesting news item that your online friends would appreciate. More sophisticated forward options include resend, which allows you to take mail that you have already sent, make a few editing changes to the content or recipient list, and then send it; and redirect, which allows you to forward a message you

have received and have it look as if it came from you, in case anonymity has been requested for some reason.

- **Archive** Many mail systems make the distinction between an overview of mail received and subdirectories where mail can be filed for future reference. You can archive status reports on a project, log customer support bug reports or feature requests, or add to a collection of amusing political quips.

E-mail can send more than text messages. One problem that developed as people started using GUI workstations and PCs is that they wanted to send diagrams, photographs, sounds, and movie clips to each other. A second problem that emerged from international connections is that ASCII is not equipped to handle many foreign character sets. MIME, or Multipurpose Internet Mail Extension, addresses both of these problems. MIME serves two main purposes:

- It provides a means for mail applications to communicate what type of data is in the mail.
- It provides a method for encoding non-ASCII data being sent through Internet mail systems.

The MIME standard, developed in 1992, supports text, binary files, graphic formats, PostScript, video, voice-messages, and other message types.

You can find specialty client-server packages, such as Lotus Notes for Windows and Macintosh, or Slate for UNIX systems, that specialize in the exchange of multimedia e-mail, if your needs require this capability on a regular basis. However, so long as your e-mail client software (as well as other software that might encounter multimedia data, such as World Wide Web clients) is MIME-compliant, your system should be able to handle these special cases.

When the post office sends mail, the mail can get lost, damaged, or be delivered late. Not only is e-mail cheaper and faster than using regular mail ("snail mail") but it is also more reliable in many ways:

- Host systems, for instance, will verify that the domain name of the addressee on a piece of e-mail can be looked up in the domain name system before sending it.
- If you have accidentally added a space or mistyped a character in the domain name, your mail will come back with a message telling you so right away.
- If certain links along the way to your recipient's host are down, the path will be rerouted.
- If the destination host machine is unavailable, your host will probably queue the mail and try to resend it for a period of days before giving up and letting you know it encountered a problem.
- If the recipient does not exist on the host machine, that host will return your message with an explanation.

Netnews

Netnews refers to the online conference discussions that are transmitted across the Internet. The term *Usenet* refers not only to the discussions but also to the special networks and protocols designed for this activity. Over 8,000 discussion groups on almost every topic imaginable can be found in Netnews groups. Participants can post questions, information, opinions, and even binary files for downloading. Where else can biophysicists find colleagues so easily? Fans of every U.S. sports team can find their buddies. Different groups serve K-12, community college, and university educators. Using Netnews, you can discuss the latest business books, and you can even look through the help wanted ads posted by employers looking for both full-time and contract workers. Browsing news groups can be a real education, given the variety of topics.

News groups have been around for a long time and have taken on many different formats, from text-only DEC Notes to fully-GUI Lotus Notes. The majority of news groups are text-based. In general, news groups are meant to be participatory and open. However, universities and larger organizations often take advantage of the Netnews format and create restricted news groups that can only be read and posted to by their own members. Using secure wide area networks (WANs), organizations can develop valuable records while restricting access to project information from competitors. When a router is configured to act as a security device and filter out certain types of network traffic, it is called a *firewall* (see Figure 4-5). Organizations can allow their employees to read and post information to the Internet-wide discussion groups and prevent outsiders from viewing any proprietary information or accessing system security files, such as passwords, kept inside the firewall.

Netnews groups are structured in a hierarchy of categories to help readers identify relevant information. There are major categories, such as comp (computer hardware, software, and protocol discussions), news (topics related to Usenet announcements), and sci (chemistry, physics, mathematics, etc.). There are also the controversial but popular categories such as the alt (alternative discussions about lifestyles, hobbies, and philosophies).

News groups rely on NNTP (NetNews Transfer Protocol), which is usually supported by your TCP/IP connection. The way news groups work is that when a message gets posted, it is passed from one NNTP server to another, making messages available worldwide in a very short period of time—from a matter of minutes to several hours. When you check a news group from your workstation, the client software requests a list of the most current messages posted to the group. You can then browse or post to a potential audience of 30 million people.

With so many topics, it becomes a nontrivial matter for system administrators to maintain news groups. Accepting all Netnews groups, called a full feed, can require over 10MB per day of bandwidth and storage. And as one system administrator confided, "Nothing generates more complaints from users than having one of the news feeds go down." People get very attached to this interactive pipeline of information.

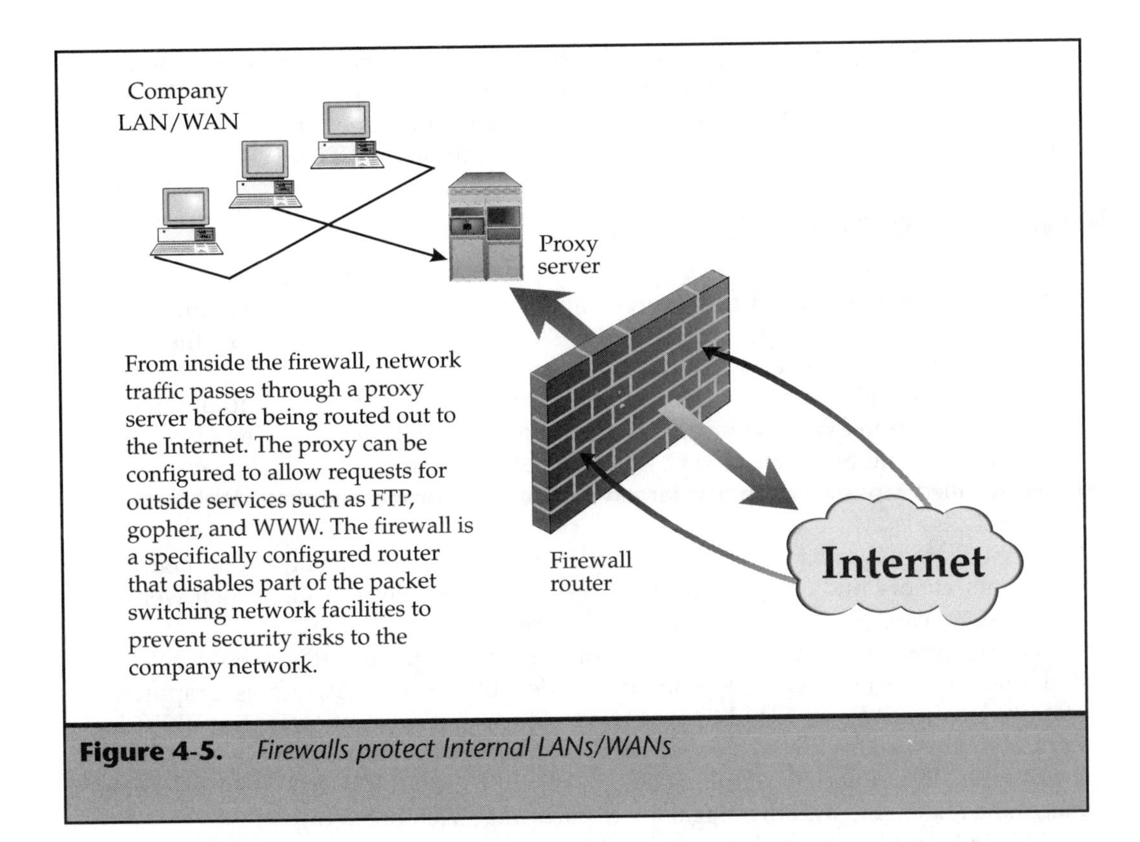

Figure 4-5. *Firewalls protect Internal LANs/WANs*

Check with your ISP to find out if it provides a netnews feed, since some providers do not; if so, find out if it is a full feed. If not, check out what groups are not available, so you can determine if they are critical to your needs.

FTP

FTP stands for file transfer protocol. The abbreviation stands for both the network protocol as well as the service it provides. With FTP, you can search for and download public domain, shareware, and private files from all over the world. The ability to transfer both text and binary files across different computer platforms and networks makes FTP a significant addition to your Internet toolkit.

To access an FTP site, you enter a full domain name. MIT sponsors an FTP site that has archives of Netnews discussions as well as current FAQs on a wide spectrum of topics. A FAQ is a document of frequently asked questions (and answers) maintained by a volunteer who keeps the information up to date; enter the address ftp.rtfm.mit.edu to access this FAQ site. As with many public sites, this one allows *anonymous login,* which means that anyone can have access to this server as long as there is a free port to the host available. FTP servers can restrict access by requiring a nonpublic user ID and password. As a courtesy to the system administrator, enter "anonymous" as the user ID and your full e-mail address as the password on all public FTP servers.

World Wide Web (WWW)

The World Wide Web (WWW) was created by physicists who were looking for a better way to cross-reference scientific research papers at the Swiss European Laboratory for Particle Physics (also known as CERN) in 1993. Since that time, it's become the fastest-growing Internet service.

Web users crisscross information resources by following hyperlinks to different information resources and services. The most popular Web client is called Netscape, which has versions available for PC, Mac, and UNIX platforms. An earlier Web browser called Mosaic, which was largely responsible for increasing the Web's popularity, is also widely used.

Organizations are using the Web for innovative ways of reaching out to their customers, clients, and constituents. Here are a few examples as well as their web page addresses, in case you want to view them for yourself.

Wired magazine (http://www.wired.com) has a Web page that is as bold, colorful, and interactive as its monthly journal. It includes full-text articles, quotes, graphics, and photos. Instead of the lead time of two to four months common in the publishing industry, *HotWired* (as the Web page is called) updates some sections of its content every week. By using this complementary strategy, it uses the Web page to encourage magazine readership and the magazine to encourage Web browsing.

Silicon Graphics (http://www.sgi.com), a California manufacturer of graphical workstations and software, recently held a contest using the Web, asking for new and innovative ways to use Silicon Graphics products. Over the course of several weeks, the company received over 5,000 entries from 43 countries. Not only did the company get terrific exposure using this technique, but it received over 5,000 ideas on how to further develop and explain the products, not to mention over 5,000 self-qualified sales leads who gave their e-mail address to be notified about the contest prize. Not bad results for a self-running Web page that was set up for less than the cost of a four-color glossy brochure!

Legislative information and full-text descriptions of bills are available on the Web through the Library of Congress (http://thomas.loc.gov). Legislation can be searched by keywords and viewed in full. Also, both the Senate and the House have gopher servers that are accessible through this Web page. While it makes some information available about the government, it goes one step beyond accessibility to interactivity. Congressional members who have expressed an interest in receiving e-mail from their constituents have listed their e-mail addresses in a document accessible with the single click of a mouse.

By entering a Web address, known as a Uniform Resource Locator (URL), a Web client can provide hypertext links to text, graphics, sounds, and movies. URLs are given in a format that lists the service first and then the host name and path. Services include http (for WWW), FTP, mailto (for e-mail), and gopher. Because the transition is so subtle from one service to another in many cases, most users do not realize that Web documents can direct a Web browser to act as an e-mail client, browse gopher hierarchies, or download a file using FTP, as well as read and post to news groups. Beyond that, Web pages offer the ability to collect information from participating users through the use of forms.

Creating basic Web pages can be done on any platform using a simple text editor. The lingua franca of WWW is called HTML (hypertext markup language), and it is characterized by simple *tags* bracketing text and giving it special properties when read by a WWW browser. For instance, to the "source" text might look like <B> Read This </B>; when read by a Web browser, it might look like this: **Read This**. Of course, the syntax for creating these tags can be generated by various Web editors. Many editors for both the Mac and Windows PCs exist. For an up-to-date list, look on http://www.yahoo.com/Computers/World_Wide_Web/HTML_Editors/.

Creating your own Web page is the first step. The next step is creating a means for others to view your Web page and interact with it. When people speak of "publishing" their Web page, they are talking about putting it on a server so that others can access it via a URL. A whole host of software packages support this activity for direct IP connected nodes. There is the shareware MacHTTP or commercial WebStar application for Macintoshes, and SerWeb for Windows. Go to http://www.charm.net/~cyber/ for a comprehensive list of support software for launching a Web server.

Entry-level Web servers are often UNIX-based, but it is not necessary to have a UNIX background to set up a Web server—especially with Internet software solutions for Power PC and Windows. The most popular entry-level Web server solutions today are shown below, and each includes a software bundle to facilitate Web page development and server administration.

Server	MSRP	OS	RAM	CD-ROM	Hard Drive
SGI Webforce	$10,995	IRIX 5.2	32MB	No	1GB
Sun Netra Server	$6,149	Solaris 2.4	16MB	Yes	535MB
Apple Internet Server	$2,909	Mac OS	16MB	Yes	700MB

Use this information for comparative purposes only. All of the servers listed are rated to handle 3,000 hits (access requests) per hour, which should be more than adequate for many initial Web servers.

Conclusion

Prior to and during the 1970s computers were regarded as centralized mainframes with which the select few would interact. In the 1980s the personal computer led to a great decentralization of computing power and facilitated the rise of many an entrepreneur and small business. The 90s is the decade in which the focus will become one of interdependent interconnections, and the Internet will continue to grow to support this demand as well as define its means.

With Internet access, you have the unprecedented opportunity to meet people and learn from them, access documents that will help you solve problems and answer questions, shop outlets on the Internet, and conduct business on a global scale. You can exchange messages, files, pictures, sounds, and video in addition to text. More and more organizations will find a Web server to be a key interface to their business community. E-mail and Web browsers will become popular alternatives for communicating and tracking down information at any time of day, from anywhere you can plug a laptop computer into a phone jack.

Described as an Information Superhighway, the Internet is a good beginning to achieving that idealistic end of high bandwidth, universal access to vast, well-organized information; it is very useful, but it is far from perfect. In the same way that automobile roadways are made up of a combination of dirt roads, paved streets, boulevards, and yes, superhighways, so too will the informaton infrastructure be composed of a combination of copper wire, coax cable, fiber optic cable, wireless communications, and other means.

Though it is challenging to keep up with the rapid changes which take place on the Internet, the early adopters of this technology who take the time to do so will have a clear edge in understanding this new environment and enjoying its advantages. Organizations can use the Internet to inform the world about their products and services. Sophisticated Web pages can bring a whole new meaning to the concepts of marketing and customer service. Information exchange with other divisions of your organization, outside technical support, government agencies, and current and potential customers can be facilitated through the use of the Internet.

Remember that the Internet is not just about hooking up computers, but allowing people to make connections in new and interesting ways. The Internet is a network of networks for computers, people, and information. It is becoming as standard a part of business equipment as the telephone and fax machine. It is a new way of researching, learning, communicating, and working.

PART TWO

Telemanagement: The Focal Point of Telephony

Chapter Five

Telemanagement

Telephony is an extremely broad marketplace. With all of the telephones, faxes, voice mails, computers, modems, bridges, routers, etc., that fall into the generic category of "Telephony Related," it is amazing anyone understands it all. Trying to allocate, project, manage and account for telephony is almost impossible and the complexity of the task is increasing practically every day.

Tele(communications)*management* has been one of the more underused tools of telephony. Telemanagement is the combining of computer technology and techniques to manage the telephony environment and especially the related expenses. Telemanagement is the managing of all telephony-related products and services. If that sounds fairly broad, well, it is. Telemanagement can be responsible for planning, implementation, tracking, and managing every component of telephony, including the PBX, ISDN, T-1s, local and long-distance telephone lines, telephone sets, authorization codes, cable pairs, LANs, fax machines, voice-mail systems, automated attendants, multiplexors, routers, interactive voice-response systems, modem pools, and the list goes on and on. In fact, every topic covered in this book and a few others can potentially be the responsibility of a well-run telemanagement system (TMS). The reason the words "can be" and "potentially" are used is because *you* have the power to apply telemanagement to as many or as few telephony tasks as you see fit.

The vastness of telephony makes it very difficult to plan for and to control. A telemanagement computer system is a major part of a smooth-running telephony operation. But behind every well-implemented telephony installation and computer system lies a human mastermind who must construct and orchestrate the telephony project. The TMS is the tool through which telecommunications managers should work their magic. The first rule of telemanagement is to plan. Even before a single piece of equipment is purchased, you must have a clear strategic plan of what you want to accomplish in the telephony arena, so that the optimal solution is found to your telephony needs.

Strategic Planning

There are a number of elements that go into strategic planning for telephony. You want to do the best job that you possibly can, so you don't want to leave anything out. You need to set out the steps that you want to go through in order to make the right decisions for all parties concerned with telephony. Uh-oh, that seems to include just about everyone. When you are putting together your plan, the best you can hope for is to please all of the people most of the time. You probably won't find utopia on earth, so remember to keep your expectations realistic. Figure 5-1 gives you an idea of how many areas of the business environment can affect telephony. You must decide what things to concern yourself with and either delegate or forget the rest. When forming your strategic plan you should keep in mind the following:

- Current Telephony Environment
- Business Objectives

- Management Information Systems Requirements
- Financial Considerations
- Operational Environment
- Technological Market
- Customer Environment
- Market Environment
- System Environment Planning
- Expansion Planning

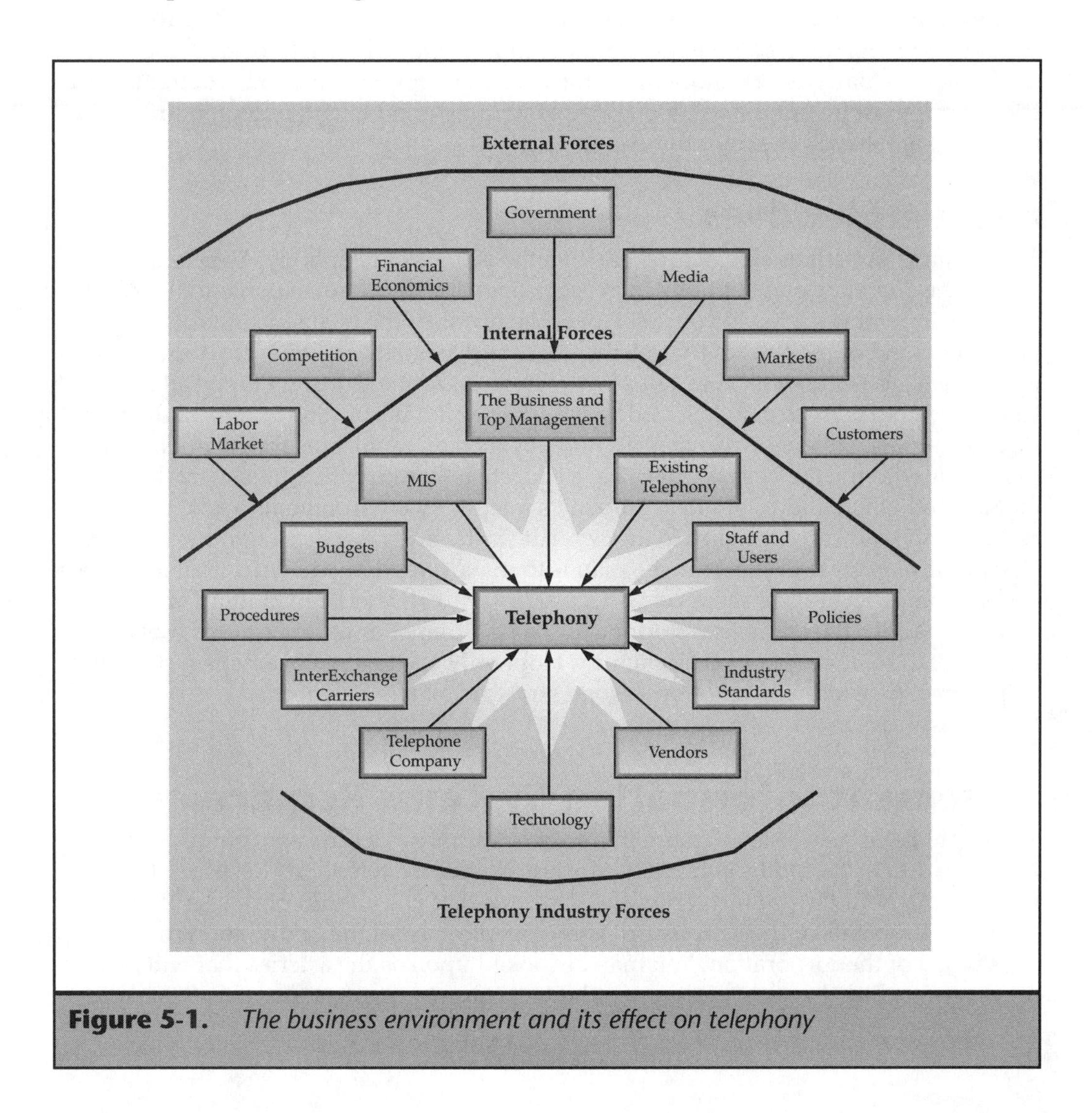

Figure 5-1. *The business environment and its effect on telephony*

Current Telephony Environment

It is imperative that you know as much about the current telephony environment as possible. There are a number of telephony-related publications, such as "Computer Telephony Magazine" and "Teleconnect." More telephony publications can be found in the recommended reading section at the back of the book. Unless you are starting from scratch, you will have to build onto existing equipment to attain your goals. The organization is dependent on the telecommunications and information processing systems that are already in place. Disrupting the fabric your company is built on is a dangerous task to undertake. You may find that you will need to discard or upgrade certain elements of the existing environment that do not fit into the strategic plan you are developing. Don't be the proverbial pack rat and try to save everything. You should try to plan for a combination of old and new. One of the jobs of strategic planning is to make the transition from the current telephony environment to the new telephony environment a smooth one. Implementing good telemanagement from the beginning is a step in the right direction.

Business Objectives

The business objectives almost always drive the actions of telephony. As business grows, the demands on telephony increase proportionally, if not exponentially. When forming your strategic plan, you *must* meet the needs of the business. In fact, it is your job to be a kind of psychic, to foresee the needs of the business. Odds are, the people in upper management haven't read this book. They undoubtedly don't know all of the technologies that are available in today's market, and it is your job to inform them. Let them know what telephony in the '90s and beyond can do for the organization. Once the people who run the business have been educated, they can make intelligent decisions on how and where they want the company's telephony environment to take them.

It is also your responsibility to know what the business needs as a going concern. One of the objectives of the strategic plan is to make telephony do what it already does better, more efficiently, or more cost-effectively. Try to always keep in mind the concept of optimization. Telephony is notorious for having tremendous amounts of waste and fraud, not to mention poor accountability. Hopefully, the telemanagement system will help a great deal in these areas, but a solid strategic plan is also needed for success.

Management Information Systems Requirements

You might have been able to ignore this section entirely a decade ago, but today the MIS department should be extremely concerned with the telephony decisions that are made throughout the organization. Which means that you should pay close attention to the requirements of the MIS group. These people control the information that is the life's blood of the corporation. You may be looked upon as the arteries that will carry the corporate blood to new locations, at higher volumes and faster speeds than were ever possible in the past. Also keep in mind that you will be generating information from the telemanagement system that might need to be incorporated with databases for which the MIS department is responsible.

When you are working with the MIS department, you are working on two fronts: the technical requirements of the MIS department on telephony and the data processing aspect of the proposed telemanagement system. In today's market, you will find that it is advantageous to have as much integration between the TMS and the MIS computer systems as possible. Since MIS must also plan for telecommunications changes, both the MIS and telecommunications strategic plans must be developed with the business considerations kept in mind to produce a cohesive implementation plan. Figure 5-2 shows the interrelationship between business objectives, MIS, and telephony.

Financial Considerations

Obviously, you must take into account the financial position of the corporation. You should try to get a rough estimate of the amount of money the company wishes to spend on the acquisition of new telephony equipment and services. Until you have made a complete strategic plan and obtained some rough estimates of the costs of the telephony components that you require, it is very difficult to actually budget for

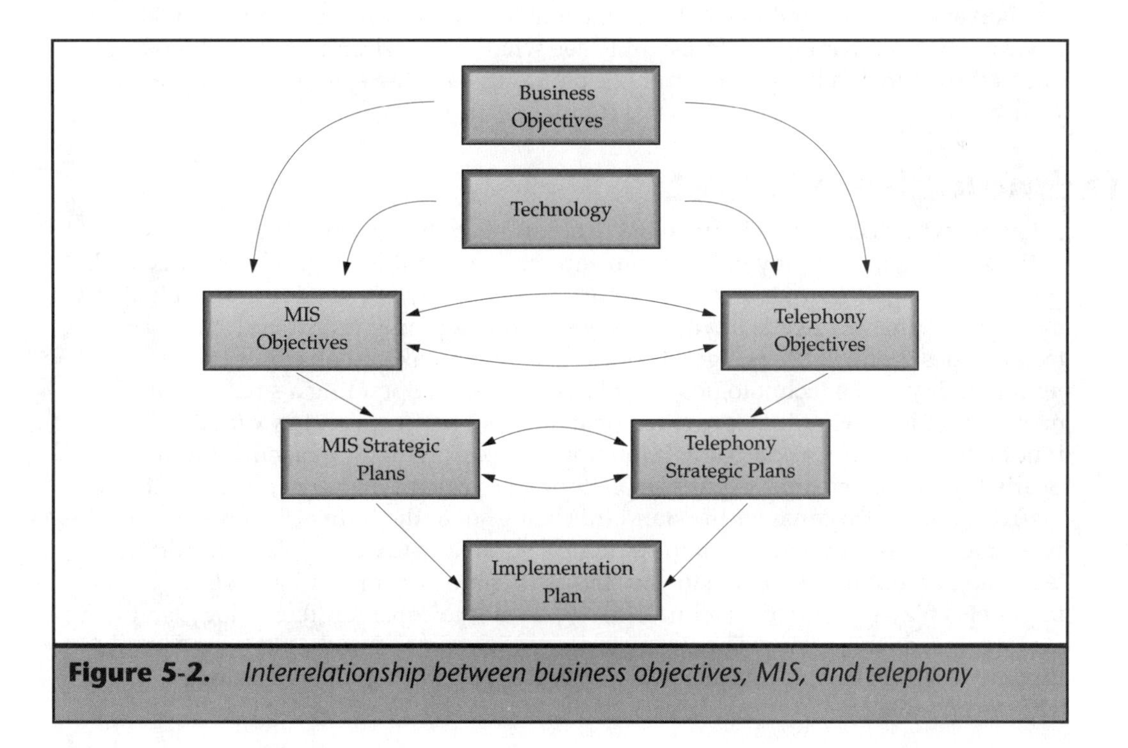

Figure 5-2. *Interrelationship between business objectives, MIS, and telephony*

telephony projects. However, if you know in advance the beginning figures, you may be able to rule out certain components that don't fit into the current financial plans.

The business considerations and the financial considerations go hand in hand. You cannot meet the needs of the business without proper financial planning. Unless you are wearing several hats in your organization, it is not your job to budget for the necessary funds. Your job is to create a comprehensive strategic plan that includes financial information and to present your findings to those who are responsible for financial planning. Finances and budgeting quite often require that you create a strategic plan for telephony that is implemented over long periods of time. Try to stay focused on the job at hand and provide the best strategic plan possible based upon all of the information you gather.

Operational Environment

In telecommunications, you must always be sensitive to the needs of the people you serve, and that is just about everyone. The best way to survey the operational environment is to go out and walk among the masses. Ask the people you serve what they think of the existing telephone system and how telephony in general helps them or could help them be more productive. Your operational environment might entail multiple locations networked together to form a web of information. As any spider will tell you, webs break and need to be repaired. They also have voids—some should be filled and others should be left alone. Understanding the current operational environment and having the foresight to see what that environment might look like in the future is a big part of strategic planning. Like the spider, you must tend to your web.

Technological Market

Knowing what technology is available to accomplish the goals you set forth is vital to the success of telephony in the organization. You might have purchased this book specifically to gain an overview of the current technologies available in the telephony market. Gauging the needs of your business before learning more about the specific technologies you will use is a good course of action to take. Look closely at the upgradability of the technologies you plan to invest in. Spending a small percentage more for similar technology provided on a different platform and by different vendors is not always an unwise decision. If the upgrade path is higher reaching and more clearly laid out on the more expensive platform, the added expense might be justifiable.

You should also consider the standards being set in the industry. Throughout your internal environment, certain standards might already be in place that affect the technological options available to you. Industry, state, federal, and global standards might also play a part in the technologies you want to consider. Technology is moving at warp speeds these days. The strategic plan you make today should be as forward-thinking as it can be, and the technologies you invest in should have as long a life as possible.

Customer Environment

Knowing your customers makes strategic planning decisions a whole lot easier to make. Quite often, the customer base will dictate to the organization the level of technology that should be implemented. For example, let's say that your customers are all over the world and you supply a service that requires a one-hour response time for any problems reported. This type of customer might force the strategic planning of a paging system. If your technical staff is spread out over the country and at certain times of the day or week different technicians are on call, you may need to plan for a national paging system such as SkyTel.

Customers can also come to expect different services that your competitors are offering. One of the best examples of this is the automated teller machine (ATM). First, the technology was invented. A banking institution invested in the technology, hoping to draw customers away from competitors. The technology worked, and the first banks to adopt the new technology gained more business than the technology ended up costing them. Even though ATMs were very expensive, the other banks had to invest in the same technology, or they would lose even more business to the banks that did invest in the added convenience of ATMs. Now virtually every bank in the industry offers ATM banking services. (By the way, ATMs are heavily based on telephony technologies.) The customer is always right, and when it comes to your strategic plan, you had better not forget who really puts the food on the table.

Market Environment

It is certainly no secret that you must keep an eye on the business market you inhabit. Trends don't get followed by people who aren't paying attention. The telephony market is driven by the technology and the people who embrace it or reject it. Your industry is undoubtedly driven by many different factors, but you can be certain that telephony is one of them. One of the most effective telephony tools that ever affected the marketplace was the toll-free 800 telephone number. Businesses found that if customers could call them for free, their overall business and profitability increased. The 800 number is also effective in opening up new markets that were not recognized earlier. There are many different applications of telephony that can open up new markets for business. Recognizing the ones that will work in your business environment is extremely valuable to your strategic plan and to your company's business.

Another aspect of the market environment is the competition. Advances by the competition can push an organization in a given direction, such as the ATM example given earlier. The competition's weaknesses can also be exploited by your organization. Let's face it, there are thousands of applications and technologies that have not been implemented in thousands of industries. The trick is to figure out which applications and technologies can be implemented to give your company a competitive edge.

Staying in the telephony realm, let's say that you sell office furniture both through distributors and direct. The market is very competitive, and you want to increase your market share. So you decide to implement two new telephony applications to open new markets and also to take over some of the existing market. The first application you implement is a fax-on-demand service. Customers and dealers can call an interactive voice-response (IVR) system and request information to be sent to them on any product that you sell. The fax-on-demand service will send pictures, specifications, and pricing information to the customer's fax machine almost instantly. You gain more customers by providing better service. See Chapter 16 for more on fax-on-demand.

The second application is another IVR that allows a customer or dealer to not only order but also check on availability or the current status of an existing order. The results of these telephony decisions on the organization are a greater number of customers, better customer service, and larger market share; and, as an added bonus, you have reduced the staffing requirements of the customer service department and decreased the demands on your sales force, freeing them up to make more sales. You may have to be more aware of the market your company works in, but with a little foresight and good business management, your telephony strategic plan will also be a business strategic plan. For more information on IVR, refer to Chapter 17.

System Environment Planning

If you can touch it, it takes up space. Telephony products take on some pretty strange shapes, and the strategic plan must include preparing and maintaining the environment that the components of your telephony plan require. Cabling is one of the more difficult parts of telephony to plan for and maintain. Let me suggest that if you are in the process of procuring cabling for your organization, your plan should include all the cable you will ever need while your company occupies the real estate. Cabling is *extremely* expensive. The majority of the cost is incurred not in the cable itself, but in the labor involved to lay the cable. Upgrading the cable on a piece of real estate is an exercise you don't want to have to do even once if you can avoid it. You may want to see where upgrades might be made by you and your staff. This can save significant dollars.

If your strategic plan includes a PBX switch, environment planning is essential. A switch room must be designed to the switching platform's specifications or a switch must be purchased that meets the specification of the switch room. Floor space, raised floors, power supply, backup power supply, air conditioning, fire extinguishing systems, wall space, and lighting are just some of the considerations that need to go into the environment planning of a PBX (see the next section for more on PBX). An improperly prepared or implemented environment can cost a corporation millions in a matter of seconds. All telephony equipment has its base in computers in today's arena. Environmental planning is usually expensive, and even more important, is fixed once everything is in place. Be sure your strategic plan does not fall short in the area of environmental planning.

Expansion Planning

You need to remember that every part of the strategic plan you set forth today will be in place and in use for many years to come. In certain areas of telephony, the upgrade path is easy to follow and implement. In other areas, not planning for expansion can be a near-fatal mistake. The most obvious example of the necessity for expansion planning is the procurement of a PBX. All PBX systems are expandable, but they all also have limitations.

For example, let's say you purchase a PBX with an installed capacity of 1,200 stations, and you currently have a need for 1,000 stations. That means you have a 20 percent growth factor built into the existing configuration. Let's also say that you can upgrade the PBX by purchasing either additional cabinets or station cards to bring the total number of stations up to 1,400 without expanding the physical environment. The PBX station capacity is fixed at 1,400 at the time you purchase the system. If expansion planning calls for more than 1,400 station lines over the life of the system, you are going to have problems. Sure, you could knock down a wall, upgrade the power supply, rewire, and so on, but your overall project costs would skyrocket. Remember to plan for expansion up front so that the purchase will perform admirably over the life of the system. For more information about PBX systems, please refer to Chapter 3.

Conclusion

A strong strategic plan will include a combination of all factors presented in this section. In addition, other factors in Figure 5-1 can be evaluated to provide a more comprehensive strategic plan. Part of the strategic plan that needs to be amended at the end of the process is the implementation plan. The implementation plan could be viewed as a subset of the strategic plans that the telephony and MIS departments must develop. The strategic plans that contribute to the implementation plan as depicted in Figure 5-2. The success of the project begins with the efforts put forth in the planning. Proper planning almost guarantees a successful project.

One of the more important steps in the strategic plan is making selections. Vendors, products, maintenance, and support must all be determined. You may choose to deal exclusively with existing vendors, or you may find the need to create a request for proposal, or RFP (see the next section). A well-written RFP must include every detail about the strategic plan to obtain the products and service necessary to get the job done. You will get out of your strategic plan what you put into it. Do not be negligent with the strategic plan, and you will achieve the goals that you set for telephony in your workplace.

Writing a Request for Proposal (RFP)

A *request for proposal (RFP)* is a set of detailed specifications that define the requirements of a project. Where extremely large projects are being planned, a *request for information (RFI)* and a *request for quotation (RFQ)* may precede the RFP process. RFIs and RFQs are written in roughly the same manner as an RFP.

When writing an RFP, you must take the strategic plan that was developed and subsequently create a set of requirements that a vendor must comply with in order for your business objectives to be met. At the same time you are compiling your RFP from the strategic plan, you should also be compiling a set of action items that you will have to perform in order to obtain your objectives. Depending on what your goals are, you may find that it is better for you to write more than one RFP for the different components of your strategic plan. There are three major components of telephony that usually require RFP generation:

- Computer-based telemanagement systems
- Switching platforms
- Voice-mail platforms

For the remainder of this discussion, we will assume that you are at least in the process of acquiring a new switching platform. Of course, the RFP method of acquisition can be used for any purchase. But, due to the amount of preparation that goes into RFP specifications, you may opt for choosing your vendors wisely and always preparing a detailed strategic plan for smaller projects.

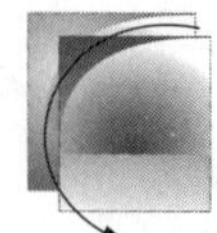

TIP: If you are planning for a major acquisition of telephony products and services, try to purchase and implement your computer-based telemanagement system first if you do not already have one. A TMS that is in place and in use while a switch installation or other telephony project is in progress will help to make for a smooth transition. You will be able to use the TMS to drive the project to a successful conclusion. In addition, the TMS should be accurate the moment the new switch or telephony project becomes active—a big time saver!

What an RFP Should Contain

Writing an RFP is a tremendous undertaking. Try to start with another RFP that your company issued in the past, if one exists. You will find a great number of sections are common to virtually all RFPs. If you have hired a consultant, ask for a copy of an RFP that is of a similar nature. For switch RFPs, the prospective vendors are also a good source of information. They will provide you with an RFP that they feel was well written, but you can be sure that they also won the bid specified. In the field of telemanagement, Select Source, a telemanagement firm based in southern California, will even provide you with a sample RFP on floppy disk. Pick and choose the things that apply to you. Don't be afraid to add items or sections that are unique to your specifications or needs. Dice and slice, cut and paste, until you have an RFP you are proud of.

An RFP will not be comprised solely of specifications of the project. Your RFP should begin with a summary statement of the scope of the project detailed in the RFP. Let the vendors know how you will be evaluating the RFP. You will find that their responses will change accordingly. Make a statement of proprietary information nondisclosure. You don't want your competitors getting insider information about your future plans. Inform the vendors of the proposal format that they should adhere to. Make sure that the vendors supply all supporting documentation necessary to verify the responses given in the proposal. State the number of complete copies of the response you will require the vendors to supply. This is your big chance to get additional copies of the vendors' documentation. Documentation can cost hundreds of dollars per copy, so be reasonable. You should also clearly state a schedule of events that will inform the vendors of the following:

- RFP issued date
- Vendor conference dates (mandatory or not mandatory)
- Proposal due date
- Evaluation of RFP response date
- Anticipated award date
- Equipment order date
- Installation schedule and due dates

A vendor qualification form, such as the one depicted in Figure 5-3, should also be included. It is always wise to ask for proof that your vendors are properly bonded and insured for the contract they may be entering into. State in the RFP any and all permits, licenses, regulations, and ordinances that may be pertinent to the project. The list of action items you have prepared for yourself along with action items for the vendor should be compiled. Responsible parties and completion due dates of project milestones should also be listed. If ongoing maintenance and equipment upgrades will be needed, make the vendors commit to you in writing the prices they will charge for their products and services in the future. There is nothing worse than making a low bid award and then being stung on the back end by inflated prices. Lastly, state the terms of payment you will agree to.

There is at least one other thing you should keep in mind when preparing your RFP. Make the bidder supply you with an itemized quotation and separate quotations for different sections of the RFP that you may wish to award to separate vendors. That way, you can divide your award between vendors if you wish. Sometimes a single source vendor is not the best solution.

Once you have written your RFP, all prospective vendors are given a copy and asked to respond to every section in detail. Bidders must state precisely how their

Vendor Qualification Form

Name of vendor: ______________________________

Address: ______________________________

City: ______________ State: ______________ Zip Code: __________

Telephone Number: ______________ Fax Number: ______________

Name of president or CEO: ______________________________

Name of account executive: ______________________________

Type of business (corporation, partnership, subsidiary, etc.): ______________

If you are a subsidiary, identify your parent company and the location of their nearest offices: ______________________________

How long has your organization been in business?: ______________

Primary form of business (manufacturer, consulting, distributor, etc.): __________

Description of above: ______________________________

Who is the manufacturer of the product you are proposing?: ______________

Where is the nearest distribution center for the manufacturer?: ______________

How many systems of this size, type, and model have you installed in the past year?: ______________

How many systems do you currently have installed in this area?: ______________

How long have you been in business?: ______________________________

Where is your nearest support location?: ______________________________

Number of personnel located in the nearest support location: ______________

Marketing:__________ Hardware support:__________ Engineering:__________

Admin.:__________ Software support:__________ Other:__________

What is your Federal Tax ID?: ______________________________

Please provide a list of **ALL** customers that you have sold to in the past five years. Please include the customer's size, equipment purchased, contact person's name and phone #.

Please attach one or more of the following:

- Dunn and Bradstreet report
- Annual report
- An audited financial statement

Please provide any additional information about your organization that you feel might be pertinent to this RFP response.

Figure 5-3. *Sample vendor qualification form*

proposed solution will meet the requirements of the RFP. If deviations are made or bidders cannot meet certain specifications, in the response to the RFP they must state the deviation or the inability to comply with the specification. In addition to the responses to the specifications, the bidders must also submit pricing information along with any and all pertinent documentation to support the claims made in the response. All bidders who respond to the RFP are then evaluated for award of the project.

RECOMMENDATION: Send a copy of the RFP to the vendors on floppy disk and request that they incorporate their answers directly into the original specifications. You will find that this is an excellent way to receive and evaluate RFP responses, and vendors like it too.

Conclusion

Writing an RFP is extremely time-consuming and difficult. If you have never written an RFP before, try to get help from someone who has. While preparing to write your RFP, ask questions to everyone about almost everything. Make up a questionnaire if you like, but always keep in mind who you are there to serve. If there are several sections to an RFP that can be awarded separately, stipulate that you want the proposals to be made with pricing breakdowns that detail each item affecting the overall purchase price. Try to avoid using consultants if at all possible. The good ones are few and far between, and always cost an arm and a leg (maybe worth it if you can find and verify their expertise). In addition, they are usually not as tuned in to the specific needs of your organization as you are and are not as willing to invest the time and effort into understanding the newer technologies that are emerging in the telephony industry almost every day. Telephony in the '90s and beyond does not stand still for anyone. Don't leave anything out of an RFP that you can think of. Vendors will not add specifications to the RFP, so make it as comprehensive as possible.

Evaluating RFP Responses

Proper evaluation of the RFP responses you receive is just as important as the writing of the RFP itself. You should create a set of guidelines that you will use to evaluate the responses you receive. RFP evaluation is usually performed by committee. It is important for you to know the members of the evaluation team and to understand the value each individual brings to the table. Many a war has been fought among evaluation committee members. If your strategic plan is comprehensive and you form your evaluation committee wisely, you will find that the members should be a bit more hospitable. Remember that you are all working towards the same goal. Your goal is to contract with the best qualified vendor(s) that respond. Success is not measured by the price one pays for products and services but by the amount of value that one gets for the dollars spent.

Higher Quality Will Always Cost Less in the Long Run

Steps to Evaluating RFP Responses

There are several steps you should take when evaluating RFP responses. You will find that organization of the evaluation process can save a tremendous amount of time and effort. Try to follow a predetermined path to your goal (a successful project implementation). Try to optimize your RFP response efforts along the way by:

- Creating a set of criteria that must be met
- Divide and conquer
- Vendor evaluation
- Technical evaluation
- Life cycle evaluation
- Financial evaluation

Creating a Set of Criteria That Must Be Met

The first step in evaluating RFP responses is to make a set of criteria that must be met. The committee members can all participate in creating the list of mandatory criteria, throwing in the most important features that they feel must be included. One requirement might be a maximum cost that will be considered, another might be that the nearest service location be no more than two hours away, and a third might be that the firm be minority-owned. You and your committee must decide what is important. The set of criteria that must be met does not necessarily have to be the same criteria that were stated as mandatory requirements of the original RFP. At this stage of the acquisition, you are trying to separate the cream from the fat. There can be only one award given at the end of the evaluation process for each section of the RFP.

Divide and Conquer

You may receive a dozen or so responses to your RFP. Each response can be hundreds of pages long and can contain thousands of pages of reference material. Once you have created your set of criteria that must be met, you should try to divide up the responses and ask each member to verify that the RFP response meets or exceeds the criteria set forth. You will find that a good percentage of responses can be immediately removed from the prospect list. You may find that your expectations were too high and that every bid does not meet the selection criteria. In this case, you may have to reevaluate your criteria, change your strategic plan, or even leave out certain items of the project until a later date. Remember, you must be realistic. If you were diligent in creating your strategic plan, odds are that you will have several vendors who will progress to the next stage of evaluation.

Vendor Evaluation

Now that you have narrowed the field a bit, it is time to scrutinize the vendors with whom you have the opportunity to do business. To be quite honest, there may be several bidders you simply do not want to do business with. You may notice in Figure 5-3 that the form asks for *all* customers the organization has sold to in the past five years. Anyone can come up with three to five customers who will absolutely rave about their vendor and what a wonderful job they are doing. Why should you let the vendors choose the references you will contact? You may not call every reference given, but you can select the references rather than having the vendor pick them for you. Vendors' references are their most valuable asset. All other things being more or less equal, the customer's satisfaction with the vendor's maintenance and support is the most important factor when choosing a vendor.

There are other things about the vendor that you should keep in mind. The vendor's overall reputation in the industry should be known. The size of the firm and its financial stability are other factors that should be considered. If the vendor has passed the acid test, then it is time to review the installation plan, training schedule, product warranty, and maintenance plan. These factors may not exclude a vendor from consideration, but they should be weighed in the evaluation process. You will have a relationship with the vendor that is awarded the contract for a fairly long time. Make sure you feel comfortable with the vendors you may choose to do business with.

Technical Evaluation

Some people will say that you should perform the technical evaluation before you perform the vendor evaluation. Technical responses are probably the most difficult to evaluate and verify. If performing the vendor evaluation first can reduce the number of prospects further in a shorter period of time than technical evaluations, then performing the vendor evaluation first is prudent.

Now that you are down to a select few respondents, it is time to sharpen your pencils and get down to business. The technical evaluation is, quite often, the deciding factor of an RFP award. If certain bidders respond to every specification with "The vendor understands and will comply" with no further explanation, BEWARE! If you did your homework, no one will be able to measure up to the high standards that you have placed in your RFP. Caveats, exceptions, and substitutions are expected when reading RFP responses. If you find none, odds are that the vendor didn't do the necessary homework and the responses are inaccurate. Bidders should respond with a direct yes, no, or an exception and support their response with documentation, examples, and references. The technical specifications should be weighted based upon their effect on the strategic plan. Categories such as mandatory, optional, and desired are also helpful when performing the technical evaluation. The technology being proposed and the *mean time between failure* (MTBF), which put simply is the amount of time you can expect your telephony product to be inoperable over a period of time, should be taken into account as well when evaluating RFP responses. Delivery schedules and the vendor's ability to meet those schedules are additional concerns that should be evaluated.

Life Cycle Evaluation

Different products and services have different life cycle expectations. You may recall that the RFP specifications ask for extended maintenance pricing and expansion equipment pricing. The initial system has an anticipated life cycle, and the system growth potential to increase the longevity of the system is just as important. Your decision must also be favorable to the future of the organization. If you choose a platform that meets the needs of today and is more cost-effective in the short run, you may find that later on you have made your investment into a black hole. Make sure that whatever telephony item you are purchasing, particularly a switching platform, is capable of being upgraded to the prevailing technologies of the industry for the life of the system. The organization has expectations of the life cycle that must be met by the selected product(s). Environmental concerns such as square footage of real estate, air conditioning, and electricity consumption over the product's life cycle can also be a swaying factor. Don't forget to look at the whole picture.

Financial Evaluation

When you are making a major acquisition of any kind, the financial aspects of the transaction always weigh heavily on the decision-making process. In the case of telephony, the variables are much greater than with other capital investments. If you are in the process of acquiring a switching platform solution, you will have to tackle the pros and cons of a Centrex solution vs. a PBX. In short, a Centrex solution costs less in the beginning, and as time goes on it becomes more expensive. You have relative certainty that the Centrex will be kept up to date with technology, but there are no guarantees beyond what is put in the RFP specifications.

A PBX acquisition may cost more initially, but over the life of the purchase should be more cost-effective. A PBX gives the organization greater control over its telephony environment, which has an immeasurable value. If properly acquired and implemented, the PBX solution is almost always the better solution.

At the end of a telephony product's life cycle, a PBX will have a net present value that can bring the cost of the overall project way down. You may also want to consider options such as lease vs. purchase. The cost of money may also play a role in the way you make your telephony acquisition. You should have a person on your committee who is fully versed in the art of money to help make these decisions.

Conclusion

Successful evaluation of RFP responses should be a combination of art, science, and financial wisdom. The art comes into play when you are trying to separate the truth from the salesmanship. Knowing how to read between the lines is an asset in any undertaking, and RFP evaluation is no exception. Also getting a human feel for the vendors you may be working with is extremely important. Don't underestimate gut instinct when it comes to choosing the vendor you are going to invest the future of your organization in. The science comes into play when you are evaluating the empirical data that is provided in the RFP response. Methodically categorize, prioritize, and rationalize the results of the evaluation, and you will be able to place your vendors in the proper order for award. Finances always play their role in major acquisitions. Remember that quality always costs less in the long run. If at all possible, try to make the art of RFP evaluation the deciding factor that gives the contract award to the best-qualified vendor.

Why Write an RFP?

You may ask yourself, why should I write a gigantic RFP that could take months to prepare? The answer to the question is: you don't want to make the wrong decision. An RFP is not needed for every telephony acquisition. But when there are hundreds or thousands of details that need to be spelled out and thousands or even millions of dollars at stake, you should not take any chances. An RFP defines exactly what is being purchased, what it must do, how much it will cost, and when it must be delivered, along with several other vital pieces of information. An RFP must be extremely precise and contain every detail that you can put onto paper. Bidders will rarely add anything into an RFP specification that is missing, even if it is for the customer's own good. Why? Because the other bidders won't be adding to the RFP, and the object of the game is to meet as many specifications as possible and to be the lowest bidder so that they will have the best chance of being awarded the contract. If the bidder strays from the path set forth in the RFP, it is unlikely that the vendor will be awarded the project.

Reasons Not to Write an RFP

There are many reasons why you may choose not to go to the trouble of going through the RFP process. In some cases, the less formal RFQ can meet the needs of the project. Obviously, if the acquisition is too small to warrant the time investment, you would not go through the RFP process. You may have a favored vendor with whom you have done business in the past and been very satisfied. If the market for competition is tight, you are upgrading an existing telephony platform, or the hassle of dealing with multiple vendors is not palatable, then you may choose to avoid the RFP process entirely. If you are a diligent manager and the fundamental guidelines of telemanagement are followed, the RFP process can be avoided and other courses of action can be pursued to achieve the goals of the organization. You have the power to make your own decisions on how to plan, implement, allocate, track, and manage your telephony environment. Try to make the right decisions from the beginning, and your life will be much easier.

Conclusion

Telemanagement can be the difference between the success and failure of any telephony environment. Telephony costs can be literally in the millions of dollars on a monthly basis for large organizations. The old school of thought that telecommunications expenditures are simply part of overhead is a thing of the past. Every dollar spent on telephony should be reallocated to the specific individuals, departments, and divisions that caused the expenditure in the first place. The goal of telemanagement as a whole is to provide the highest overall quality of service, features, and performance to the organization at the lowest cost. An educated telemanager equipped with a good computer-based telemanagement system should be able to optimize the telephony environment, provide a high quality of products and services to the organization, and at the same time allocate all capital expenditures back to their responsible parties while at the same time cutting costs. The variable dollars saved or made through good telemanagement can more than pay for the hard dollars spent on other telephony products and services. Try not to overlook the value of good, solid telemanagement.

Chapter Six

Computer-Based Telemanagement Systems

A computer-based telemanagement system (TMS) is the focal point of all telephony equipment and applications that are used throughout an organization. As the number of components of telephony continues to rise, the need for computer-based telemanagement systems becomes greater. No one person in an organization has a complete understanding and control over all telephony components. A TMS should be the central location where all telephony activities can be viewed, tracked, and managed. *Single point of entry* is another reason why more and more organizations are implementing, looking for, or building new computer-based telemanagement solutions. Many database elements with regard to telephony are duplicated on several platforms. A single point of entry removes the need to enter the same information over and over again.

The need for control over the telephony giant is growing rapidly. There hasn't yet been a telemanagement system built that can perform all of the functions that a TMS should in today's world. In Figure 6-1, the operations of a computer-based telemanagement system are shown. As the demands on telemanagement from business continue to rise, more and more software development is being done in the area of telemanagement to try to fill the technological gaps that exist in the industry today. The task of making TMS the single point of entry for all telephony will be never-ending, but the value of telemanagement systems is here right now in the present.

Traditionally, telemanagement has been the computer tool of the businessman. The controller was interested in knowing where his multimillion-dollar telecommunications budget was going and who was spending it. Without telemanagement, a corporation has little, if any, internal accountability for its telecommunications expenses. Money is the driving force behind almost all telemanagement functions. And in the telecommunications business, we're talking about big bucks. Hundreds of thousands, even millions of dollars can be spent, lost, squandered, or stolen if a telecommunications system is not properly administered, monitored, and managed. The telephony resources of a company must be controlled just like any other resource a company has. The hardest part about telemanagement is that, to some degree, *every* person within the organization has a stake in what goes on in the telecommunications department.

Telemanagement is where computers and telephony collide. The reason for this collision is that all current telecommunications equipment is grounded in computer technology but is completely designed around a different application. Telemanagement is the one arena of telephony that is truly a computer-based application in the more standard form of computing. The applications are computer based, but the information is primarily telephony based. And there you have the collision.

The reason there is so much conflict in the area of telemanagement is that too many departmental organizations have a stake in what a telemanagement system does. By default, telemanagement systems have traditionally been run by the telecommunications department. In the beginning, TMSs were used strictly to perform the task of call accounting. Managers wanted to know who was calling where and how much it was costing them. In order to perform this task, a computer was used to collect call detail records (CDRs), also called station message detail records (SMDRs), from the PBX and then price, sort, and perform an accounting function on the CDR data

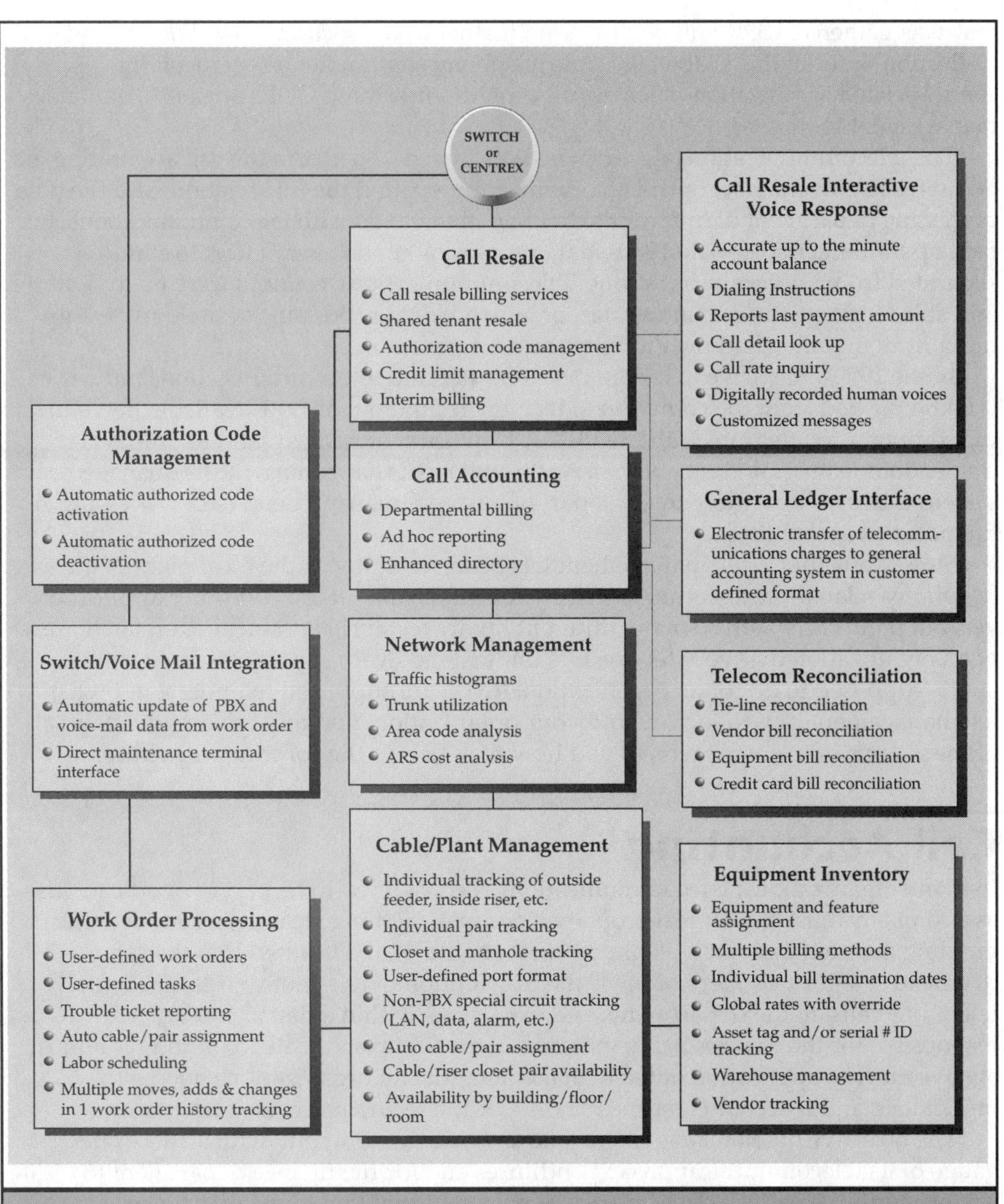

Figure 6-1. *Operations of a computer-based telemanagement system*

that was gathered. CDR will be discussed further in the section "CDR Format and Call Collection" later in this chapter. In some telephony environments—particularly service-related organizations such as call centers—incoming CDRs are also a valuable management tool.

The telecommunications department was forced to perform the call accounting job, because the powers that control an organization wanted the information, and the data processing department didn't want to have anything to do with telecommunications. In fact, up until the mid-to-late 1980s, data processing made every effort to stay separated from telecommunications. Telecommunications vendors were ready, willing, and able to merge the two industries long before data processing wanted to give up the control that they had over a company's information.

In the 1990s, the drive to merge the computer and telecommunications industries is being pushed from the computer side. New technology is now available that allows the computer to integrate with the PBX or Centrex switching system to unleash tremendous features and capabilities. As this rollover is happening, the telemanagement system is becoming a more integral part of business management, data processing, finance, and accounting.

Slowly but surely, telemanagement has been emerging as the focal point of all telephony-related products and services. Telemanagement is a computer application, yes, but it interacts with so many different highly technical and business-related areas of an organization that very few people understand every aspect of what it can and is supposed to do for an organization. If you are supposed to understand what telemanagement can do for you and your organization, you must first understand all of the aspects of telemanagement and how they fit into the corporate structure.

Call Accounting

In many organizations, telecommunications is considered to be part of overhead, just as a company must pay for real property it occupies, lighting, heating, water, electricity, building maintenance, etc. Telephony costs should not be lumped into the same overhead category. *Call accounting* is the distribution of telephony costs, particularly telephone calls, to the departments and extensions within a department that are responsible for the charges being incurred in the first place. Call accounting is almost always the first application of telemanagement that an organization chooses to implement, and it accounts for the majority of the recurring costs of telephony.

The objective of telemanagement is to provide the company with an accurate gauge of its telecommunications expenditures and for the finance department to create a telecommunications budget for each department that they must manage and are responsible for. The telecommunications department can then become a zero-based budget department. A zero-based budget department is a department that has no funds allocated directly to it. Instead, all of the funds needed by the department come from billing other departments for the products and services provided. Every penny it spends is funded by the departments that cause the expenditure in the first place. By accurately distributing the overhead costs of telecommunications into the chargeback system, a true zero-based telecommunications department can be achieved.

When your organization chooses to perform call accounting with its TMS, you must start by gathering the pertinent information with which to populate the database. (For this discussion, it is assumed that you already have a TMS acquired either through the process described in Chapter 5 or by some other means.) You need to gather information about your organization's accounting practice, personnel, and the rest of your existing telecommunications environment in order to perform the call accounting function.

Since telephony chargeback probably wasn't done in the past, it may be necessary to develop a method of chargeback. Start with the controller of the organization and ask how exactly chargeback should be performed. The controller is responsible for the accounting and financial functions throughout an organization that should have already been laid out in the call accounting function of the TMS in the strategic plan (the process of creating a strategic plan is discussed in Chapter 5). What are some of the ways that telephony costs can be allocated? The following is a sample list of hierarchical breakdowns and basic forms of chargeback that your organization may choose to use.

EXTENSION This method simply bills the individual extension back for the telephone call charges it is responsible for. The extension is usually assigned back to an individual within the organization. This method is only used in the most basic accounting environments. Very little control is offered using this method, and therefore it is rarely used.

DIVISION/DEPARTMENT, EXTENSION This is the most common form of chargeback used in business today. Most organizations are comprized of divisions and departments within divisions. Extensions are assigned to personnel that work for a particular department. When the chargeback function is performed, the call charges are billed back to the originating extension and then subtotals are generated for each department and each division. Budgets are usually issued at either the division or department level, but the TMS should have the capability to track budgets all the way down to the extension level.

TENANT, DIVISION/DEPARTMENT, EXTENSION This method of chargeback is similar to the method mentioned in the previous section. The primary difference is that a tenant (also referred to as a location) subtotal is added, usually for corporations that maintain multiple sites. One of the more powerful functions of a TMS is to gather telephony data from all locations of network, so that it may be compiled into a comprehensive telephony management tool.

FORCED AUTHORIZATION CODE (FAC) OR PERSONAL AUTHORIZATION CODE (PAC) The FAC, also referred to as a PAC, is a code ranging from 2 to 14 digits in length. For security purposes, FACs should be seven digits or longer, but the decision is up to you. A FAC is assigned to an individual. In most cases, the individual is associated with an extension, which is associated with a department, which is associated with a division, which is associated with a tenant. In order to use FACs, the switching platform that you are using must support the authorization code function. Telephones can be assigned what is called a class of service. A *class of service* is what allows certain telephones to make long-distance calls and other telephones to only make interoffice calls or local calls (a hall phone for instance).

A FAC can also have a class of service assigned to it. When a person is using a telephone with a class of service that does not allow long-distance calling, the switch can prompt the user for a FAC. The FAC then temporarily raises the class of service of the telephone so that the call can be placed. The FAC is a popular tool in telecommunications today. A person can roam all over the installation and still have easy phone access. In addition, the FAC is assigned back to an individual, so no matter where the code is used, the proper party is charged. For more information on FACs, refer to the "Toll Fraud Detection" section later in this chapter.

GENERAL LEDGER General ledger is another popular method of chargeback. A general ledger number can be assigned at virtually any level of the chargeback hierarchy. Quite simply, the TMS must subtotal the telephony charges to the appropriate general ledger level. General ledgers can become a little more complex depending on how they are applied. For instance, a general ledger number may be comprised of several embedded codes, such as a location code, a division/department code, a fund number, and a charge type code. Your controller may wish to see the telephony charges broken down by local versus long-distance charges, equipment versus work order charges, overhead versus one-time charges, and the list goes on and on. The base general ledger numbers for each of these categories do not change from extension to extension; only the embedded codes will change. This method of chargeback is very powerful, but it is also more complex to implement and to maintain. Your TMS should be able to perform this valuable subdivision of telephony-related charges.

FUND NUMBERS Funding is a concept that business people are very familiar with. The idea of having funds that all expenditures made on behalf of a particular undertaking are charged against is not new. Funds are usually assigned a fund number to make chargeback easier. The TMS gives telecommunications the ability to charge back its expenditures to the appropriate fund number. Knowing not only who is spending your telephony dollars but for what purpose is a valuable tool for an organization. Through analysis of the accounting generated by the TMS, you can establish patterns of telephony expenditures in certain areas. For example, let's say that you are funded for $100,000 of marketing. You can easily tally up the money spent on marketing literature, advertising, employees, etc. But now, the variable costs of telemarketing can also be added to the funding pool. You will then be able to directly attribute all expenditures back to the proper funds.

Fund numbers are also sometimes assigned to types of charges. Local, long distance, cabling, equipment, labor, and overhead are just some of the types of telephony charges that may be assigned fund numbers. A TMS can be designed to group all or certain users' telephony expenditures by the type of charge being incurred.

PROJECT CODE A *project code* is either a special form of FAC or a department-like code that is assigned to individuals working on a specific project for a period of time. If an organization is contracted to perform certain functions for an external third party, a project code may be the desired method of chargeback. Probably the best example of project code chargeback would be in a law firm. Each client or case could be assigned a project code, and every time an employee of the law firm makes a telephone call on behalf of the client or for the case, they punch in the appropriate project code. The TMS then automatically groups and charges for the calls and potentially for the lawyer's time on the telephone. Project codes are also very popular with the government and with corporations that work on multiple projects for the government. Another unique function of the project code chargeback method is that the codes are usually temporary and are added and removed as projects are completed and new ones are begun.

All of the forms of chargeback given here, and most definitely a few others, can be combined in any way, shape, or form that your organization chooses. Your strategic plan should reflect the chargeback method that you have chosen to use. Make sure that your TMS can allocate the telephony expenditures back the way your organization chooses. This one function can be the difference between success and failure of the TMS that you implement.

Administrative Database Population

Based upon the chargeback method that was chosen, you must now gather all of the pertinent data necessary to populate your TMS databases. If your TMS is integrated with the MIS financial records system (FRS), this task may be very simple. The greater the integration that you can achieve between the TMS and the FRS, the better. You will find that integration of computer systems in the 1990s is almost always a symbiotic relationship.

Personnel records in a TMS are driven by the extensions that the personnel are assigned to. This information forms the directory database. The TMS needs to know the following information if applicable about each telephone or individual in the organization in order to perform the call accounting function: extension number, person assigned to the extension, division/department number, and general ledger number. For additional information about what might be tracked by a TMS pertaining to an extension, please refer to the "Directory Function" section detailed later in this chapter.

There are several other databases that must be populated for the call accounting function to be performed. The division/department database must be built and maintained inclusive of, at a minimum, the division/department numbers, their descriptions, and, if applicable, their assigned general ledger numbers. A database of the tenants that will be tracked needs to be defined and maintained. If authorization codes are going to be used, a database of the codes and their responsible parties must also be generated and maintained. Sometimes this database may be a simple cross reference to the directory database. A database of all fund numbers and their definitions must be built and maintained if fund numbers are to be used. If project codes will be used, a database of the project codes along with a description needs to be built and maintained. If FAC, fund, or project codes are used, the switching platform must also be populated with these codes and the appropriate class of service assigned.

Telecommunications Database Population

There are other telecommunications-related databases that must be established in order to perform the call accounting function. The first and most important is a definition of the telephony circuits, called *facilities*, that connect your switch to the rest of the world. These circuits have many different names, such as route, trunk group, trunk, tie-line, FX lines, LEC lines, T-1s, WATS line, and the list goes on and on. See Appendix B for a definition of these terms.

You must define all of the circuits in a TMS database. The TMS wants to know how the switch will identify that these lines are being used. It wants to know how many of them you have, how much to charge for their usage, what grade of service you want to supply on these lines, and a host of other things. A definition of all telephony circuits must be obtained through whomever maintains the switching platform. The switch technician should be able to supply you with this information.

CDR Format and Call Collection

The call detail record (CDR) format is one of the most important pieces of information to gather when you are implementing a TMS. A *call detail record* is a data stream of alphanumeric characters that contain the calling information. The definition of the CDR, data dictionary, record layout, whatever you want to call it, is what links the telephony switching platform to the computer-based TMS. Telemanagement systems are passive devices that only know what they are told. The CDR definition combined with the actual CDR data, the telecommunications databases, the administrative databases, and the rate tables gives the call accounting module the ability to produce the information expected.

In order for the TMS to collect the CDR data, you must also define the communications connectivity between the TMS and the switch's CDR port. This is done by specifying the baud bit and parity that will be used on the communications ports and any handshaking software or hardware that might be required. For more information on communications ports and connectivity, refer to Chapter 15. Generally speaking, switches do not require any handshaking—they simply send the data stream out the port, and if there is a piece of computing equipment listening, then fine, but if there is not, then the data is thrown into the proverbial bit bucket where it can never be retrieved.

One feature of your switching platform that you should be aware of is what is called answer supervision. *Answer supervision* is a signaling method by which a switching platform is notified that a telephone call was answered. The switch tells the TMS in the CDR that the call was answered or was not answered. Some switches provide some kind of flag in the CDR that can be read by the TMS to determine if the call record being reported is for a call that was answered or one that was not. Other switches only generate a call record if the call was answered. However, the majority of PBX switches still do not provide answer supervision. This means that a call record is generated for every call that is placed through the system, whether it was completed or not. The TMS has no way of knowing if the call was actually completed or not, because it is passive and only knows what is reported to it.

The TMS may have to be programmed to discard call records of short duration under the assumption that the call was not completed. You may have to use a trial-and-error method to determine the right minimum duration setting for the TMS. A number somewhere between 15 and 45 seconds should be correct for most domestic traffic, and 60 seconds may be more appropriate for international calling. Your TMS should be able to determine if the call is domestic or international by the number dialed or by the long-distance facility used.

The amount of information contained in a CDR varies from switch to switch, both Centrex and PBX. Look at a single call record that you see on your telephone bill. At a minimum, all of the information that you see on a single telephone record is contained in the CDR with the exception of the call cost and the location that the call was placed to. These two pieces of information are added to the processed CDR to create a call record that you can understand. You should also be aware of the information that is contained in a CDR that is not seen in the printed call. A switch is connected to the outside world through facilities. The CDR tells the TMS what facility was used for that specific call, and subsequently how to price the call record. Analysis of the facility information is performed by traffic management and network optimization, as discussed later in this chapter.

Automatic number identification (ANI) information can also be included on incoming CDRs. ANI lets you know who is calling your organization. If your switch supports ANI recognition and your telco supplies the ANI information, your company, especially your marketing department, will be interested in knowing who is calling them and from where. CDR can also contain information about the port, the telephone, signaling used, feature used, time to connect, and so on. The amount and content of information included in the CDR will vary from switch to switch. Find out what information you are interested in and then make sure that your switch and the TMS can support and track the information.

In addition to the standard outgoing call record that you should be familiar with, a switch can also generate additional CDR information. CDR information can be produced for a number of specialized functions. Incoming call detail information can be supplied. Under most circumstances, incoming call records are summarized by extension during end-of-day processing and then thrown away, because there is little need to store every individual incoming call record. If ANI information is included by your switch in incoming CDRs, you may want to keep the detail information for further analysis. Incoming call volume is about equal to the outgoing call volume. If you plan to store incoming call information, make sure that the TMS is configured to process and store the additional call volume.

Station-to-station calling can be reported. This will show all calls placed through the switching facility to other extensions connected to the switch. This type of CDR can help you determine interdepartmental calling patterns, and combined with the other CDR information, the total time an extension is on the phone on a daily basis. Station-to-station traffic is extremely useful to the security department when trying to track harassing phone calls or other unauthorized access to the system. Station-to-station traffic can generate up to three times the number of CDR as outgoing calls alone. The TMS can be programmed to discard the calls at the end of the day, but if you desire to keep the station-to-station CDR detail, you must make sure that the TMS is configured for the additional volume of call records.

Special CDR can be generated for conference calls, feature usage, 911 calls, ISDN calling, call attempts, and so on. Make sure that your switch provides the TMS with any and all information that you feel is pertinent to the proper management of your telephony environment. The TMS can only work with the information it is supplied with. Try to keep this in mind while you are searching for new ways to put your TMS to good use.

Call Buffering Devices

Call buffering devices supply an added level of security to your TMS. A *call buffer* is either a personal computer or a solid-state unit, such as the one depicted in Figure 6-2, that is designed to collect, store, and pass CDR on to another computing device, usually a TMS. Call buffers range in price between $1,000 and $5,000, depending on the buffer quality and the storage capacity of the buffer you purchase. Buffers can be

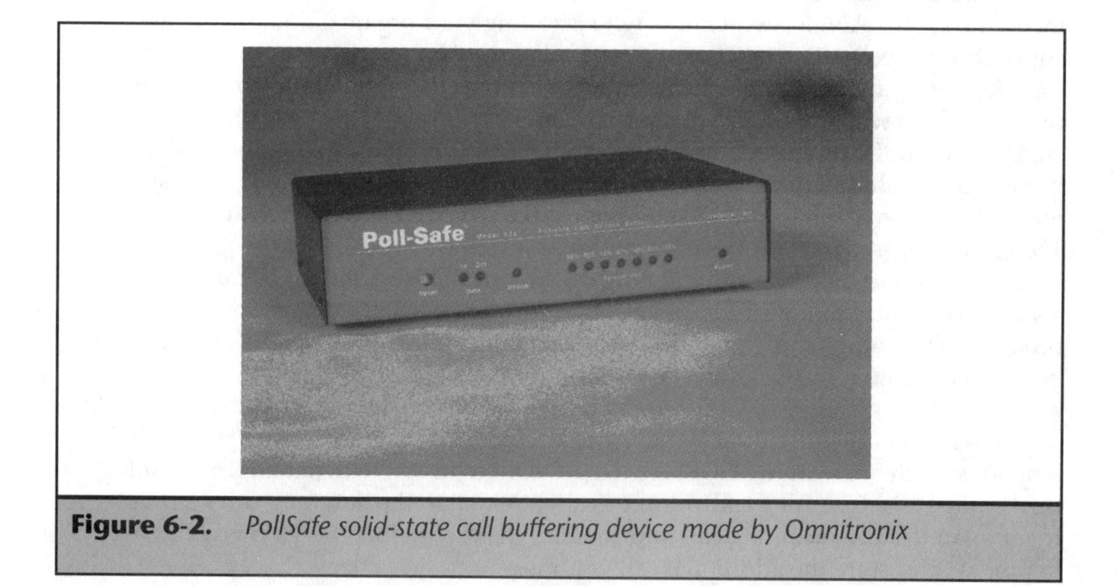

Figure 6-2. *PollSafe solid-state call buffering device made by Omnitronix*

configured to hold between 4,000 and 175,000 call records. Some buffers can be chained together to provide even greater storage capacity.

A call buffer is necessary when a TMS is required to collect call records from multiple sites in a polling method rather than in a direct connect mode. The buffer collects and stores the CDR until the TMS calls the buffer, at which time the CDR is downloaded to the TMS. It is *highly* recommended that you acquire a solid-state buffer that is capable of storing between three and seven days worth of calling volume from your switching location. The solid-state call buffering devices have no moving parts to break and are extremely reliable, in comparison to the PC call buffer, which can hold more call records but is simply another computer waiting to break. Computers are prone to down time; the TMS, a computer, is no exception to the downtime rule.

With the exception of a few switching platforms that supply internal call buffers, if the TMS is not operating properly for whatever reason, call records will disappear into oblivion unless you have a call buffer in place. Even if you are collecting call records in a direct connected, real-time mode, a call buffer is recommended to add integrity to the system and to provide handshaking to assure against call loss. If you are using the TMS as a chargeback vehicle, losing calls for one day, or even one hour in some cases, can cost more than the price of the call buffer. Do not cut corners when it comes to your call buffers—this is money well spent.

Call buffers provide a number of other features and functions. They can put out both visual and audible alarms to warn you when they are filling up. Call buffers can be programmed to call a certain telephone number, such as a pager, to warn someone when they are nearing capacity. They can call the TMS and say it is time to be polled. Some call buffers have some basic toll fraud detection that they can perform. They can also provide handshaking between the TMS and the buffer where the switching platform potentially could not. This assures that data is not lost during collection due to processing lag. Some call buffers, such as the one depicted in Figure 6-2, have a built-in UPS supply that will allow the buffer to collect CDR for up to 15 hours after power to the unit is lost. The call buffer can also store CDR that has been collected for up to 45 days with no external power. Call collection is the heart of call accounting. Without accurate call collection, your TMS will not supply accurate information.

TIP: Always purchase a call buffer for every switch that the TMS will be collecting calls from. The money spent on call buffers is very small in comparison to the peace of mind and the integrity that they add to the telephony environment. If your TMS call collection module fails even once, the cost of the buffer will be recovered.

Pricing

One of the most powerful functions of a telemanagement system is that the computer gives you the ability to define the pricing method that you will use to charge back the telephone calls made through your switching network. Now, your switching network may contain only a single switch, but it is a network nonetheless. You probably have negotiated some specialized pricing for all of the long-distance calling that your organization does. This pricing is significantly less than standard long-distance rates. This is because of the extreme volume of calling that the long-distance vendor is expecting to obtain, the dedicated long-distance connection that you probably have, and the additional switching function that you provide instead of the regional Bell operating company (RBOC). See Figure 6-3 for a list of the seven major Bell operating regions of the United States.

You will probably be able to negotiate somewhere between 30 percent and 65 percent off of your long-distance calling charges based upon your geographical location, calling pattern, and size of your organization. Since January 1, 1995, your long-distance vendor can even carry your intralata traffic. *Intralata* refers to a call that is placeable by your RBOC without switching the call to another carrier, usually within 100 miles of the call origination. These calls used to be the most expensive domestic calls your organization would make, because the RBOCs had a monopoly on carrying this traffic. Now the RBOCs only have a monopoly on local measured traffic. Look for RBOC rates to rise in the near future. Since long-distance vendors can now also carry this traffic, the intralata market is now competitive and prices have fallen. Refer to the "How to Choose a Long-Distance Vendor" section later in this chapter for more information.

As a telecommunications administrator, you and your business must decide on how you will redistribute these costs back to the others in your organization. Remember

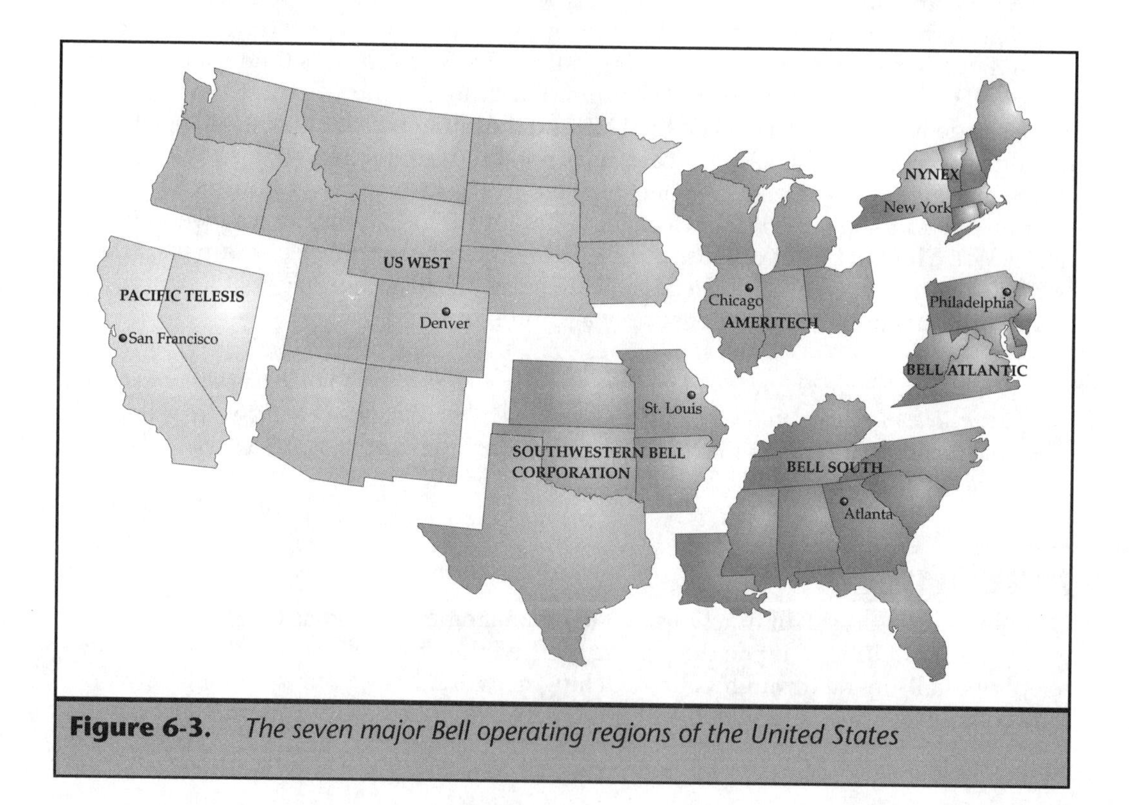

Figure 6-3. *The seven major Bell operating regions of the United States*

that the goal is to make telecommunications a zero-based accounting department. In order to attain the reductions in long-distance costs mentioned above and the reductions in line fees that you get when you own a PBX, you incur other, hopefully smaller, costs. These costs should now be factored back into the costs that you redistribute. The net result of all this effort should be a significant overall reduction in cost and increased quality of service to the organization.

Your TMS should give you the power and control to create or modify the rates that are charged for the long-distance calls made. Normally, you will start with the standard long-distance rates used by a long-distance carrier, for instance AT&T DDD (direct distance dialing) rates. All of the local and long-distance carriers publish tariff rates, which can be obtained in a database format. Your TMS should have come with at least a one-year subscription to this tariff database. CCMI, a subsidiary of United Communications Group, Inc., is probably the nation's leading provider of tariff databases.

The tricky part of tariff databases is that they must be created so that each database reflects your specific location as the center of the calling universe. All calls must be priced based upon origination at your specific location(s). The database uses the NPA (area code) and the NNX (the exchange—the first three digits of your seven- digit local telephone number) of your place of business to center in on your location. These are the first six digits of your long-distance telephone number. The TMS can price the calls your organization makes based upon the rate table or any other method you choose. Some of the features you should expect out of a strong TMS pricing module are:

- Percentage surcharge increase or decrease
- Flat surcharge amount increase or decrease
- Minimum cost assignment to calls
- Minimum per-minute cost assignment
- Flat rate override of rate table charges
- Flat per-minute override of rate table charges
- Flat tenths of a minute billing (six-second increments)
- First and subsequent segment billing increments
- Evening and night discount calling percentage
- Local calling exception pricing
- Discount or surcharge pricing based upon the type of user of the phone system
- Multiple pricing computation and storage for the same call record (three or more)
- Pricing based upon facility used
- Network facility pricing
- Private network billing, flat, per minute, and tenths of a minute
- Private network surcharge addition to LD charges flat, per minute, and tenths of a minute

These are just some of the pricing methods that should be available to you in your TMS. The ability to price the same call record multiple ways will enable you to charge back one rate and at the same time calculate the actual rates that you should be charged by the local and long-distance vendors. This is extremely useful if you are planning to perform vendor bill reconciliation, which is detailed later in this chapter. Being able to price a call record yet a third way is useful when you would like to play a what-if scenario. Trying to find the correct pricing method to make telecommunications a zero-based accounting department is not always easy. Experimenting with your pricing algorithm before implementing it is a useful tool.

If you have sublet space, have contractors working on-site, or have different categories of users, such as faculty, staff, students, doctors, nurses, patients, concessions, and/or personal users, then having the ability to price call records at different rates based upon the type of user of the system is a very important feature. Sometimes you may be reselling the telephony back to the user. Other times you may just want to distribute the burden of overhead disproportionately. If the choice is to be yours, the TMS must be able to assign charges based upon the types of users of the system. For more information on this subject, refer to the "Call Resale" section later in this chapter.

Reporting

Telemanagement systems generate information for use in a number of areas of business. The most obvious reporting of a TMS is the call accounting function listed above. Due to the numerous variations of call accounting report formats, this book will not provide examples of the various chargeback reports. A description of the minimum reporting functions that should be supplied by your TMS is provided below.

- Call detail chargeback reports should look very similar to your personal telephone bill that you receive at home. The extension, call date, time, number dialed, location, duration, and cost should be included at a minimum. The call detail information should be summarized at the bottom of the call detail chargeback report. The primary difference between your personal telephone bill and the one your TMS will generate is that the amounts charged back and what is charged back are determined by you. Sorting and level of detail are also determined by you.
- Summary call accounting reports should reflect at a minimum the following items at each subtotaling level: total number of calls, total number of minutes, total cost, average number of minutes, and average cost.
- Budgetary reports should be available, in the event your organization chooses to track budget variance. The budgetary report should be summarized at the extension or FAC level. The following information should be included both for the period-to-date and the year-to-date: number of calls, total time, total cost, budgeted amount, and variance percentage.

- Ad-hoc reporting allows the user to make inquiries on the call record database based upon any field stored in the call record. The searching fields provided in ad-hoc reporting should include, but not be limited to: tenant, division/department, extension, authorization code, call date, time, number dialed (including all of the subfields of the number dialed), cost (all three costing fields), facility used, and trunk within facility used. Ranges of the search fields should also be specifiable. The ad-hoc feature is very powerful, and it can be used under dozens of circumstances to provide valuable information. It will provide you with quick feedback about the usage and performance of your telephony environment. Ad-hoc reporting is also used to assist in toll fraud detection. A strong ad-hoc reporting module is a *must* for any TMS.
- Number dialed detail or summary reports can show you every telephone number that was dialed through your telephony network. The report should be able to tell you how many calls were made to each number, who was making those calls, how much time was spent on the calls, for how much money, and averaging information. This information is useful to the business, but it is also valuable in conjunction with network optimization, which is discussed later in this chapter.
- An area code summary report can help the higher-level individuals understand where the long-distance telephony dollars are being spent. This report summarizes the calls placed to each area code or country code to show you: how many calls were placed, total duration on the line, total cost, average time per call, and average cost per call. This report is especially handy when you are choosing a long-distance vendor. A long-distance vendor's account representative can be handed this report, and can then accurately estimate the amount of traffic the vendor can anticipate receiving from your organization. Subsequently, the vendors can provide you with accurate proposals for your long-distance business.
- Exception detail and summary reports should be provided and configurable. The user should be able to tell the exception reporting programs what qualifies as an exception. Exceptions should be definable by cost, duration, time of day, and number of calls. The exception reporting is used to establish patterns of calling as well as track abuse of the telephony network. The information generated from exception reports can be combined with other information to help with general management, network management, and fraud detection.
- Watch dog reporting is a useful reporting tool that your TMS should provide. The watch dog report can be set up to report when any telephone number that you specify is dialed. As many numbers as you desire can be reported. The local radio stations, pizza parlors, home telephone numbers, and any other numbers that you determine to be abused can be added to the watch dog list. From that moment on, the TMS will find the telephone calls for you.

- Incoming call analysis can provide your organization with some vital information about its operation. You will be able to analyze the performance of your customer support staff, marketing department, operator consoles, and any other group of people that takes incoming calls to your organization. Of course, individuals can also be monitored. Total number of calls answered, total time on the phone, average time on the phone, and cost, if inbound 800 charges are assigned, can be reported. Statistical information about groups of individuals can also be generated to show performance and efficiency of individuals within a group. If ANI information is provided for incoming CDR, then additional detailed and summary information may be generated to obtain demographical information vital to sales and marketing.

These are just some of the reports that you should expect from a good TMS. If the TMS has the data available to create other valuable information for you, then the structure of the TMS should be flexible enough to provide you with the information you need. Just keep in mind that the TMS can only report on information it is provided. Switching platforms still do not provide a great deal of information to the TMS via the CDR port. In the future, look for switching systems to provide more information to TMS for technical and managerial reasons. Patches and custom programming are also available from some switch vendors to provide more information to the TMS. If you are searching for a TMS or require more information about call accounting and the reporting function, please refer to Appendix C.

Toll Fraud Detection

Toll fraud detection is the ability of the TMS to detect unassignable long-distance usage and to determine who is responsible for the long-distance usage in the first place. A TMS should be able to identify and assign almost all toll fraud in the telephony environment that it oversees. Toll fraud detection entails investigation of incorrect postings, such as an authorization code that is in use, but not defined in the TMS, or unauthorized use of someone else's authorization code. Toll fraud can also appear when DISA (direct inward system access) lines are used by your organization. DISA allows an employee to call directly into the telephone switch and to access outgoing lines through the use of FACs and unique dialing patterns.

Unfortunately, DISA is extremely volatile and can also open your telephone system up to illegal usage. The TMS may be able to detect the fraud, but to help prevent the fraud from happening, the switching system must be designed with the telephone hacker in mind. *Never underestimate hackers!* These people have nothing better to do with their time than to sit around and try to break into your phone system. Without toll fraud detection in place on your TMS, you might not even know that your phone system's integrity has been breached until you get $100,000 added to your monthly phone bill.

A TMS collects the CDR information directly from your local switching system. The RBOCs and the long-distance vendors do not know what extension placed the call, only that the call was placed from your installation. When it comes to toll fraud, knowing the origination point of the call is a vital piece of information. Whether your telephony system is Centrex- or PBX-based, you can be certain that toll fraud detection is a major concern that should not be treated lightly.

How to Choose a Long-Distance Vendor

Choosing a long-distance vendor today is not nearly as difficult as it used to be. In the past, each long-distance carrier had certain areas of the country in which it was very strong. Their pricing might have appeared very attractive to certain areas, but other areas of the country may have been weak, and the pricing reflected that weakness. Over the past five years, there has been a move to monopolize a company's long distance by offering favorable rates based upon mileage rather than the area code and exchange that you are calling.

The area code summary report described earlier will give the long-distance vendor an accurate picture of your calling pattern. Print this report out and supply it along with your request for bid. It will make the long-distance procurement process much easier, but if you don't have a TMS yet, you may have to compile the information by hand or supply your phone bills to the long-distance vendors if they require it. However, you probably do not want to give them that much detail about your organization. The more attractive rates are contingent upon the long-distance vendor getting all of your long-distance business.

In addition to the lower overall rates that are proposed by the long-distance vendor, the circuits needed to carry the traffic are many times thrown in for free. A T-1 circuit that can carry up to 24 voice communications channels simultaneously costs somewhere between $750 and $1,750 a month per T-1, depending on your installation's relative distance from the long-distance vendor's hub location plus installation fees. (See Chapter 10 for more information about T-1s.) These fees alone can add up quick, so remember to negotiate your best overall deal. Sometimes a free T-1 with slightly higher long-distance rates, after analysis, can be more cost-effective.

One other thing to write into your long-distance contract is penalties for lack of performance, down time, and/or response time. Although you will find that long-distance vendors are pretty responsive, having a penalty clause can reduce your down time even further and may put a few extra dollars back into the company's pocket.

TIP: When requesting bids for long distance, ask for all of the costs broken out and then ask for all of the fees included in the long-distance rate quoted. Also ask for discounts for signing two-, three-, four-, and five-year contracts. Take the proposals made to you and analyze your detailed calling pattern. Don't forget to incorporate your anticipated telephony growths to help determine which vendor, method of pricing, and term of contract is best for you.

Network Tie-Line Reconciliation

Network tie-line reconciliation is when a TMS collects call detail record information from multiple telephony nodes that are connected to each other. The TMS then combines the multiple call records together into a single consolidated call record that contains all of the legs of call transmission and their associated costs, as depicted in Figure 6-4. The configuration will undoubtedly be different, but the model remains basically the same.

What is a tie-line anyway? A *tie-line* is a dedicated circuit (usually a T-1) that connects two locations, typically connecting two branches of an organization. Tie-lines (among other things) allow you to make connections by dialing shortened extension numbers rather than entire telephone numbers to reach other branches of your organization. Tie-lines can be used to carry voice, data, and video information, depending on how the circuits are configured.

The call passed over a network will produce two or more call records, each having information that the other does not. However, there is some information that is common between the call records. The TMS must be able to combine the calls based upon the common data. The date, time, number dialed, and durations should be

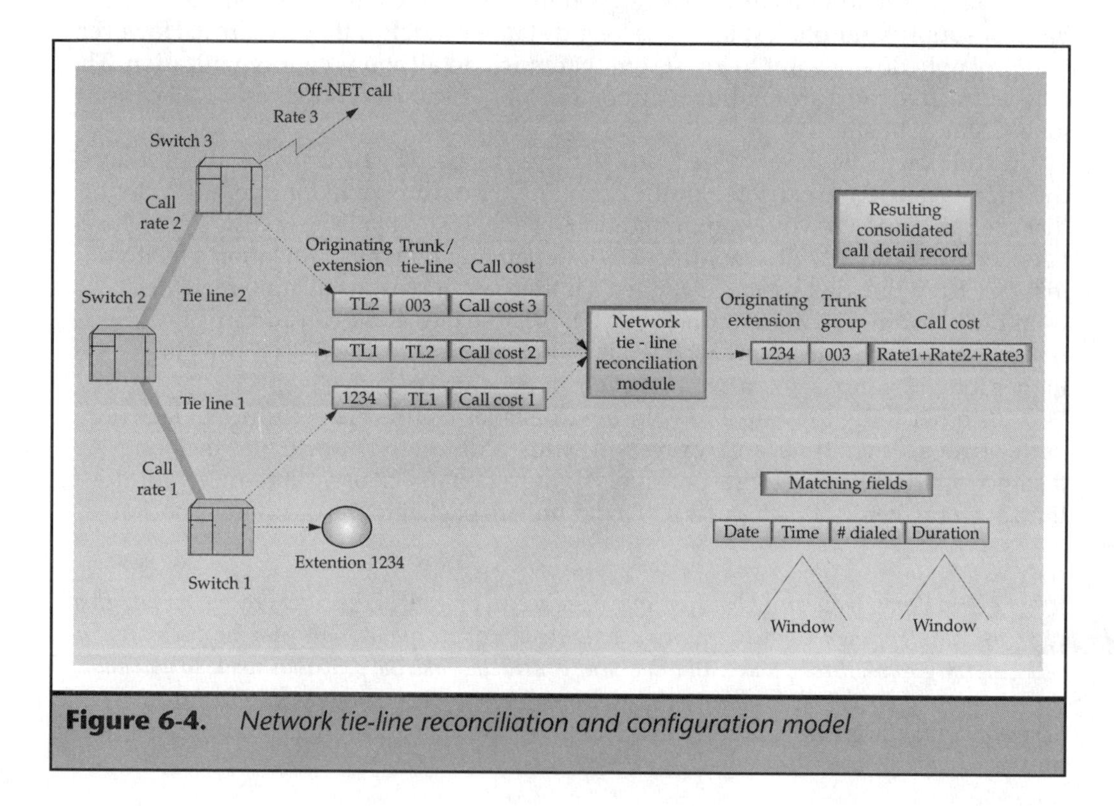

Figure 6-4. *Network tie-line reconciliation and configuration model*

roughly the same for all records. The switch clocks, time zones, and the connect time through the network can make the data not 100 percent equivalent. The network tie-line reconciliation module allows you to adjust the margins of error, the windows so to speak, to obtain an optimal hit-to-miss ratio when combining the call records. Unmatched call records can be matched manually or the windows can be adjusted and the data reprocessed.

Synchronizing your telephony network is extremely helpful when performing network tie-line reconciliation. Once the call records are matched, the extension and authorization code from the originating record must be combined with the trunking information found in the terminating record. The reconciliation must then combine the call charges to obtain an accurate call cost. Once the reconciliation is complete, the call record can be passed to the call accounting system for further processing.

Vendor Bill Processing

Vendor bill processing is the process of validating the amounts that your telephony vendors are charging you. Without a TMS, this process is impossible. Telco vendors realize that they make errors in billing, almost always in their favor, but they do not have a way to fix the discrepancies unless you, the consumer, point out the errors. Vendor bill processing is performed by downloading some form of electronic media that contains your local and long-distance telephone bills onto the TMS. A conversion utility will be needed to convert the telco call records to the same format as the TMS call record. Every vendor supplies their vendor tape in different formats and on different media. A common media must be selected between the two systems.

Once the telco detail is downloaded and converted, the TMS sorts the telco's call detail records and the TMS call detail records in a chronological order. The TMS then performs a call-by-call comparison of each and every billable item. If the call records do not reasonably match in duration or cost (which is definable by you), they are included in a discrepancy report. This process is very similar to network tie-line reconciliation in function. With the discrepancy report generated, you can then go back to the local and long-distance vendors and request refunds for incorrect billing in the telco's favor.

The telcos almost always provide the refunds prescribed by the vendor bill processing module. Why? Because they realize that their switching systems can report incorrect information to their billing system just as easily as your switching system can supply your TMS with incorrect data. However, your TMS is dedicated to information tracking for your organization and your organization alone, while theirs is responsible for millions of customers. Both systems will make mistakes, but your TMS is undoubtedly more accurate than the telephony vendor's.

Vendor bill processing has been known to save companies tens of thousands of dollars every month, and it can easily pay for the entire cost of a full-blown TMS. Long-term savings can be in the millions of dollars, paying for further telephony enhancements. Your TMS must be configured specifically for this task, because approximately twice the number of call records must be stored online to perform the process. If your organization has over 500 extensions, odds are that vendor bill processing will be a major cost saver for you.

General Ledger Interface

General ledger interface is an add-on module to telemanagement that allows financial information gathered by the TMS to be shared with the financial records system (FRS) being maintained by the data processing department for the accounting, business, finance, and controller's offices. The general ledger interface module can reduce the amount of paper generated by the TMS, reduce the amount of redundant data entry, and create a feeling of continuity between the two departments.

Step number one in the process of integrating the two systems is to find common ground. Usually the FRS has been in place long before a TMS was in the picture. Business accounting is a very personal thing. Every company has finely tuned their FRS to perform around the way they do business. Since the TMS must perform an accounting function as well, the general ledger interface portion of the system must be tailored to the specifications of the FRS. Department codes, general ledger numbers, project codes, fund numbers, and a dozen other methods of grouping and allocating costs must be accommodated by the telemanagement system.

Integrating the TMS with the FRS can be an easy task, or it can be virtually impossible. When you developed your strategic plan to procure a TMS, you should have verified that the TMS could support the accounting structure that the accounting department required. If you did your homework, the TMS database structure should be able to support the accounting structure that you desire. Now you must integrate the two systems. If you are really fortunate, planned properly, and were lucky enough to find a TMS with all the bells and whistles that you wanted, the TMS might be able to use the same database engine that your FRS does. Oracle, Informix, Sybase, and Ingress seem to be the more popular relational databases of the day. Your FRS may use one of these databases or some other method of database storage and retrieval.

If the FRS and the TMS use the same database engine and the two systems either coexist on the same platform or can be connected via ethernet or some other networking protocol, your job may be done. The MIS department, which usually runs the FRS, may even choose to download every telephony detail record directly to the e-mail system, so that no paper needs to be generated for internal chargeback of telephony expenditures. This level of integration is extremely desirable, but rarely achieved. Remember to set realistic goals for your TMS and the general ledger interface module. If your systems are not fully integrated, you may have to do a little work to interface the TMS with the FRS. Your TMS vendor should be able to provide you with conversion utilities to integrate the two systems. If the hardware platforms cannot be integrated, it may be necessary for the TMS to generate a diskette or magnetic tape of the general ledger information. Whatever it takes, as long as the information is presented properly to the FRS, your TMS has done its job.

Network Optimization

Analyzing all of the components that make up an organization's telecommunications network to determine where performance can be raised and/or costs can be cut is the goal of network optimization. Voice, data, and video traffic can all be optimized either separately or together.

Network optimization is a process of gathering traffic information from all pertinent sources connected to the network and then making additions, deletions, and changes where necessary to obtain an ideal configuration of local, long-distance, FX, tie-line, and ISDN facilities. Telephony network optimization is performed in almost the same way that a LAN supervisor might optimize the data traffic through the environment. The LAN administrator has advanced computer network tools to help with the optimization process, not to mention the network protocols, bridges, and routers that assist with the ongoing optimization of the LAN environment. The telemanagement network optimization application is the best tool that a telecommunications administrator has.

The primary difference between a LAN administrator's optimization job and a telecommunications administrator's job is the amount of information that needs to be gathered and the diverse sources of information that must be combined to provide a comprehensive telephony network optimization plan. Oh, and let's not forget about *money*. Network optimization for LAN administrators is primarily a performance and routing issue. Network optimization for telephony can span the globe in hundreds of different ways and can involve billions of dollars per year. Poorly managed networks, not to mention individual PBXs, cost users *big*! As much as a 30 percent to 70 percent inefficiency in network and external telephony connection sources is common, and up to 100 percent cost reductions in local area calling costs where network nodes are located. If your company is large with locations across the country and you have tie-lines between locations, you can conceivably call from coast to coast for no additional cost.

If you are large and/or growing, you may be trying to determine where it is best for you to put tie-lines between your locations. Tie-lines aren't free, so you must spend where your savings will be greatest or the telephony performance increases the most. With the greater demands of data and video traffic added to voice traffic, tie-lines are needed more than ever. By examining your actual calling patterns from the call accounting reports available, the traffic management information described below, and the organizational plans for telephony growth, you can determine where tie-lines are justifiable.

FX lines should also be examined for potential savings. An *FX line* is a special service that, for a fixed fee, will allow you to call a toll area for free. There is a break-even point for purchasing FX lines, but if your calling traffic dictates that there is a cost savings to be had, you may as well take advantage of it.

Traffic Management

Traffic management is the process of collecting all telephony traffic to determine the state of the network. By analyzing the traffic management information gathered by the TMS, the calling patterns of your organization are more clearly defined. The network optimization module is dependent on traffic management.

Telephone systems have what is called an ARS (automatic route selection) pattern. ARS tells the switch which outbound line(s) to use to make the connection that the caller has just requested. If the caller is trying to make a long-distance call, the ARS pattern might connect the caller to the outbound WATS (Wide Area Telecommunications Service) line. Or, if they are trying to call over to corporate headquarters, the ARS pattern might be programmed to use the network tie-lines to complete the call. ARS patterns also contain second and sometimes even third and fourth choices. Management can readily see how the ARS is programmed and also determine if there are any problems with the existing network through this detailed analysis.

Take a look at Figure 6-5. This figure shows an organization's typical calling pattern for a specific day, over a specific trunk group, broken up into time slices. Analyzing this type of report over, say, a three-month period, the user will be able to identify areas of waste and places for growth in the telephony network.

What information must a traffic management module generate anyway? Briefly, here is a description of the reports you should expect:

- Calling traffic by trunk and time of day, usually represented in graphical as well as spreadsheet format (see Figure 6-5). Information about erlangs and/or centi-call seconds (CCS) is provided for added analysis. (See Appendix B for definitions of these terms.) Outbound, two-way, and incoming trunks can be analyzed with this report. Optimization recommendations can be made based upon the traffic and a desired grade of service. A *grade of service* is a desire to have a minimum of x percent of all call attempts be completed successfully with limited traffic overflow to other facilities. Typical traffic goals are 95 to 99 percent. Analysis of several months of reporting is necessary before conclusions can be drawn.
- Trunk utilization by trunk group provides a more specific picture of the individual trunks within a trunk group. Outbound, two-way, and inbound trunks can be reported on. Traffic totals are provided for individual circuits that include total calls, total time, average time, and overall percentage of traffic carried by each individual trunk within the route or trunk group. Period-to-date totals are also presented. By looking carefully at these statistics, you can see where there are facilities problems as well as over- and under-configuration of trunking facilities. Refunds for vendor-provided facilities

that are not functioning properly can also be obtained through proper use of this report.

- Exchange and route traffic analysis, both detail and summary, brings traffic management to a new level of detail. Exchange and route reporting generates a list of all area codes and exchanges dialed within the network and the route or routes that carried the traffic. Call traffic overflow is easily seen in this report. Through proper analysis of this one report alone and some pricing information gathering from your RBOC, you can determine if FX lines are justifiable. The ARS pattern is also well-defined by this report.
- Route and exchange traffic analysis, in both detail and summary, shows you precisely what you are using your facilities for and how much your facilities are being used. Waste and room for downscaling are more easily seen through analysis of this report over time. The potential for ARS modification is also visible using this traffic management tool.

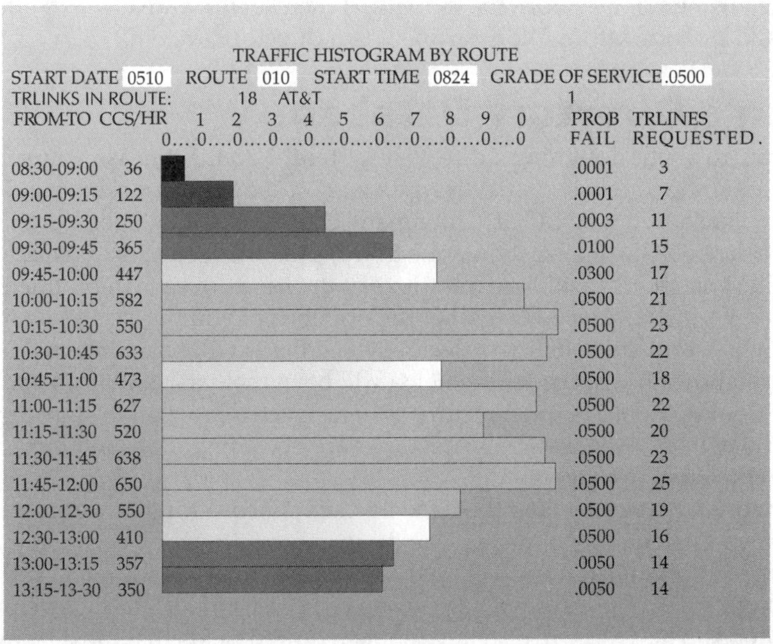

TRAFFIC HISTOGRAM BY ROUTE

START DATE 0510 ROUTE 010 START TIME 0824 GRADE OF SERVICE .0500

TRLINKS IN ROUTE: 18 AT&T 1

FROM-TO CCS/HR 1 2 3 4 5 6 7 8 9 0

0....0....0....0....0....0....0....0....0....0

FROM-TO	CCS/HR	PROB FAIL	TRLINES REQUESTED.
08:30-09:00	36	.0001	3
09:00-09:15	122	.0001	7
09:15-09:30	250	.0003	11
09:30-09:45	365	.0100	15
09:45-10:00	447	.0300	17
10:00-10:15	582	.0500	21
10:15-10:30	550	.0500	23
10:30-10:45	633	.0500	22
10:45-11:00	473	.0500	18
11:00-11:15	627	.0500	22
11:15-11:30	520	.0500	20
11:30-11:45	638	.0500	23
11:45-12:00	650	.0500	25
12:00-12-30	550	.0500	19
12:30-13:00	410	.0500	16
13:00-13:15	357	.0050	14
13:15-13-30	350	.0050	14

Figure 6-5. *Sample traffic histogram report*

Traffic management can become extremely difficult when your telephony network is large. Decisions made with regards to your telephony network should be made with painstaking detail, analysis, planning, and creativity in mind. Network changes can be very costly, and you don't want to make any mistakes.

Directory Function

The directory function of telemanagement is necessary because the information required to perform the call accounting function is a subset of the information that is needed to perform a directory lookup and printing function. Therefore, it is a normal progression to take the directory function and place it on the TMS. That way, as soon as the information is entered for a new directory entry, that same information can be made available to the call accounting module and other modules that reside on the TMS. You should look for a directory system that has a strong phonetic search feature, so the lookup function will be user-friendly and efficient.

The directory function is a curious animal. Information that can affect the directory database can come from a number of sources. Your personnel department, the security department, the accounting department, the telecommunications department, the housing department for schools, the individual themselves, and a host of other areas can have and provide information pertinent to a directory system.

Designing a Directory Database

How do you go about designing a directory system? You need to ask yourself and your staff a number of questions: What information should be tracked in a directory field, and what should it be called? How many characters of data need to be stored for the field? Where will the information come from? Is there a need to do database lookups by this field? You should ask these questions of the heads of each department and everyone on your staff. Once you gather the opinions of your associates about the directory, it is your responsibility to decide which fields will be maintained in the directory database, in what order the fields will be in the database, and which fields will be key lookup items. Some directory systems even allow you to store a graphical image of the person being tracked. Directories are becoming more and more advanced as computer technology grows.

Do not forget that every field that is to be maintained through the directory must be maintained by you and your staff. The design of the directory system is a unique task in that your organization will choose to store and maintain different information from various sources. The directory system you use must support the field definitions that you choose. The TMS will also have some required fields that must be maintained in the directory database. Try to remember who will be using the system. First and foremost, the operators must feel comfortable with the module and how it operates. After all, they are the ones who will use it the most.

Since directory information comes from so many sources, you must plan ahead for how the information will be gathered. Usually the majority of the directory information already exists in other departmental databases. The MIS department may be able to supply you with a directory database to get you started, but you must coordinate a method of sharing your directory information with the MIS department's databases. A form can also be developed that new users of the telephony system must fill out, so the information can be manually entered. If you will be using a work order processing module, you will find that the majority of work order information can be incorporated into the directory system automatically. For more information about work order integration, refer to Chapter 9.

Directory Database Security

A complete directory database stores a great deal of proprietary information. You certainly don't want anyone to be able to view your authorization codes, employee home telephone numbers, names, and addresses. You may be tracking dozens of pieces of information that are either personal or proprietary, which only select individuals should be able to maintain, look up, view, or print. Security is usually supplied through operator IDs, passwords, and security clearances. Your TMS must be able to supply the security you deem necessary for your valuable directory information.

Directory Import, Export, and Printing

A directory system should be able to import existing directory information from a number of other sources. There must be a conversion media available between the computer systems and some common fields that the TMS can use to merge multiple databases, but importing directory information should not be a problem. Exporting and printing your directory database are similar. The export database will need to be created in a format specified by the receiving system or the requesting party. A directory print is roughly the same process, except the information usually has page header information and a specialized page layout. The directory system should give you the power to define the record layout, selection, and sort criteria. Security should also be in place for the export and printing functions. If your TMS database is integrated with the MIS system, you will find that the process of maintaining, importing, exporting, and printing your directory becomes much simpler.

Call Resale

Call resale is the process of turning your organization into a telephone company. Local and long-distance calling purchased in bulk is far less expensive than it would be if purchased by an individual. Organizations can negotiate drastic reductions in local and long-distance rates when the volume of traffic is known. Reductions of 40 percent, 50 percent, or even 60 percent on toll charges can be negotiated based upon the site's geographical location, the volume of calls, and the calling pattern of the organization. A *calling pattern* is when you analyze precisely where and how often you call. Your calling pattern can tell the educated local or long-distance salesman precisely how much business can be expected and how profitable that business will be.

The telemanagement system is already collecting, pricing, and tracking every telephone connection that is made through the switch. A good TMS can price the same call several different ways. By adding on the call resale module to a TMS, the subscribers to the service can be offered a substantial reduction in the cost of placing calls, while at the same time the organization can make a nice little profit to pay for other telephony expenditures and upgrades. Call resale has paid for the cost of more than one phone system in the past and is popular with both users (employees, customers, contractors, doctors, patients, students, etc.) and organizations.

How does call resale work? When people are at your location, they are what we call a captive audience. Because the telephones on your site/installation/campus are owned by you and not by the visitor/contractor/employee/patient/student, you have the opportunity to sell these guests their long-distance service at a lesser rate than they would have to pay if they picked up one of your telephones and used a credit card. Rates typically charged to subscribers of a call resale system range from 20 percent below AT&T DDD rates to 10 percent above AT&T DDD rates. Additional services can also be offered (for a fee), such as: a monthly line fee, voice-mail, call forwarding, call waiting, calling plans, etc. In essence, you can offer and charge for any service that your switch or your staff can provide. Figure 6-6 shows a call resale account inquiry screen, including a portion of the telephone bill that has been accrued. Call resale is really a complete accounts receivable/accounts payable system. Everyone can save or make money and at the same time enjoy savings on their telephone bill if call resale is implemented correctly.

Call Resale Interactive Voice Response (IVR)

Call resale interactive voice response (IVR) is one of the newer operations added to a telemanagement system platform. The call resale IVR works about the same way that banking by phone does. The user is greeted by a pre-recorded human voice which instructs them to enter their account number and then typically their authorization code for verification. Once the user has been verified from the TMS database, information about the account, such as the account balance, payment information, cost of any call that was placed, dialing instructions, and call rate information can be retrieved from any touch-tone phone. Don't you wish your phone company would offer this convenient feature? Look for more features to be added to IVR systems in the near future, such as integration to other telemanagement modules. For more information on IVR systems, please refer to Chapter 17.

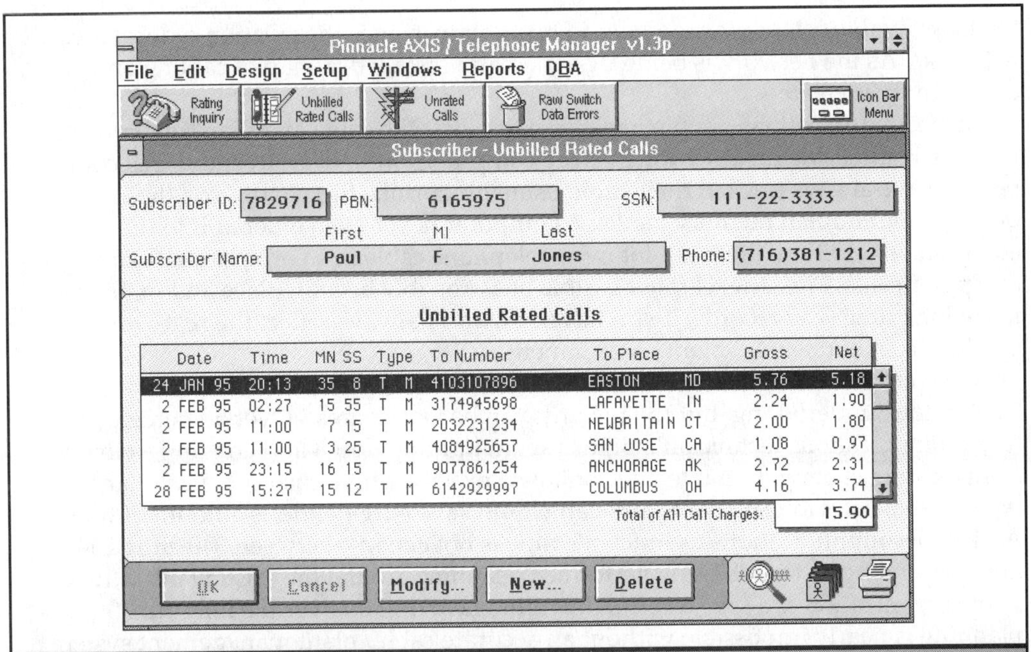

Pinnacle AXIS / Telephone Manager v1.3p

File Edit Design Setup Windows Reports DBA

Rating Inquiry | Unbilled Rated Calls | Unrated Calls | Raw Switch Data Errors | Icon Bar Menu

Subscriber - Unbilled Rated Calls

Subscriber ID: 7829716 PBN: 6165975 SSN: 111-22-3333

Subscriber Name: First Paul MI F. Last Jones Phone: (716)381-1212

Unbilled Rated Calls

Date	Time	MN	SS	Type		To Number	To Place		Gross	Net
24 JAN 95	20:13	35	8	T	M	4103107896	EASTON	MD	5.76	5.18
2 FEB 95	02:27	15	55	T	M	3174945698	LAFAYETTE	IN	2.24	1.90
2 FEB 95	11:00	7	15	T	M	2032231234	NEWBRITAIN	CT	2.00	1.80
2 FEB 95	11:00	3	25	T	M	4084925657	SAN JOSE	CA	1.08	0.97
2 FEB 95	23:15	16	15	T	M	9077861254	ANCHORAGE	AK	2.72	2.31
28 FEB 95	15:27	15	12	T	M	6142929997	COLUMBUS	OH	4.16	3.74

Total of All Call Charges: 15.90

OK Cancel Modify... New... Delete

Figure 6-6. *Sample call resale account inquiry screen*

Courtesy Pinnacle Software Corporation

Facilities Management System

Facilities management is a term used in the industry that covers three main functions of telemanagement: cable/plant management (CPM), equipment/feature inventory management (EFI), and work order processing. In order to track the assets of an organization properly, the facilities management databases must be accurate and kept up to date. If used to its fullest ability, the facilities management system can be the single point of entry used for all telemanagement applications. In the future, you will see that the facilities management system will be the single point of entry for almost all telephony-related products.

Cable/Plant Management

Cable/plant management (CPM) is the formidable task of trying to track and manage every single piece of wire and cable that is in use, available, bad, and pending connection within the entire organizational structure. LAN administrators have taken the task of cable/plant management to heart almost since the creation of local area networks. As the network is being designed and installed, a map of cabling and connectivity is created.

Telecommunications people have been tracking the cable and plant information they are responsible for since long before computers were ever invented. Telecom people have always tracked their cable using blueprints, flip charts, and the ever-popular human memory banks. A computer never even entered into the equation. In addition, there is a lot more telephone cable in a corporation than there is computer cable. Plus, the telephone cable has probably been in place and in use for a fairly long time. The records, if you have any, don't always reflect the real-world conditions of your cable/plant environment.

Converting cable/plant information onto a computer-based TMS can be a tremendous undertaking. But the rewards can be just as great. Over 50 percent of a telecommunications technician's time is spent hunting down and creating a circuit from existing cabling. If that technician knew exactly where each cable went and which cable pairs to use, that technician would save a tremendous amount of time. And a telecommunications technician's time is not cheap—between $40 and $250 an hour depending on the job being performed. Planning around the existing infrastructure of the organization can save a company millions of dollars over the long run. Such planning is nearly impossible without an accurate cable/plant management system in place. See Chapter 7 for more information on cable/plant management.

Equipment/Feature Inventory Management

Equipment/feature inventory management (EFI) is the allocation, accountability, and management of all telephony-related equipment and feature services that an organization wishes to track. Every telephone set, speaker phone, fax machine, modem, desktop video, ISDN link, voice-mailbox, call forwarding or conference calling feature, and just about anything else you can think of is an asset that a corporation might want to track. The assets that EFI tracks range in value from just a few bucks
to thousands of dollars.

Items can be tracked generically by a description or they can be detailed all the way down to the serial number or the asset tag. Knowing where corporate assets are being used and by whom is valuable information that any good manager should have. When no one is held accountable for the expenditures that are incurred by making various telephony-related decisions, there is no reason for planning or to be frugal. The main reason to have a telemanagement system in the first place is to distribute the costs of telephony back to the departments and individuals that caused the expenditures to be made. With all of the new technology that is associated with telephony today, the need for equipment/feature inventory management is greater than ever.

EFI is also responsible for tracking the availability of items. Warehouses can be stocked with available physical items, while nonphysical items, such as voice-mailbox ports, can be stored in a logical warehouse for tracking purposes. The vendors of equipment and supplies can be tracked along with product lead times, costs, and warranty periods. Purchase order numbers can also be tracked by the EFI. Refer to Chapter 8 for more information on EFI.

Keysheet Telephone Maintenance

Keysheet telephone maintenance is the process of defining every single button on a digital telephone. Keysheet telephone maintenance is a subfunction of EFI that helps further define the features being tracked by EFI. It is extremely valuable to know precisely where every telephone extension appears within the telephone system. The same telephone line could appear on hundreds of telephone sets, or just one. The telephone switch tracks this information internally, but the information is difficult to obtain and is usually encrypted in such a way that an administrator can find no value in the information. The TMS, however, is user-friendly and can bring up desired information about telephone programming in an instant. The data can be represented graphically or in a standard report format. The information gathered and maintained in keysheet telephone maintenance is used by the administrators of the switching system and also quite extensively by the people who are using the work order processing system.

Work Order Processing

Work order processing (WOP) was created to provide a single point of entry for telemanagement systems. Almost every related file and field within a telemanagement database can be updated through the work order processing system. Telemanagement deals with such a broad number of applications, and yet they all seem to revolve around the user. The user is usually assigned a telephone, and from this point on the telemanagement system branches off to form the various modules that are described in this chapter.

Work order processing sits over the telemanagement system like an umbrella. If it is used properly, virtually no other database maintenance needs to be performed to keep the system up to date. As work is performed, a technician's time is used, equipment and features are allocated, and cable/plant management is used, the work order processing module automatically updates all related databases. In addition, the accountability for the work being performed within the work order is passed on to the general ledger interface module of the telemanagement system. Refer to Chapter 9 for more information on work order processing.

Switch Integration

Switch integration is a major step in the direction of a single point of entry. One of the major goals of any well-designed computer system is to cut down on the amount of redundant data that is entered into the system. One of telemanagement's primary goals is to do just that. Most telemanagement systems have one to three external connections. The first is to gather SMDR data from the switch. The second might be for collecting SMDR data from remote locations, and the third might be to supply the general ledger interface function detailed earlier in this chapter.

One additional external connection is required in order to provide switch integration. The majority of PBX systems use a simple TTY (teletypewriter) interface port expecting VT100 terminal emulation (see Chapter 15 for more information). In order to provide switch integration, a connection between the TTY port and the TMS must be made. The TTY port is the gateway to programming the PBX. If the correct command sequence is known by the TMS, all PBX functions can be programmed in an automated fashion through the TMS.

Where does all of the information come from that is necessary to program the PBX? The TMS provider must gather the information necessary for programming your specific switching platform and then make the TMS issue the proper commands and incorporate the appropriate variables from the facilities management module. The task of gathering all of the commands to program a switch is very difficult. In fact, the command structure will vary from revision to revision of a specific switch.

At this time, different telemanagement systems appear to support certain revisions of certain switches. No one telemanagement system supports all switching platforms. Most switches at this time are not supported by any TMS, but this should change in the near future. When choosing a TMS you must decide if this level of integration is desired. The cost for this add-on module ranges from $5,000 to over $30,000, depending on the TMS vendor, the switching platform that the TMS must integrate with, and the switch commands that the TMS must support. When you factor in the cost of the technician who would have to perform the tasks of the switch integration module manually and the reduction in human error, you will readily see the long-term benefits of switch integration. If properly implemented, switch integration will prove to be a tremendous time saver and valuable asset to the work flow of any organization.

Voice-Mail Integration

Voice-mail integration is the process of creating, modifying, and deleting voice-mailboxes through the telemanagement system. When a person's telephone is created or someone wishes to add the voice-mail feature to his or her telephone, a work order will be entered. Upon completion of all physical tasks on the work order, the administrator will close out the work order on the TMS. At such time, the voice-mail integration module would open up a connection from the TMS to the voice-mail system and then create the boilerplate voice-mailbox.

At present, a very limited number of voice-mailbox platforms have been integrated by the major telemanagement systems. Octel and Centigram seem to be the most popular voice-mail platforms for TMS integration at this time. If you are just now creating your telephony environment and are going to implement telemanagement and voice-mail, you may want to look into telemanagement and voice-mail systems that integrate with one another. In addition, TMS vendors are starting to write their voice-mail integration as open-ended modules where the command structure for the creation, modification and removal of mailboxes can be defined by the user.

Try to accurately gauge the value of this function for your organization. If you are spending significant time maintaining your voice-mail system, this add-on module for TMS can be a tremendous time saver. For more information on voice-mail systems, refer to Chapter 14.

Conclusion

The computer-based telemanagement system is a continually evolving component of telephony. As telephony technologies are developed and implemented, the telemanagement system must by nature change to incorporate the new technology into its structure. The potential benefit to businesses of using computer-based telemanagement systems—now and in the future—will continue to grow. By centralizing the management of telephony onto a single computerized system, the term "computer telephony integration" takes on new meaning. No serious user of telephony should be without a telemanagement system.

Chapter Seven

Cable/Plant Management

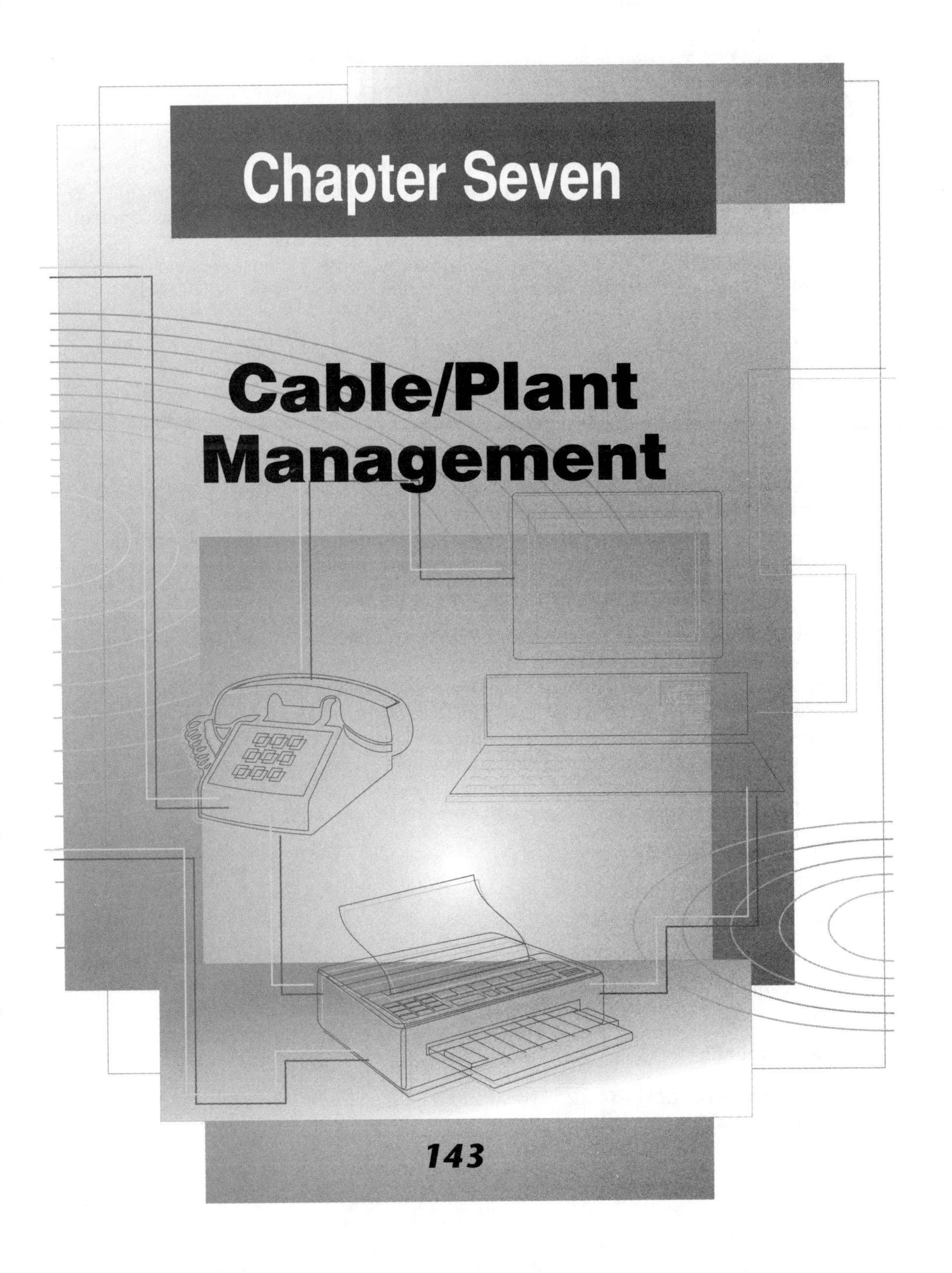

Cable/plant management (CPM) is one of three areas covered by the term "facilities management," along with equipment/feature inventory management and work order processing (see Chapter 8 for more information on equipment/feature inventory and Chapter 9 for more information on work order processing). Cable/plant management, the focus of this chapter, is the process of mapping all existing cabling and circuitry equipment used throughout the telephony environment. This includes all available and in-use physical plant used as well.

To understand CPM, let's consider a computer network. When a person is given control of a computer network, he or she is expected to know, at the very least, precisely where each connection to the network is, what the network is connected to, and how the connections are made. Computer networks are designed, tracked, and managed by technicians using computer technology. A logical and graphical map of the computer network is usually created in advance of the actual cabling work or, at the very least, the cable definition is made while the cable is being put into place. CPM performs the same function for the telephony environment that computer network management does for the LAN. Unfortunately, tracking and managing telephony is infinitely more complex.

The infrastructures of most buildings are designed by architects on drafting boards and eventually end up in a set of blueprints. The contractors take those blueprints, build the buildings, make provisions for the conduit, and install the prescribed cables. The technicians then come along and connect the telephone wires to one another. And that is it. As circuits are created, there is little, if any, documentation of the connectivity. The switch database might hold some clues to the definition of your physical plant. However, tracking is sometimes not done at all. Thus, the primary resources of information for cable/plant management have traditionally been technicians' memories, flip charts, and original blueprints which do not show connectivity.

Now ask yourself this question: How much money does it cost to run your telecommunications department? Don't think small—think of ALL your telephony expenditures. Implementing CPM in a telephony environment can help you gain control of the following resources, so you can start saving money:

- Technicians
- Local circuits
- Long-distance circuits
- Real estate
- Utilities
- Personnel
- Accounting

Many of these costs can be reduced by managing your circuits and physical plant. Your planning ability can make your company's efficiency rise in almost every area of business. The time needed to perform telephony tasks is greatly reduced due to the additional information that is available to you and your staff through CPM.

LAN and Telephony Circuitry

Unlike a LAN, where terminals are usually connected via a daisy-chain cable connecting one terminal to the next, phone wiring has at a minimum a separate, dedicated pair of wires for each and every piece of telephony equipment. Depending on the telephone equipment used, additional pairs of wire may be needed for a single telephony device. That means if you have thousands of telephones, then you have thousands of pairs of wire connecting each one separately back to the switch room.

LANs can use twisted pair, ethernet, coaxial, fiber, wireless, and several other media to make their connections. Telephone systems use almost the same variety of media. As you review this chapter, keep in mind that CPM for telephony can also be applied to the LAN environment. All of the cabling and circuit definitions are flexible and can be tailored to the users' needs. If you partition the CPM database (which most CPM systems allow for), you can use the same platform to track and manage your LAN as as well as your telephony environment.

Cable Circuitry Model

What exactly does a cable circuitry model look like and what does it help you accomplish? Let's begin by taking a look at the whole picture and then breaking it down into the components that make up the telephony environment. Figure 7-1 shows a building in a typical telephony wiring configuration.

Wire in a building where a switch resides passes through the main distribution frame (MDF) to the riser cables or inside feeder cables to the intermediate distribution frame (IDF) and eventually to station wire. Wire in a building where a switch does not reside passes through a building distribution frame (BDF) instead of an MDF and is connected in the same manner. The wire in the BDF must then be cross-connected to an outside feeder cable to be transported back to a switch room (see also Chapter 3).

How can the cabling plan in your environment vary from the one shown in Figure 7-1? Figure 7-1 begins and ends with your primary building and does not show the potential connectivity between buildings. Actually, cables are also coming from your telco providers and cables are going to (potentially) other locations that your switch services throughout the organization. These cables are for the most part copper twisted pair, shielded and unshielded. An increasing number of installations are converting to a fiber-optic backbone or coax cable.

This discussion will focus on copper wires and cabling, because the vast majority of telephony circuitry is copper. Channels/circuits within fiber or coax can be managed in the same manner as copper wiring. The labeling of the cable type may change, but a circuit is still a circuit. Management of copper is more complex because

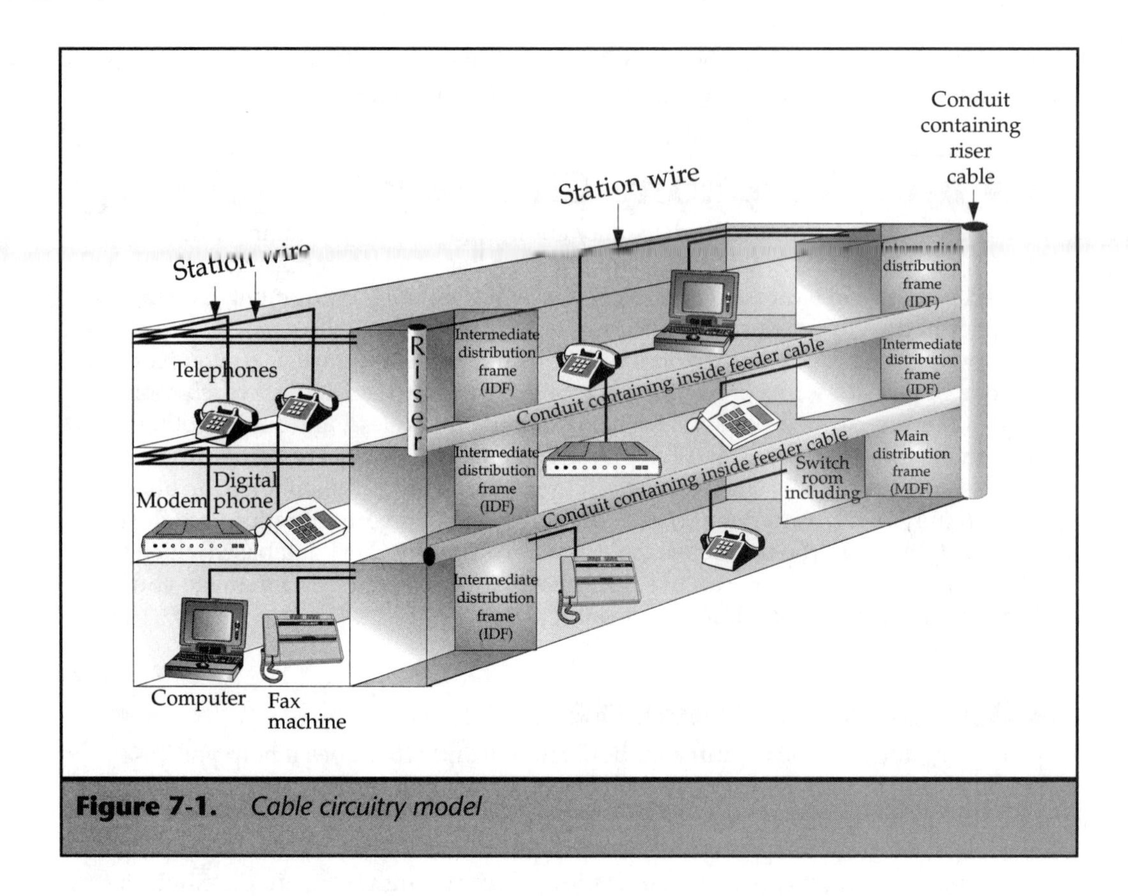

Figure 7-1. *Cable circuitry model*

of the number of physical strands of wire that must be tracked. To a CPM system, the description of what the circuit is physically made of is just a description. CPM is concerned primarily with tracking the circuits, so don't be too concerned about the cable itself. Rather, focus on the mapping aspect of CPM.

For now, let's begin by analyzing the path that your cable takes within the Main building where your PBX resides.

Switch Room

Starting from the switch room in the lower right hand corner of Figure 7-1, your cable definition begins. The switch room is where your PBX will reside. All of the telephony circuits throughout your building(s) begin at this location.

A circuit begins at the switch port and ends at the telephone jack. Switch ports are connected to an MDF (main distribution frame) cable that takes several ports over to the MDF block. Your actual cable mapping begins here. The MDF is one of the walls of wire that you have in the switch room. Every port can be found on the MDF, usually by a label, such as 120308A. The "12" in the label may denote the cabinet of the switch where the card is located, "03" may denote the row or card within the cabinet, and "08A" may denote the actual port. This is just an example of how a cable is referenced back to a specific port. Every switch has a different configuration for its equipment, so the nomenclature will vary.

Station Wire

Several other sets of cable and wire are present in the switch room that you should be concerned with. Notice the lines in Figure 7-1 that go out, around, and down. These lines depict station wire. Every telephone jack that you have in every room is connected to station wire. If you need another telephone jack installed in a room, the telephone technician must use existing station wire in the room to create the telephone circuit. If station wire is not available, the technician must cut a hole in the wall, install the jack, then run a pair or more of wire (usually, a four-pair cable will be used, but this will vary from location to location) from the room back to either the MDF, BDF, or the closest IDF, where the station wire can be cross-connected to another pair of wires that takes the circuit back to the switch room and eventually to a port. That station wire now makes a circuit from the telephone to a port on the switch so that telephony services can be delivered.

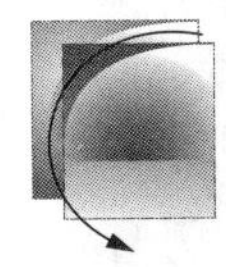

TIP: A technician's time is extremely valuable. Avoiding the cost of installing station wire in one location by using a secondary location that is suitable is highly desirable. Cable/plant management will show you where your wire is and where it is available.

Riser Cables and Inside Feeder Cables

Another set of cabling you can identify are riser and inside feeder cables. Both riser and inside feeder cables travel from one location to another inside conduits, also called riser shafts. A *conduit* is simply a pipe or sometimes just a hole or shaft that cables are inserted into to go from one location to another. The actual thickness of the conduit is also important and will be discussed in the "Conduits" section later in this chapter.

Riser Cables

Riser cables are used to provide circuit connectivity throughout the different floors in your building. The riser cable is used to take a large number of cables from one location up through conduits or shafts to another floor where there is usually an IDF (also called a closet). By taking a large number of pairs of wire over to a remote location on one larger cable, the cost of providing service to the various locations in a building is greatly reduced.

Inside Feeder Cables

Inside feeder cables will only be present if you have more than one IDF on each floor. In fact, depending on your building's configuration, you may have dozens of IDFs connected by inside feeder cable.

Of course, your building(s) may not look like the one in Figure 7-1, but the concepts remain the same. The number of floors, square footage of each floor, and shape of the building will obviously vary the number, location, and placement of each component. The switch room sometimes may be located in the basement or on the top floor. Your inside feeder may run through several locations, and the wiring plan may not make much sense to you at all. The architect of the building should have taken into account the distances and costs associated with the design and configuration of the infrastructure. You must learn how to manage and deal with the legacies you have been left by the architects, contractors, and technicians who came before you.

Conduits

Conduits are quite often overlooked when tracking and managing cable/plant. The pipes, holes, tunnels, conduit—whatever you want to call it—has limitations, and if your site is older, the available conduit space may be limited. CPM systems have the ability to track the *gauge* (thickness) of the wires and cables that are running through the conduit. By adding the gauges of the wire to the size of the conduit, you can determine where additional cable can be installed in the existing conduit, reducing the need for major contracting. Of course, the main goal is to avoid adding cable to the infrastructure in the first place. But if cabling upgrades are required, knowing the available conduit area is extremely helpful.

Cross-Connections

Each length of cable must be identified to provide a comprehensive cable/plant picture. It is sometimes necessary to define the intermediate connectivity points along with any peripheral circuit-related equipment being used. You must decide on the level of detail that you will define throughout the CPM system. CPM can be performed as an end-to-end operation, or you may define every piece of pertinent information throughout the circuit.

All of the different segments of cable or wire discussed so far must be connected to form a circuit. The circuits are formed by making cross-connections, also referred to as *bridges* or *jumpers*, as depicted in Figure 7-2. Cross-connects can be made between two or more blocks that separate out the individual pairs of wire, or cross-connects can be made between two entire segments of cable.

Figure 7-2 also shows the station wire frame for each IDF. When a phone is activated, the station wire is cross-connected to the riser cable or the inside feeder cable. Depending on the current connectivity of the riser and/or the inside feeder cable, additional cross-connections may be needed to create a complete circuit to a

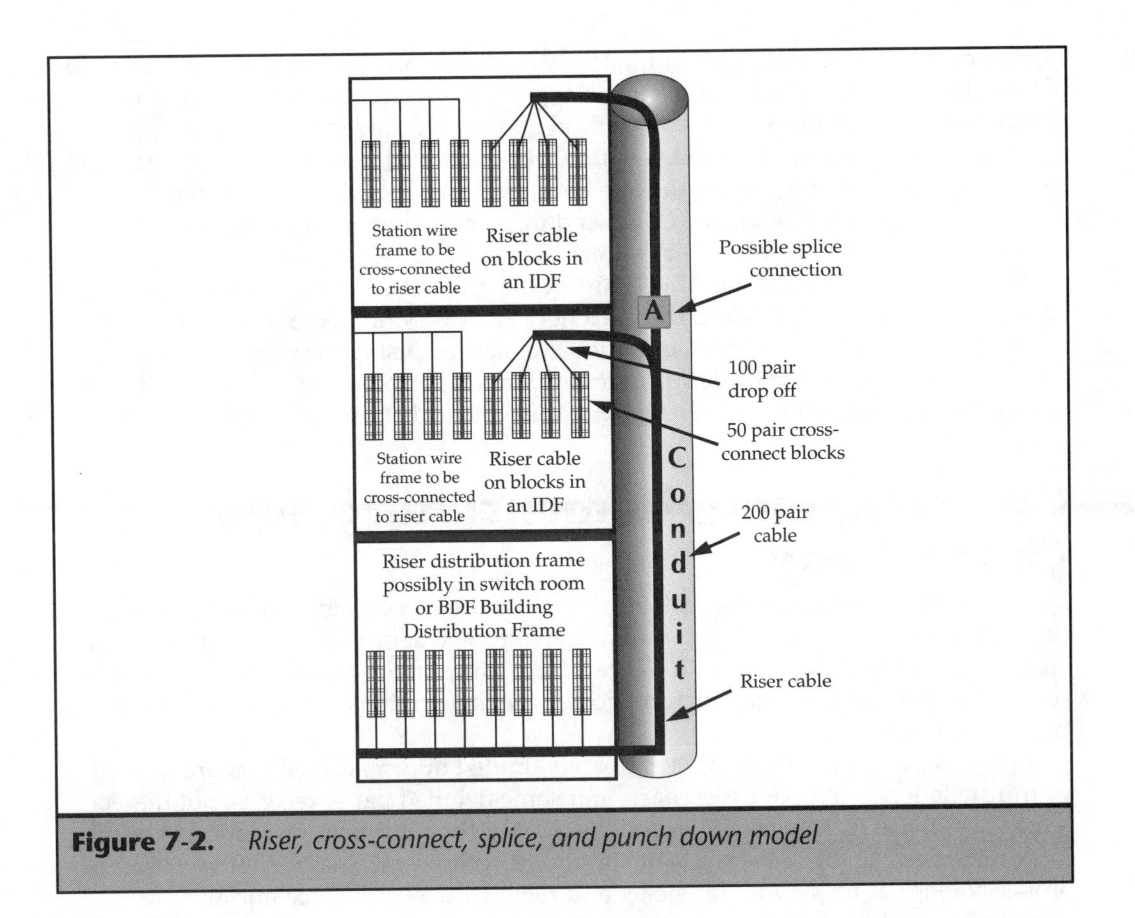

Figure 7-2. *Riser, cross-connect, splice, and punch down model*

switching node. The CPM module is designed specifically to manage and track all circuits. It will show you where wires and cross-connects are made as well as where cable is available.

Cross-connecting entire segments of cable together is called a *splice*. Splices can also be used to connect fractions of a cable together with another cable. The remaining pairs are usually punched down or wire-wrapped on blocks for use in the IDF.

Cross-connections may be daisy-chained with multiple *drop-offs*. This is how most homes are wired. In the PBX environment, daisy chaining is almost never performed. The PBX is programmed to make the same telephone extension ring on multiple phones. Understanding how, where, with what, and why cross-connections are made is a large part of understanding CPM.

In Figure 7-2, a 200-pair cable is punched down (connected) in the first floor switch room. "Punch down" gets its name from the device used to connect the wires together: you punch the wires with the punch down tool to connect and cut the wire in one physical action.

The cable(s) travels up a conduit to the second and third floors. One hundred pairs of the cable are dropped off on the second floor, while the remaining one hundred pairs continue to the third floor. The method of drop-off used will vary based upon the unique circumstances of the installation, but the model remains basically the same. Depending on the method used, the one hundred pairs of cable that continue to the third floor could also be punched down and then cross-connected to another riser cable going up to the third floor, or the technicians could have spliced the cable together. In some cases, one hundred pairs of the original cable may be left intact, allowing the cable to continue to the third floor with very limited manual labor.

Inside feeder cable is connected in the same fashion, using the same equipment and basically the same terms. The primary difference between inside feeder and riser cable is that the conduit goes horizontally for inside feeder and riser cable is vertical.

Cable Mapping, Nomenclature, and Connectivity

Cable mapping is the formidable task of trying to track every single pair of wire or circuit from beginning to end. You will need to know where all cables reside, not just the circuits that are in use. If you have decided to take on the massive task of mapping your entire cable/plant infrastructure, now is the time to think about nomenclature and connectivity.

Odds are, you are starting with extremely limited information. If you are lucky, your installation might have flip charts and some detailed paper records, but the accuracy of those records is in question. Invariably, the information is coded to be cross-referenced against blueprints or simply the technician's memory. If you use proper nomenclature, cross-referencing and relying on a specific technician's recollection can be avioded.

Determining the level of connectivity that you will track is another decision that you must make. Just as you should create a strategic plan for telemanagement, as described in Chapter 5 , a strategic plan should be developed for the implementation of CPM (although not quite as detailed). Figure 7-3 shows an example of how cable mapping and nomenclature is provided.

Cable Mapping

Cable mapping can be accomplished using two basic methods: graphical or database. A graphical CPM system is three dimensional, showing every room and piece of wire throughout an organization. Graphical CPM is extremely cumbersome (even more so than a database CPM system) and requires a greater amount of setup and maintenance time. The amount of additional information that is provided in a graphical cable mapping model does not necessarily make your telecommunications staff more productive. The primary reason for performing CPM is to make your staff more productive, and even though a graphical CPM system is nice to look at, it does not make your staff a

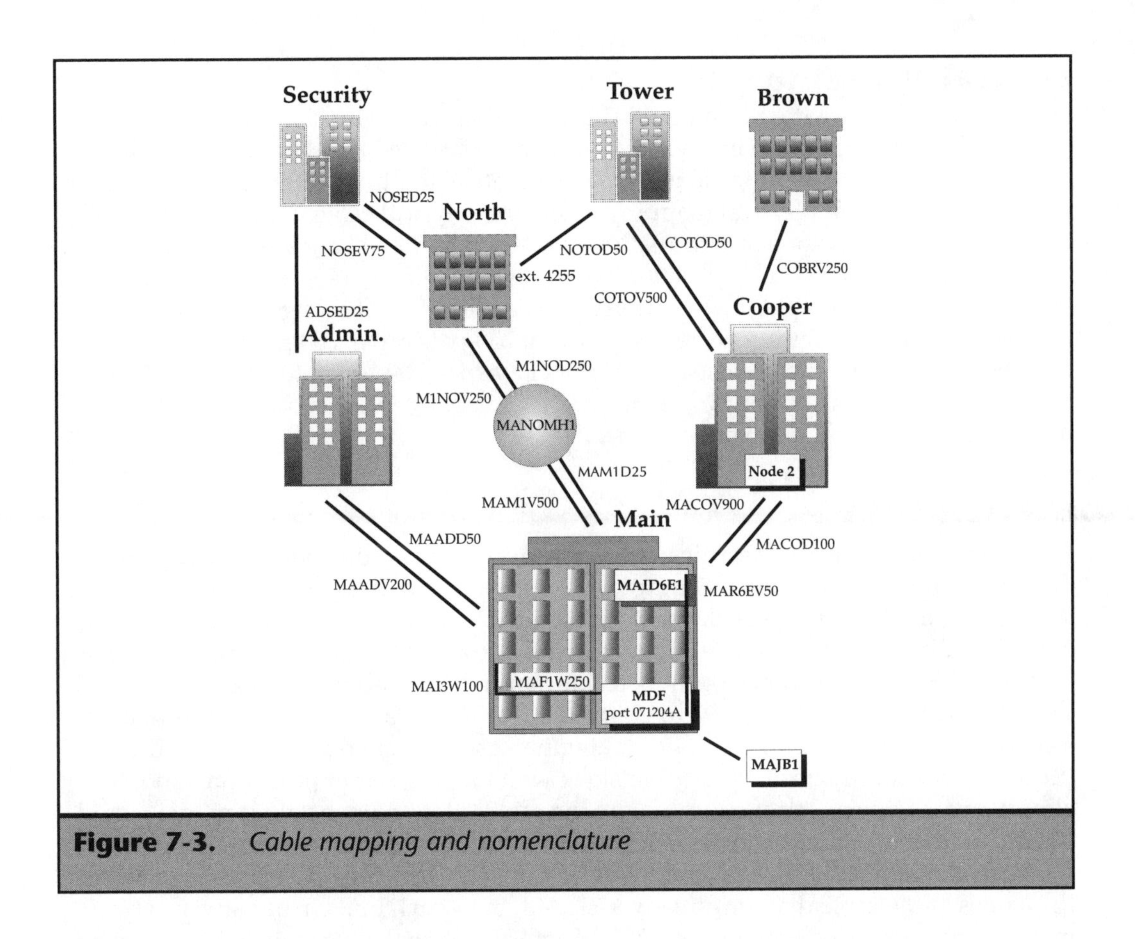

Figure 7-3. *Cable mapping and nomenclature*

meaner, leaner productivity machine. Therefore, graphical CPM systems are not recommended.

Database CPM systems require a little less data gathering, but the amount of effort needed to perform database cable mapping is still very great. Database cable mapping can be constructed in one of two ways: physical mapping or logical mapping. Both methods are accurate, but depending on how you wish to present the cabling information, you may find one method preferable to the other.

Physical Mapping

Physical mapping is done by literally taking the physical size (meaning the number of pairs that the cable contains) of each cable and then documenting exactly where the cable begins and ends. The actual pairs within the cable are mapped to a specific location on a frame or to a splice or cross-connect. Locations are detailed in the same manner.

Logical Mapping

Logical mapping is a little bit more flexible and artistic than physical mapping. In certain areas where a cable simply goes from one location to another and the entire cable has one specific use, logical mapping is identical to physical mapping. Logical mapping differs from physical mapping when one physical cable is logically divided up to carry different kinds of information, voice versus data being a common example, or a cable may be spliced at one location and sent to two different locations on separate, smaller cables. For example, the two cables between the Main building and the Administration building in Figure 7-3 are one physical 250-pair cable. 200 pairs of the cable have been assigned for voice use and 50 pairs have been assigned for data use. Only one physical cable exists, but for organization's sake the cable has been logically defined as separate cables to show its use.

Nomenclature

At the beginning of your strategic plan, you must decide on the method of naming the cables, frame, locations, and equipment used in the connectivity process. Typically, the existing method of nomenclature is pathetic: "Oh, just go to the second floor of the North building, look for room 207, and you will see a small door on the right. The IDF is in there." This is an example of how technicians tell each other where to find an IDF. Can you imagine what they have to go through to find an actual pair of wires to make a circuit? (Actually, the entire circuit-making process is detailed in Chapter 3.) If you use a properly constructed CPM system along with the work order processing module, almost anyone given a brief education on the cable/plant nomenclature model would be able to find any component monitored by CPM.

Nomenclature applies to almost every element of CPM. You may be forced to use the names of the cable that are already assigned, but watch out! On any given cable map you may find five cables called "C1." That is because they are labeled based upon where they are rather than where they fit into the macro picture of the cable/plant environment. You will also find that the cables might be labeled something like C1, C2, C3, C4, 250W, 500N, etc. This tells you almost nothing about the physical location or state of the cable unless you cross-reference the blueprints that show where the cables go and what they connect. If you must use already assigned names, or you believe that changing your cable names is a bad idea, then you will find that most CPM systems can accommodate your needs. But if you want to truly bring your CPM system into the computer age, you may want to use a method of nomenclature similar to the one detailed below and shown in Figure 7-3.

Naming Buildings and Locations of Cable/Plant

It is time to start naming the buildings and locations to be managed by the CPM module. Let's begin with the various buildings you are responsible for. You may already have names or numbers for the building(s) on your campus or complex. By all means, you want to use names that you are familiar with. In the example given in Figure 7-3, all of the buildings have names. You want to define every location pertinent in the environment, so the names of the buildings are perfect for the application.

In the cable/plant module, you want to be able to define not only buildings but also locations. An example of a location might be a BDF, IDF, closet, junction box, or manhole. Any location that you can send a person to, you want to have defined. Buildings and locations can be assigned complete addresses, and if necessary a description of the location can be provided to further detail where to find that location. For example, the following description could be provided for the junction box in the lower-right corner of Figure 7-3: "Junction box 1 (MAJB1) is located on the southeast corner of the Main building, approximately 20 feet from Mission Road."

Names of other locations should be standardized as well. The name you assign to your IDFs should practically tell you where the IDF is located. The location MAID6E1 in Figure 7-3 informs the reader that the location is in the Main (MA) building, it is an IDF (ID) on the sixth (6) floor on the east (E) side of the building, and it is the first (1) IDF on the floor. The location description may include additional information, such as an actual room number or directions, but you can most likely find the location strictly by the name *if* you know the nomenclature being used.

Naming Cables

The same method of nomenclature applied to buildings and locations can be slightly modified to work for cabling. Looking at Figure 7-3, you can see several examples of cable nomenclature. Let's begin in the Main building. The riser cable represented on the right side of the building is called MAR6EV50. From the previous example, we can deduce that the cable begins in the Main (MA) building, it is a riser (R) cable going to the sixth (6) floor on the east (E) side of the building, and it contains voice (V) carrying 50 pairs of wire.

Additional information about the cable can be included in the description of the cable and the other components of the CPM module. The description might contain information such as what specific MDF or IDF it is served from and what IDF it is connected to. Information about the physical wire, such as the manufacturer, type of wire, gauge, and load might also be included (see Chapter 3 or Appendix B for a complete description of these terms).

Outside feeder cables can be named to provide this same valuable information. An outside feeder cable is any cable that goes outside of one building to provide telephony services to another building or location. Let's look at the outside feeder cable MAADD50 in Figure 7-3. The name tells us that the cable runs between the Main (MA) building and the Administration (AD) building. The cable is designated for data (D) use and contains 50 pairs of wire. The same additional information listed in the previous paragraph can be included about outside feeder cable.

Choosing a method of nomenclature that works for you and your environment is very important. If you feel that a strict numbering scheme works for you, then by all means use it. Remember, though, that the goal is to have cable/plant records that are easy to understand and that anyone can use in as short a period as possible. Maps, flip charts, and memories should be secondary sources of information that rarely need to be referenced.

Connectivity

Connectivity is the definition and tracking of all points where wire or cable is connected to other telephony items. Connectivity can be tracked between two cables, a cable and a frame, a pair of wires and a port, a pair of wires and a phone, fax, or modem, and the list goes on and on. The level of connectivity that you decide to track will greatly influence the amount of work and effort you will have to put forth in order to create your CPM database. Below you will find examples of the two extremes of connectivity tracking: end-to-end and complete detail. The examples can be *roughly* followed by referring to Figure 7-3, starting in the Main building at the MDF.

Here's an example of end-to-end connectivity:

- Port 071204A is connected to...
- Extension 4255, assigned to John Edwards.

Here's an example of complete detail connectivity:

- Port 071204A is connected to...
- MDF cable MAMDV009 pair 20, which is located on ...
- MAMDF01 column 5, row 20, which is cross-connected to frame...
- MAM1F01 in column 1, row 20, which is connected to...
- Outside feeder cable MAM1V500 pair 20, which is spliced at MANOMH1 and connected to...
- Outside feeder cable M1NOV500 pair 20, which is connected to frame...
- NOBDF01 in column 1, row 20, which is cross-connected to frame...
- NORIF02 in column 2, row 20, which is connected to...
- Riser cable NORI2V100 pair 70, which is connected to frame...
- NOIDF02 in column 2, row 20, which is cross-connected to frame...
- NOSTF01 in column 1, row 20, which is assigned to...
- Room 216, west wall, jack ID 216WA, which has been assigned to...
- Extension 4255, assigned to John Edwards.

All of the detailed descriptions assigned to the buildings and locations referenced are also available for lookup and reporting when reviewing a cable or circuit map. You will probably want to find some middle ground between end-to-end and complete detail connectivity. As long as the mapping process shown above is, there are many circuits that travel through dozens of locations and cross-connects. A circuit map can be enormous, so you want to be sure you know what you are getting into.

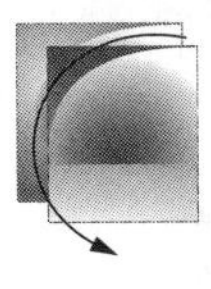

TIP: Evaluate the overall connectivity of your environment before making a decision about the level of detail you will ultimately track. At a certain level you may experience the law of diminishing returns. Remember, the goal is to increase productivity to the highest level possible.

Automatic Connectivity

When you are evaluating a CPM system, one of the more important features to look for is the ability of the system to automatically associate cables and pairs to create connectivity. Building your cable model is like creating a spider web where none of the ends actually touch. Cables are defined that have beginnings where other cables end, but the connections between the wires are not yet made. CPM helps make the telecommunications department more productive by showing where manual labor must be performed and where the work has already been accomplished. *Automatic connectivity* is the process of taking the ends of two or more cables that reside in a specific location and logically cross-connecting the individual pairs of wire to continue the circuit path. These connections should already be made in the real world.

A strong CPM system will allow you to logically cross-connect hundreds, even thousands, of pairs of wire by simply defining the cable, the starting and ending pair you wish to auto cross-connect, and specifying a second cable, starting and ending pair. The two rules that must be followed are that the two cables must reside at the same location and the number of pairs specified must be the same for each cable. Once the cable circuit is extended, the user has the ability to see directly from one end of the circuit to the other and also view all points in between if desired.

Near and Far

The concept of near and far was developed in telemanagement to help the technician find the shortest path from a specific location back to a switching point. As cables are put into the CPM definition and the locations of the ends of the cables are defined, one other piece of information should be included: how close that link is in the chain to a switching location. The nearer you are to the switch room, the closer you are to your goal. If a cable is three cross-connections away from a switching hub, then that cable would be assigned a number four, because there are three additional links of cable nearer to the switch that must be traveled to return to the switch room.

When you have finished defining your web of cabling and connectivity, you will find many strands of cable that have not been completely connected to make a circuit. Knowing which cables bring you closer to your destination and which do not is very valuable. The work order processing module (discussed in Chapter 9) uses this information to help configure for new service. Without the near and far information, all the intelligence needed to make a new telephone connection would have to be provided by the technician configuring the telephony circuit.

Building the Cable/Plant Database

Once you have defined all of your cable and the cross-connects in the TMS, you begin the formidable task of defining every single telephone circuit that is currently in use. After all this work has been done, the TMS can account for all physical plant and cabling for future lookup and reporting. This information is also made available to the work order processing module for circuit addition, modification, and deletion. (Work order processing is discussed in detail in Chapter 9.)

If you have decided to perform CPM on your TMS, you have to invest a great deal of time and effort in the beginning to get a very limited return on your investment. As mentioned in Chapter 5, the best time to implement CPM is just before you have a new switching platform installed or you are upgrading your infrastructure. Why? Because all of the effort put into defining the CPM database will save a greater amount of time in the installation and implementation of the new PBX and/or the infrastructure upgrade. Another valuable reason to implement CPM just before procurement is because the database should be extremely accurate during and after the installation process. By using the TMS as the managing tool to set your work force in motion, you will achieve new heights in efficiency.

Where do you begin defining your CPM database? You need to determine what information about your infrastructure is currently in place. You will probably find that the switch database, flip charts, and blueprints are the only physical records of the current infrastructure. The information will be encrypted, difficult to read, and is quite often inaccurate. Here is your big chance to do it right, so don't hold back.

You probably come from a computing background. The data that you gather should be created in a form that is user-friendly and compatible with typical computing techniques. Your goal is to paint a clear picture of the connectivity throughout your physical plant so that anyone can understand it.

Sometimes gathering the final pieces of information to make a complete circuit path is extremely difficult. Ultimately, the goal is to build complete circuit paths and then assign them to extensions or groups, which are then cross-referenced to the telemanagement system for charge back and/or reporting purposes. When you defined your cabling environment, you probably did not create complete circuits—only cable paths were defined. The final legs that show the actual connection to the telephony device and the port on the switch were probably left off. The final goal is to have a completed circuit and to have the CPM system know every pair of wire that is in use, available, or bad. An example of a complete cable circuit path in an actual cable/plant management system can be found in Figure 7-4. You can achieve the goal of CPM by performing a complete physical inventory or by continued use of work order processing.

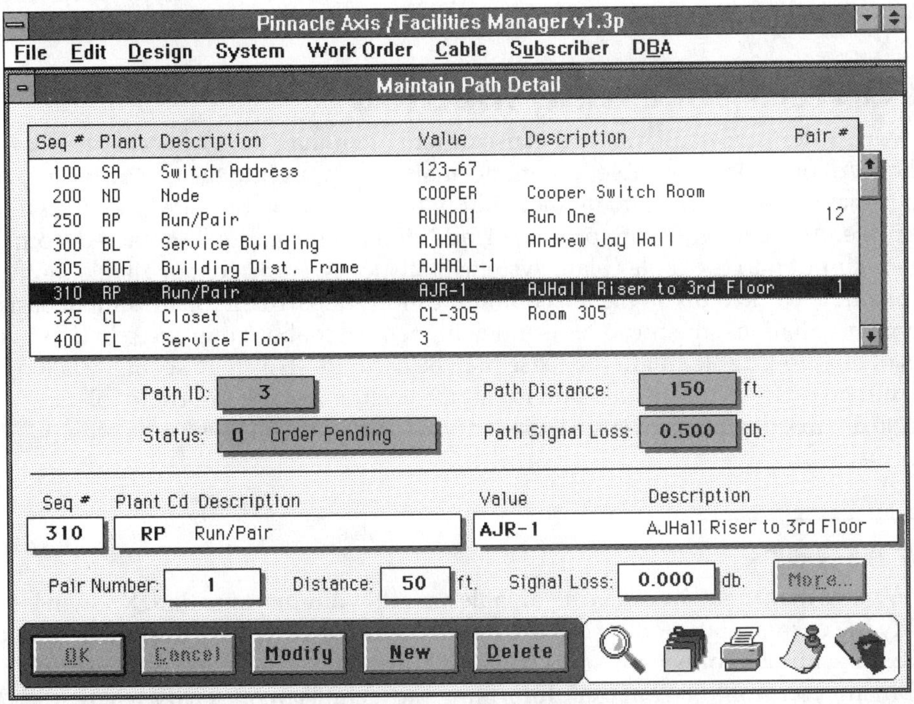

Pinnacle Axis / Facilities Manager v1.3p

File Edit Design System Work Order Cable Subscriber DBA

Maintain Path Detail

Seq #	Plant	Description	Value	Description	Pair #
100	SA	Switch Address	123-67		
200	ND	Node	COOPER	Cooper Switch Room	
250	RP	Run/Pair	RUN001	Run One	12
300	BL	Service Building	AJHALL	Andrew Jay Hall	
305	BDF	Building Dist. Frame	AJHALL-1		
310	RP	Run/Pair	AJR-1	AJHall Riser to 3rd Floor	1
325	CL	Closet	CL-305	Room 305	
400	FL	Service Floor	3		

Path ID: 3 Path Distance: 150 ft.

Status: 0 Order Pending Path Signal Loss: 0.500 db.

Seq # 310 Plant Cd Description RP Run/Pair Value AJR-1 Description AJHall Riser to 3rd Floor

Pair Number: 1 Distance: 50 ft. Signal Loss: 0.000 db. More...

OK Cancel Modify New Delete

Figure 7-4. *Cable/plant management circuit path entry screen (courtesy of Pinnacle Software Systems)* *Courtesy of Pinnacle Software Corporation*

Physical Inventory Method

Physical inventory is the labor-intensive task of going out to every single telephone and then defining the entire circuit path with all of the detail that you have chosen to track and manage. As data is collected, it can then be entered into the facilities management modules. A physical inventory not only covers the cable/plant portion but also the equipment/feature portion of facilities management. Every phone, fax, modem, etc. along with every cable pair and circuit must be mapped in the physical inventory process. This is the primary reason why facilities management is best implemented at the time a new switch is procured or the infrastructure is upgraded. Many of the steps required to bring a new switch online or upgrade the infrastructure are duplicated in the physical inventory process.

Performing a physical inventory is expensive and time-consuming. You must attack this task with full force and a detailed strategic plan. While you are gathering your physical plant information, new changes are still being made. Both speed and accuracy are required to create a timely and useful CPM module database.

Work Order Processing Method

The work order processing method is used more often than the physical inventory method. All of the known cable/plant information is input into the CPM module. Any circuits that can be defined are defined. But in the beginning, the CPM module will be inaccurate. As work orders are processed, additional information is gathered from the field and input into the cable/plant system. Statistically, 50 percent of all telephony equipment is worked on by the telecommunications department on an annual basis. In some installations, the turnover is much higher. After about one and a half to two years of use, your cable/plant database may be accurate. This may seem like a long time, but how large your installation is and the accuracy of the pre-cable/plant module database can have a large impact on how rapidly your system is brought up to speed.

Summary

Cable/plant management can be a valuable asset to your organization. Knowing quickly the current status of the cabling infrastructure is extremely important. If you counted the number of wasted steps that a technician takes when servicing telephony equipment, you would be amazed. Informing the technician of exactly where to go and which pairs of wire to use increases his or her productivity threefold. Planning for new telephony services is also made easier through lookup and reporting. Waste and misappropriations can also be identified. If a nomenclature scheme is devised that is easy to use and understand, turnover of employees will be less of a discomfort. Knowing where your assets are is the goal of every organization. The cable/plant infrastructure of your organization should not be overlooked.

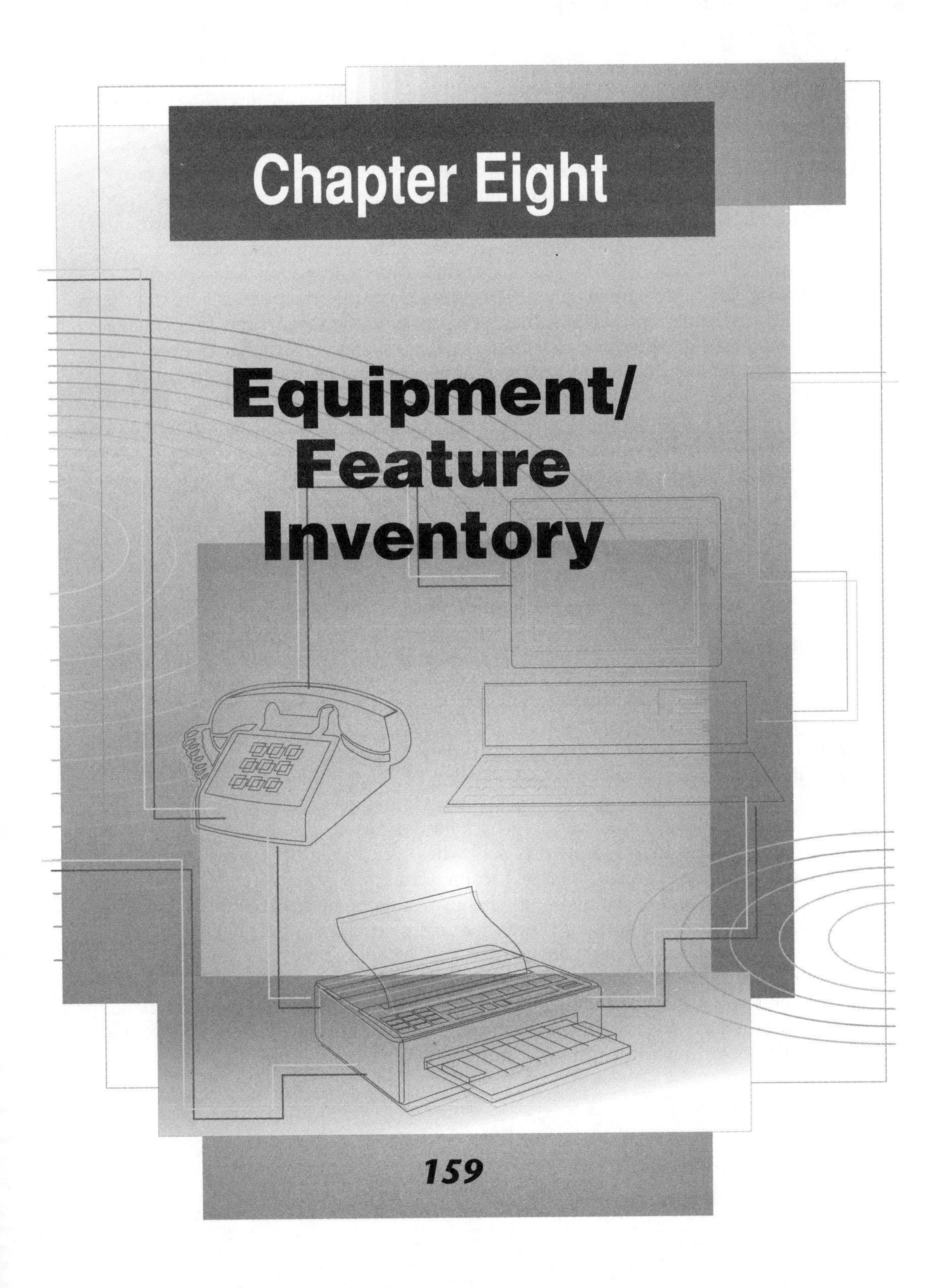

Chapter Eight

Equipment/ Feature Inventory

Equipment/feature inventory (EFI) is the second of three components that make up facilities management. (Chapter 7 covers cable/plant management, and Chapter 9 covers work order processing.) If your goal is to know where your assets are, then EFI is the vehicle for you. Telephony equipment and services are typically distributed freely throughout an organization. Computers, databases, and highly specialized programming make it possible for EFI to effectively perform this detailed tracking of the almost countless number of telephony components.

You might ask yourself why EFI exists. The moment an organization purchases a PBX, it turns itself into a micro-phone company. How does the phone company charge for all of the products and services that they supply to their customers? The answer is that they are running a very large telemanagement system (TMS) that includes EFI. When you order an additional line, a special telephone, a custom calling plan, or a special feature such as call waiting, call forwarding, or conference calling, a record is entered into the EFI module; the phone company then itemizes the equipment and/or feature to you and bills you for it. The amounts the phone companies charge for their products and services are tariffed and filed with the FCC (Federal Communications Commission). The method of chargeback that is used is based simply on a telephone number, which is assigned to an individual or organization. That individual or organization has a mailing address where the invoice is sent, which is then presumably paid to the phone company.

Think of the telecommunications department as a small telephone company. Even if your organization is serviced in a Centrex environment, it is the organization, not the individual person or department that caused the expenditure, that is billed for the products and services ordered. You must coordinate with the accounting department to create a method by which chargeback can be accomplished within the organization.

Right now, when a department or a person asks for a new phone, fax machine, voice-mailbox, etc., the telecommunications department simply fills the request. No accounting or tracking is performed at all. With EFI, that is all about to change.

Keeping track of where your assets are, both in use and available, is just one of the valuable functions of EFI. Internal chargeback for those assets is also extremely valuable. You probably would like to know precisely which vendors supply what equipment and features you use. The cost, volume discounts, lead time, and warranty period offered for the equipment and features you use are also valuable pieces of information. You should know where you have outstanding or open purchase orders available for acquisitions of additional equipment and services. Without EFI, there is no accountability for your company's telephony assets or the vendors that supply you with the telephony components you use. The goal of equipment/feature inventory is to track, manage, and reallocate the cost of all the telephony assets of your organization and to provide you with a powerful tool to optimize your telephony expenditures.

Equipment/Feature Inventory Planning

You may want to implement EFI for a number of reasons, but undoubtedly the major reason is money. The assets of a company represent money, and the users of those assets represent expenditures. The first step you will need to take when implementing EFI is to determine what the financial and accounting goals of your organization are. If you are implementing a complete telemanagement system, the accounting structure to be used by telecommunications may already be established and known. If at this time you are only implementing EFI, you may have to work closely with the accounting department to determine how chargeback should be accomplished. For more information on accounting chargeback methods and telephony, refer to Chapter 6.

Before you begin to track your assets, you must decide what you will be tracking. EFI can be used to track virtually every asset that your organization has. The software design of most EFI systems does not restrict you to tracking only telephony-related items. Anything that you wish to monitor is fair game. The item does not even have to be physical. A voice-mailbox, a hunt group, or a particular service are all items that can be accounted for by EFI. You could also track light fixtures, desks, chairs—even paper clips, if you wish. EFI can also track physical items generically, by serial or asset tag numbers. You choose the level of detail. The task set before you in the planning phase is to decide what assets are worth the trouble to track, and at what level you shall track them.

After you have determined what items are worth the effort to track, the next step is to identify all of the vendors that supply you with those items. All of the details about your vendors and their history with your organization should be gathered. If your company issues open purchase orders to vendors, then purchasing information must also be gathered.

You must also ask yourself this question: Do I want to track EFI assets that are not in use? If the answer to this question is no, then you really do not have to worry about warehouses or stockrooms. You can simply define some phantom location where all available stock is kept and have the system track only that there is a quantity of the item available. If the answer to the question is yes, however, you will need to define the locations where you will store your available physical equipment. In addition to tracking where available equipment is, you need to know about equipment that is in repair—in other words, an asset that is not in use but is also not available. EFI is designed to track and manage all of these areas of concern if proper planning is done.

This chapter will not focus on a particular software package. Several reputable vendors of complete telemanagement systems are listed in Appendix A. When you are reviewing the information in this chapter, keep in mind the things that you will expect of the EFI software module that you choose. This chapter will point out the key factors that make up a strong EFI module—it's up to you to make sure that the computing system you use can do the things you want.

EFI Accounting

Accounting for equipment and features, as with all accounting functions, is unique to every organization. Your organization may simply want to know where its assets are and who is using them. On the other hand, if the goal of your organization is to make telecommunications a zero-based accounting department, then you must determine the best way to charge back for the equipment, features, and services that you provide.

The most common form of chargeback currently used in business is by extension number. This is very similar to the method used by the phone companies. Of course, the extension is usually associated with a person and is assigned to a division/department or some other hierarchical structure. The invoices, reports, and/or general ledger transactions can then be sorted and generated at any level of detail or summary throughout the hierarchy by the EFI module.

Other methods of grouping and potential chargeback will also be made available in a strong EFI management system. A billing ID does not have to be an extension number. Many companies choose to bill directly back to general ledger numbers, group IDs, project numbers, fund numbers, and the list goes on and on (see Chapter 6 for more information on chargeback methods). Grouping and then regrouping of billable items is also very common, based upon the needs of the accounting department and the desires of management. A strong EFI module will take all items and group them as specified within the accounting structure of the organization.

A strong EFI module is very much database-driven. Any programmer will tell you that if the data is on the computer and is accessible, then summaries, extracts, and reports can be created to meet the needs of the user. Try to remember when you are choosing a TMS that you want to have integration (database information not duplicated between programming modules) and access to the database (see "Integration" later in this chapter). If the EFI software you choose as part of your TMS is running on a database that is familiar to your MIS department, then the information you want will be at your fingertips. Now that you better understand the overall reason for EFI, accounting for the money, let's take a look at the individual components that make EFI work.

Items

An *item* is a generic name for any type of equipment or feature that EFI tracks. Items can be anything that you want to define. Some specific examples of item names are 12-button digital phones, speaker phones, call forwarding, and voice mail. Yes, this is a telephony application, but there is nothing to stop you from tracking other items of concern as well. You will find that EFI is also well suited for tracking computers, office furniture, supplies, and just about anything else. Try to be focused in the beginning, but also keep in mind that the organization, particularly the MIS or computing department, might want to expand the functionality that EFI provides.

Your plan begins with a list of items that you want to track. Take the sample form shown in Figure 8-1 and gather as much information as possible about each item that EFI will be responsible for. In the beginning, you may not know the answers to all of the questions on the form. Don't worry; as your EFI definitions begin to take shape, you will be able to fill in the blanks.

ITEM DEFINITION

ITEM CODE: ____________________

DESCRIPTION: ____________________

____

COMMENTS: ____________________

____

BILLING TYPE - PRORATED, ADVANCED, FIXED, VARIABLE (P,A,F,V): ________

DEFAULT BILLING AMT: __________ PURGE UPON BILLING (Y,N): ______

MINIMUM QUANTITY ON HAND: ______ PHYSICAL ITEM (Y,N): ______

DEFAULT CLASS OF SERVICE: OPTIONAL ____ GROUP CODE: OPTIONAL ____

COMPLETE THE FOLLOWING IF THE ITEM IS PHYSICAL

ITEM TRACKING BY - SERIAL NUMBER, ASSET TAG OR GENERIC (S,A,G): ______

INVENTORY CONTROL BY - ORGANIZATION, WAREHOUSE OR NONE (O,W,N): __

PRIMARY VENDOR: ____________________

ALTERNATE VENDOR: ____________________

MAINTENANCE VENDOR: ____________________

SUBSTITUTE FOR ITEM: ____________________

Figure 8-1. *Item definition form*

A few line items on the Item definition form may need some explanation:

ITEM CODE The *item code* refers to that specific type of item for as long as you use the system. Try to devise a system of nomenclature that will be easy to understand and descriptive of the item. This should not be a problem in that most item codes are eight to ten characters in length. You may also enter a more detailed item description and any comments you might have regarding the item.

BILLING TYPE The *billing type* is usually important to the accounting department. *Prorated* is used when partial billing for a partial period (usually a month) of usage is required. This is the most common billing method. *Advanced* is used to bill for the item a month in advance. *Fixed* billing tells the system that the amount charged for the item cannot be changed when it is applied to a specific billing ID, and the system will bill for the item a month in arrears. *Variable* is used to allow for the billing amount to be modified at the time the item is assigned to a billing ID, and the system will bill for the item a month in arrears.

DEFAULT BILLING AMOUNT *Default billing amount* gives you the ability to pre-define an amount that will be charged when this item is assigned to a specific billing ID. This amount can usually be changed at the time the item is assigned to a billing ID.

PURGE UPON BILLING *Purge upon billing* is set to yes when there is a one-time charge for an acquisition of an item. This flag also makes the system more flexible for billing items such as the cable used in the circuit or a flat rate for the technician's time to install another item.

MINIMUM QUANTITY ON HAND *Minimum quantity on hand* should be defined if you intend to use the system to notify you when it is time to order more of an item.

PHYSICAL ITEM *Physical item* is an important concept to understand about EFI. Many assets, such as voice-mail boxes and conference calling, cannot necessarily be touched; however, that does not mean that they should not be accounted for. If an item is physical (a telephone, a fax machine), the system is capable of tracking additional detailed information about that type of item, such as serial numbers and locations of items in stock.

CLASS OF SERVICE *Class of service* is a description of an item specifically reserved for telephone equipment. The class of service corresponds to a programming class defined in the switching system.

GROUP CODE *Group code* is found in some EFI systems. It allows you to group items into categories for reporting. Examples of groups are telephones, modems, telephone features, and labor.

ITEM TRACKING *Item tracking* by serial number, asset tag, or generic gives you the option of how you will track individual physical items. Unless an item being defined has a high value, it is recommended that you define items generically. When new items are acquired, EFI will ask for every serial or asset tag number to be entered into inventory if the item is not defined generically.

INVENTORY CONTROL *Inventory control* by organization, warehouse, or none gives you the option to choose how reordering is monitored. The minimum quantity on hand can be cross-referenced against the inventory levels of the entire system, or at individual warehouses, or not activated at all. EFI uses this information to generate reorder reports or, in some systems, actual purchase orders.

VENDOR The *vendor* information pertains to the vendors that supply and maintain the item being described, and is used by the system to generate stocking reports and purchase orders. Maintenance requests can also be generated through the use of this information. For more information see "Vendors," later in this chapter.

SUBSTITUTE FOR ITEM *Substitute for item* can be used to indicate an item to be used in place of an item that is out of stock.

You will use the information that you gather on these forms to populate your EFI database. Every system provider has its own opinion of what should be included in an EFI. Try to select a system that is robust enough to offer flexibility in item definition. You will find that this primary capability of a good EFI system provides a great deal of the functionality.

Item Packaging

Item packaging is a function of some EFI systems that allows you to group items that are commonly used together for assignment to billing IDs. In the section "Assigning In-Use EFI" later in this chapter, you will be shown how assignment is performed. Item packaging drastically speeds up the process by allowing you to assign multiple items to a billing ID in one quick step. Figure 8-2 shows some examples of item packages. Say a new executive has just been hired, and that person is going to receive the executive package of telephony items. In one step, all six items in the executive package in Figure 8-2 can be allocated, if the EFI system you select provides this useful item packaging capability. In addition to the greater efficiency achieved when packaging is available, an additional reporting function—your item package usage—is made available.

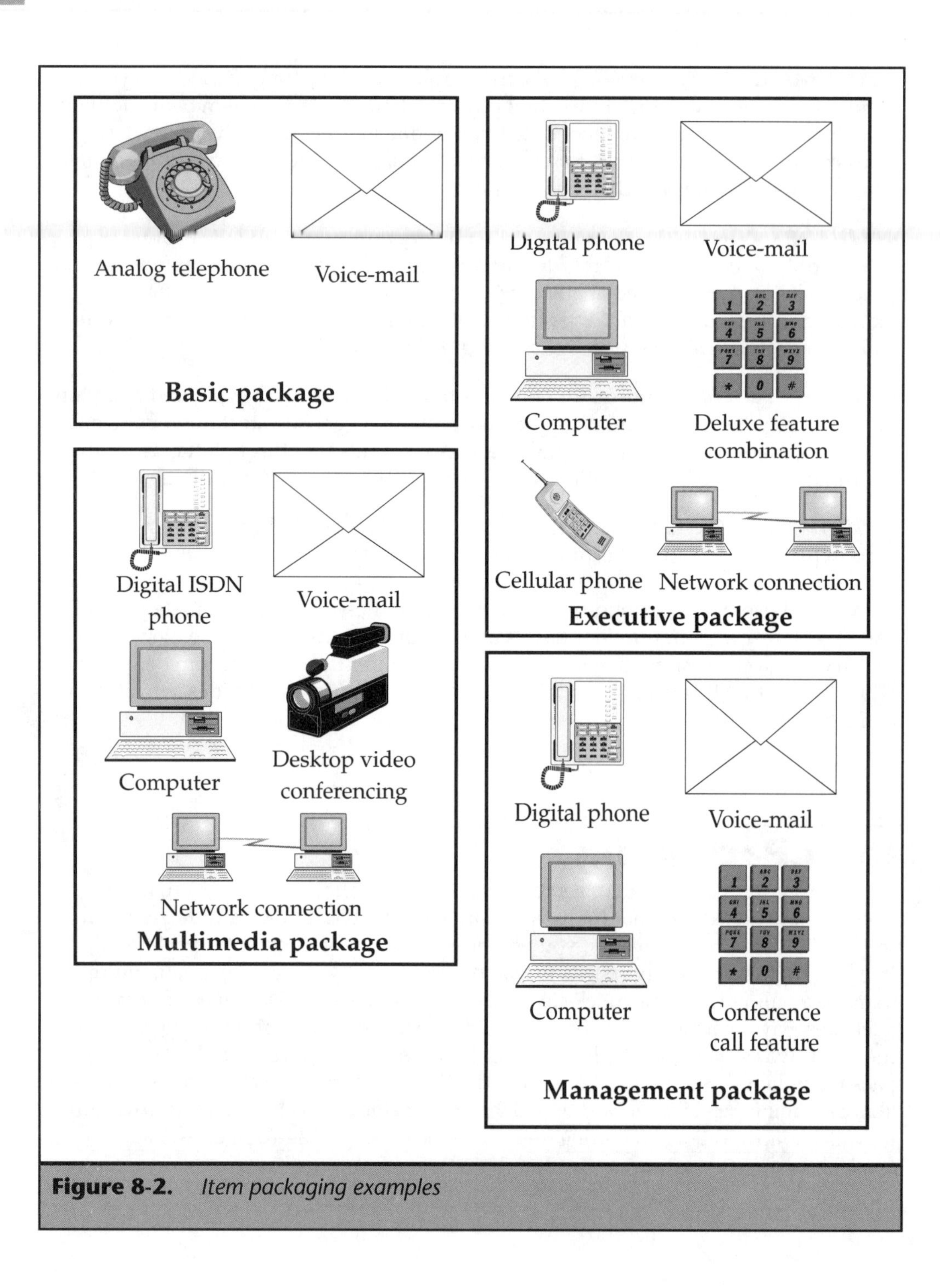

Figure 8-2. *Item packaging examples*

Vendors

Undoubtedly, your organization uses multiple vendors to supply its needed products and services. When you define what items you will be tracking in the EFI system, you are also narrowing the list of vendors that will be tracked by the system. Take the sample form provided in Figure 8-3 and fill it out for every vendor that supplies an item the EFI system will be tracking. The form is pretty much self-explanatory, but there are some components that deserve further explanation.

The *open purchase order* field is an enhanced function of some EFI systems. In addition to performing the inventory and tracking functions, EFI has the ability to track purchasing. In some advanced applications of EFI, the system will inform the user when certain items are below recommended stock quantities and, when requested by the user, will actually generate a user-defined purchase order from the information stored in the EFI database. This information usually needs to be coordinated with the purchasing department. In some network situations, the purchasing department can have direct access to this valuable information, so it may also achieve new levels of efficiency.

The *item* information at the bottom of the form should include all items, including services defined as items, that the vendor can supply you with. The warranty period, item price, quantity discount, lead time to acquire the item, and the purchase order number to be used for that specific item (if applicable) should also be defined.

If the vendor supplies more items than there are lines on the form, no problem—just fill out additional forms until all items are defined. EFI will use this information that you enter into the database to provide a wealth of information about the vendors, products, and services that you use. If you are also going to implement work order processing, the vendor information in conjunction with the product history and performance data gathered by the work order processing module (discussed in Chapter 9) will be available to supply you with additional, valuable information.

Warehouses

When most people think of a warehouse, the picture of a huge building with poor lighting, rafters, and storage racks appears in their minds. By the definition used in an EFI system, a warehouse is any place that houses items being monitored by the EFI module. Stockrooms, switch rooms, closets, and yes, even gigantic warehouses all fall into this category.

A warehouse definition will consist of simply a warehouse ID (usually eight to ten characters), a description, an address (if applicable), a telephone number, and a contact name or names. Later on, you will assign items that you have in stock to these warehouses. If you have chosen to track items generically, the entry of items into a warehouse should be a rather simple task (remember, this will vary from one EFI system to another). If you are tracking items by serial number or asset tag number, you will

Vendor Description Form

Vendor: ______________________

Name: ______________________

Contact: ______________________

Address 1: ______________________

Address 2: ______________________

City: ______________________

State: ______________________

Zip code: ______________________

Phone: ______________________

Fax: ______________________

Comments: ______________________

____

Item Code	Description	Warranty Period	Cost	Quantity Discount	Lead-Time	Open P.O. #

Figure 8-3. *Vendor description form*

find that assigning items to warehouses and removing them from the warehouse is much more time-consuming. If you order a group of 100 items and you are tracking them generically, EFI will allow you to make one entry. If you are tracking by serial number or asset tag, you will need to make 100 entries. Make sure you realize what you are getting into when it comes to item tracking and warehouses at the individual specific item level.

One type of item not handled very articulately in current EFI systems is an item that is under repair. These items are not in use, they are not available, and they are usually not even on the premises. A little creative thinking can solve this problem. Repair items are usually sent back to the vendor that supplied the item or to a maintenance vendor. Create a warehouse for every vendor to whom you send items for repair. When an item is taken out of service for repair, you can place the item in the appropriate vendor's warehouse. You will then be able to see precisely how many items are breaking down that are being supplied by a specific vendor, and you will also be able to track the history of your items by checking your vendor warehouses. This is a great way to keep an eye on item and vendor performance.

Billing IDs and Billing Groups

The concept of a billing ID is not uncommon to most of us. In the telemanagement arena, billing is usually performed by extension number. Without a doubt, this is the most common form of chargeback used today. All of the methods of chargeback described in Chapter 6 can be applied to EFI, but there are occasions where an organization may choose to group its EFI information in a different and unique fashion.

Billing IDs

Billing ID is not always synonymous with extension number. This is the most common form of telephony billing ID, but as was stated earlier, EFI can be used to track and manage things that are not necessarily telephony-related. Many EFI software packages will allow the billing ID to be any set of alphanumeric characters the user enters, typically between six and ten characters long. Billing IDs may be assigned back to an individual or they may be assigned back to a department, a contract number, a budget number—you get the idea. EFI can track any asset, and, through the creation of flexible billing IDs, you are not restricted to telephony methods of chargeback. If the computing department, physical plant, housekeeping, or any other department chooses to use the tools that are available in a strong EFI module, they may assign and group their information in any way, shape, or form that they choose.

Billing Groups

Many organizations would like to have the opportunity to group the information they manage in a number of different ways. The most common form of billing group is a division/department, but EFI does not place this restriction on the user. You may find that there are a number of groups for which you would like to produce consolidated reporting. Contractors, tenants leasing space, concessions, and other third parties are prime examples of possible billing groups. This is accomplished by defining billing groups and then assigning billing IDs to the billing groups you have created. Once you have accomplished this task, you may generate detail or summary information about the equipment and features that you have assigned.

Integration

Integration with other TMS modules is done by many TMSs in the EFI module. If you are using the extension number as the billing ID, the link to the call accounting module is in place. By combining the call accounting information with the EFI information also stored in the TMS, a completely integrated and detailed billing report/invoice, such as the one depicted in Figure 8-4, can be produced. In modern computing, this type of information can be distributed in a number of ways. The trend is to finally go paperless and to transmit telephony billing information through e-mail directly to the individual.

Security

Almost all telemanagement systems have the ability to subdivide the entire database to prevent unauthorized access to unrelated information being tracked by the same computer system. Even if the data is being stored in the same database, security is usually available to prevent unwanted eyes from looking at other people's information. Usually this is accomplished through the use of several computing vehicles. The computer system itself usually has some first level of security beginning with the physical terminal location. Next, computer systems typically have password security to help prevent unauthorized access. Many TMSs are designed using a relational database such as Oracle, Informix, or Sybase, which can provide database security. The TMS modules can usually provide the final level of security by restricting certain users from access through program restriction.

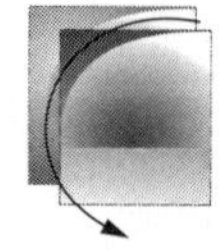

TIP: *Always implement security around any computer system with valuable information, such as a telemanagement system. If your TMS has integration to call an accounting module, the information contained in your TMS is protected by federal law. If you are primarily a telecommunications person, get help from others in the computing department to implement computer database security.*

MONTHLY BILLING REPORT
XYZ COMPANY
AUG 01, 1995 THRU AUG 31, 1995

09/01/95 PAGE: 1

DIVISION: ADM ADMINISTRATION BUILDING: 100 SPECIAL SERVICES
DEPARTMENT: SALES ADMINISTRATIVE SALES FLOOR: 2
BILLING I.D.: 2104 DRAKE, KAREN

DATE	TIME	NUMBER DIALED	LOCATION	CALL DURATION	CALL COST	USED ROUTE	REQ ROUTE	ACCOUNT CODE
08/01	09:11A	667-9500	SMITH ASSOCIATES, BILLERICA	10	1.02	001	001	
08/10	11;39A	848-5130	BRAINTREE, WA	60	0.00	001	001	
08/12	11:39A	808-372-3633	AMARILLO, TX	3	0.65	002	002	51720000
08/18	02:18P	215-667-1571	PHILADELPHIA, PA	12	2.59	002	002	51710000
08/19	02:06P	34-6138	LYNNFIELD, WA	2	0.39	001	001	
08/20	07;37A	957-8475	DRACUT, WA	1	0.08	001	001	
08/20	10:15A	1- 555-1212	INFORMATION	1	0.50	001	001	
08/21	04:28P	817-776-7336	FT. WORTH, TX	9	1.94	002	002	51720000
08/24	02:52P	213-846-9221	LOS ANGELES, CA	20	10.75	002	002	51730000
08/25	04:14P	213-846-9221	LOS ANGELES, CA	1	0.22	002	002	51730000
08/27	09:39A	416-634-0972	TORONTO, ONTARIO	15	7.42	002	004	
08/27	11:40A	515-886-3527	AMES, DES MOINES, IA	3	0.45	001	002	
08/28	12:14P	317-933-3347	INDIANAPOLIS, IN	24	6.60	002	004	
08/28	02:47P	808-372-3633	AMARILLO, TX	5	2.25	001	002	
08/28	03:23P	602-247-3379	ARIZONA	15	8.16	001	002	
08/28	03:37P	1- 555-1212	INFORMATION	1	0.50	001	001	

EQUIPMENT/FEATURE		UNIT AMOUNT	QUANTITY	TOTAL AMOUNT
CC	CONFERENCE CALLING	5.00	1.00	5.00
MODEM-14	14400 BAUD MODEM	20.00	1.00	20.00
PNN-BLTT	BLUE TOUCH TONE PHONE	7.00	1.00	7.00
PRN-560	PRINTER HP 560	25.00	1.00	25.00
SD	SPEED DIALING	3.00	1.00	3.00
SP101	SPEAKER PHONE	15.00	1.00	1.00

SUMMARY	TOTAL CALLS	TOTAL DURATION	AVERAGE DURATION	AVERAGE COST	TOTAL AMOUNT
LONG DISTANCE	14	3:00	0:13	2.89	40.51
LOCAL	0	0:00	0:00	0.00	0.00
DIRECTORY ASSISTANCE	2	0:02	0:01	0.50	1.00
INCOMING	43	2:43	0:04	0.07	3.10
EQUIPMENT/FEATURES					125.00
TOTALS	54	5:45	0:06	0.76	169.61

BUDGET VARIANCE: CURRENT: 1055.39- Y-T-D: 8500.00-

Figure 8-4. *Sample telephony detail billing report/invoice*

Building In-Stock Items

Once you have created the overall definition of your operating environment within EFI, it is time to inform the system about all assets of the organization that will be tracked. The first step is to define the available stock of all items being tracked. In the case of feature items, you need only define the number of available feature items. Some feature items have quantity limits, such as the number of voice mailboxes available. Other features, such as outbound calling privileges, are programming-related and are not limited by a physical number. Items such as labor costs also do not have to be stocked. Physical items need to be defined in the system at either the corporate or warehouse level if the system is to be able to allocate those items from stock at a later time.

Building stock is the final step you must perform before you will be able to actually construct individual telephony definitions and assign them back to a billing ID. Now all you have to do is figure out where every piece of equipment and every feature that you are going to track is and what billing ID it should be assigned to.

Assigning In-Use EFI

You have gathered all of the information to assign EFI to a specific individual or billing ID. Now you are ready to start building your in-use EFI database. The foundation was poured when you defined your items, packages, vendors, warehouses, and in-stock inventory. The final step in building a comprehensive EFI database is to assign all monitored items to the proper billing ID. Use the sample form provided in Figure 8-5 to gather the EFI information. You may want to modify the form slightly to provide information about the way you are tracking EFI and perhaps to include a summary list of the items and descriptions that you are monitoring.

One method of gathering EFI database information, which is very effective but a little demanding, is to send the form to every department and ask the department heads to have each employee fill out the form to the best of their ability and return it to the telecommunications department. Inform the department head that any telephone that is not reported will be disconnected. After about a week, the majority of the forms will be returned. At this point, you should remind everyone to please fill out and return the billing ID assignment form to avoid having their phone service disrupted.

In the meantime, you and your staff must begin to populate the EFI in-use database with the information gathered on the forms. If additional items must be added to the in-use configuration, then it should be done at this time. An extension/feature group list should be available from the switch database. Try to verify and update the information on the forms from the switch database information as you go. Remember, people don't always know what they have or what they want. You will be able to determine what they have and will hopefully be able to help them determine what they really want (with regards to telephony).

At the end of one month of information gathering and data entry, you can produce a list of extensions that are defined within the EFI system and compare it to a list of active extensions available from the switch database. A quick manual scan of the two databases will reveal the undefined extensions. If the person or persons responsible for the extension cannot be identified, turn the phone off. The person whose phone it was will contact you, and then you will be able to make your EFI database as accurate as possible. Before you do turn off any of those extensions, first make sure that the extension isn't assigned to the president or some other person who may not take kindly to this method of data collection. This method of data gathering is very successful, but make sure you avoid the pitfalls.

NOTE: This method of data collection is specifically designed around the telephony application of EFI. Adapt the data collection process to the environment in which you operate.

As you are actually entering the database information, certain EFI systems will request information about the physical location of the equipment. If you are using the extension number as the billing ID and you are also using cable/plant management, the location information should be provided automatically by the system. If you are not using the extension number for all of your billing IDs, you may be required to gather location information along with the data that is included on the form in Figure 8-5.

One final note about initial database entry and EFI. Some EFI systems have a rapid data entry mode. This program allows you to enter multiple EFI records in a user-configurable, quick, and easy-to-use form. The data entry screen does not contain all of the additional detail information and pretty formatting that is included in the standard data entry and lookup screens, but an EFI record can be entered about three times faster than in the standard data entry module. This little module found in some EFI systems is a big timesaver.

Key Telephone Maintenance

In addition to tracking all of the equipment and features of a particular station, many EFI systems allow you to map all of the programming functions assigned to every button on every telephone in the organization. This is not a financial accounting of the extensions, hunt groups, call forward functions, conference calling functions, etc. *Key telephone maintenance* is performed to know which buttons on the telephone set physically hold these functions.

How is this information useful for you? A telephone number can ring on many different telephone sets throughout an organization. Let's say that you are removing an extension from service, but aside from the telephone for which this extension is the primary telephone number, you don't know where else that extension appears. Key telephone maintenance, through the keysheet lookup function, would be able to tell

Billing ID Item Assignment Form

Billing ID: ____________________

Name: ____________________

Location: ____________________

Floor: ____________________

Room: ____________________

Item: ________	Start Date: ________	End Date: ________
Amount: ________	Quantity: ________	Serial/Asset Tag #: ________
Purchase Date: ________	Purchase Vendor: ________	Warranty Period: ________
Item: ________	Start Date: ________	End Date: ________
Amount: ________	Quantity: ________	Serial/Asset Tag #: ________
Purchase Date: ________	Purchase Vendor: ________	Warranty Period: ________
Item: ________	Start Date: ________	End Date: ________
Amount: ________	Quantity: ________	Serial/Asset Tag #: ________
Purchase Date: ________	Purchase Vendor: ________	Warranty Period: ________
Item: ________	Start Date: ________	End Date: ________
Amount: ________	Quantity: ________	Serial/Asset Tag #: ________
Purchase Date: ________	Purchase Vendor: ________	Warranty Period: ________

Figure 8-5. *Sample billing ID item assignment form*

you exactly what telephones that extension appears on and where those telephones are located. In addition to this helpful function, this module identifies all in-use and available extension numbers throughout the system and supplies this information to the work order processing module.

Lookup and Reporting

It takes a tremendous amount of effort to implement EFI. There is so much information that has to be gathered and input into the TMS system design that it sometimes doesn't feel worth the effort. You must push through the difficult data gathering and input phases to realize the benefits of the information that will be available to you through the lookup and reporting functions of EFI. In the end, if you have implemented all three components of facilities management (cable/plant management, equipment/feature inventory, and work order processing), then all of the assignment information stored in EFI will be accessible and maintained by work order processing as work is performed. The updates to EFI and cable/plant management will be automatic.

Lookup is extremely helpful to you when you need to know the details of a particular billing ID. The complete equipment, feature, and cabling information for a specific extension should be at your fingertips. Low-level diagnostics of problems can be performed without ever leaving your TMS terminal. All of the EFI information should be readily available for your review through the lookup functions.

Reporting is a major portion of EFI. The primary reason for EFI is accountability, and reports contain a great deal of the accountability information. What types of reports should you expect from a strong EFI system? Some of the reporting functions that you should expect from an EFI management system are the following:

- In-use detail and summary by billing ID and by billing group
- In-stock availability of all items by item or by warehouse
- Vendor list with all items provided by vendor, including all purchasing and performance history
- Item list with all vendors providing the items, including all purchasing and performance history
- Item usage and item package summary and detail

This is just a short list of the reports you should expect from an EFI system. The real trick is for you or your computer people to have easy access to the databases that are being maintained by EFI. That way, if the report, lookup, or extract that you want is not supplied, you can access the information directly and generate it yourself. The database structure is what makes EFI such a powerful tool.

Conclusion

EFI is a part of facilities management that is an extremely powerful tool for business. It gives you the ability to identify where all of an organization's assets are and who is using them. Waste, abuse, and under- and over-configurations can also be easily identified. The benefits gained from the accounting and chargeback functions of EFI are readily seen. The largest benefit gained through the implementation of EFI, aside from the financial aspect, is the information made available to the third part of facilities management—work order processing. Help desks and technicians can now draw directly from the information contained in EFI to optimize the customer service and technical responsibilities of a telecommunications department. The work order processing function is discussed in detail in Chapter 9.

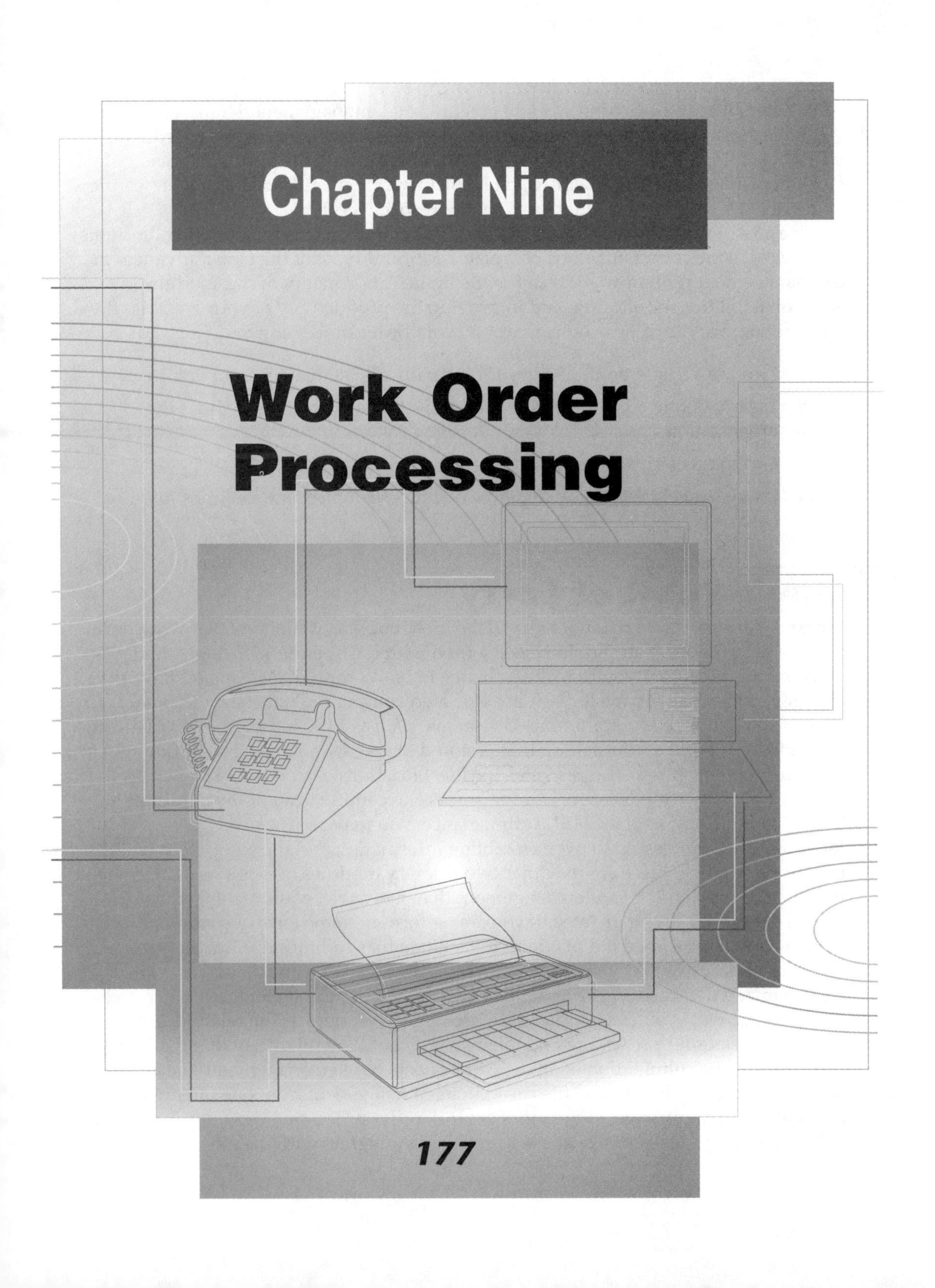

Chapter Nine

Work Order Processing

After all the painstaking work of creating a telemanagement database, you need a way to create, delete, and update all TMS-related items in one coordinated operation. That operation is work order processing, the third component of facilities management (the other two, cable/plant management and equipment/feature inventory, are discussed in Chapters 7 and 8, respectively).

The goal of work order processing is to act as an umbrella over all the other functions of a TMS. It's the cornerstone of a complete TMS, and most of the major advances in telemanagement right now are being made in the functionality of this module. In addition to billing, monitoring, and improving the productivity of your entire staff, here are some of the things you can expect work order processing to do for you:

- Provide a single point of entry for virtually all telemanagement databases
- Clearly define all telephony-related tasks that are done throughout the organization
- Group tasks into complete actions
- Provide a vehicle for you to define common troubles and solutions for your help desk

Single Point of Entry

Without a doubt, single point of entry is the most important function that work order processing provides. You should be concerned with the operation of single point of entry, because it allows you to make a change in one computer system and have that change automatically made for you in associated computer databases, and potentially also reflected in other computer-based systems, such as PBX and voice-mail platforms. Centralizing all of the information that is stored, tracked, and managed throughout a telemanagement system also provides the ideal location for updates to be performed on a global basis. Why does work order processing centralize all of the information that is stored throughout a TMS? At first glance, you may see the need for work order processing to oversee cabling information and the equipment/feature inventory. After all, these are the other two primary modules of facilities management. But, there is much more to telemanagement than just cable, equipment, and features.

All of telephony is interrelated. If you are performing some form of communication and you are using some kind of circuit to facilitate that communication, then you are using telephony. Telemanagement centralizes all the diverse technologies onto a single, computerized database platform. Work order processing ties it all together in a nice, neat package that is efficient and easy to use. Figure 9-1 depicts the relationship between work order processing and telemanagement. You might say that work order processing is the culmination of all the components of telemanagement. The ability to access and update all database files used in telemanagement from one integrated set of applications is the strongest function that work order processing provides. Let's look at how the components of work order processing integrate with the other components of telemanagement.

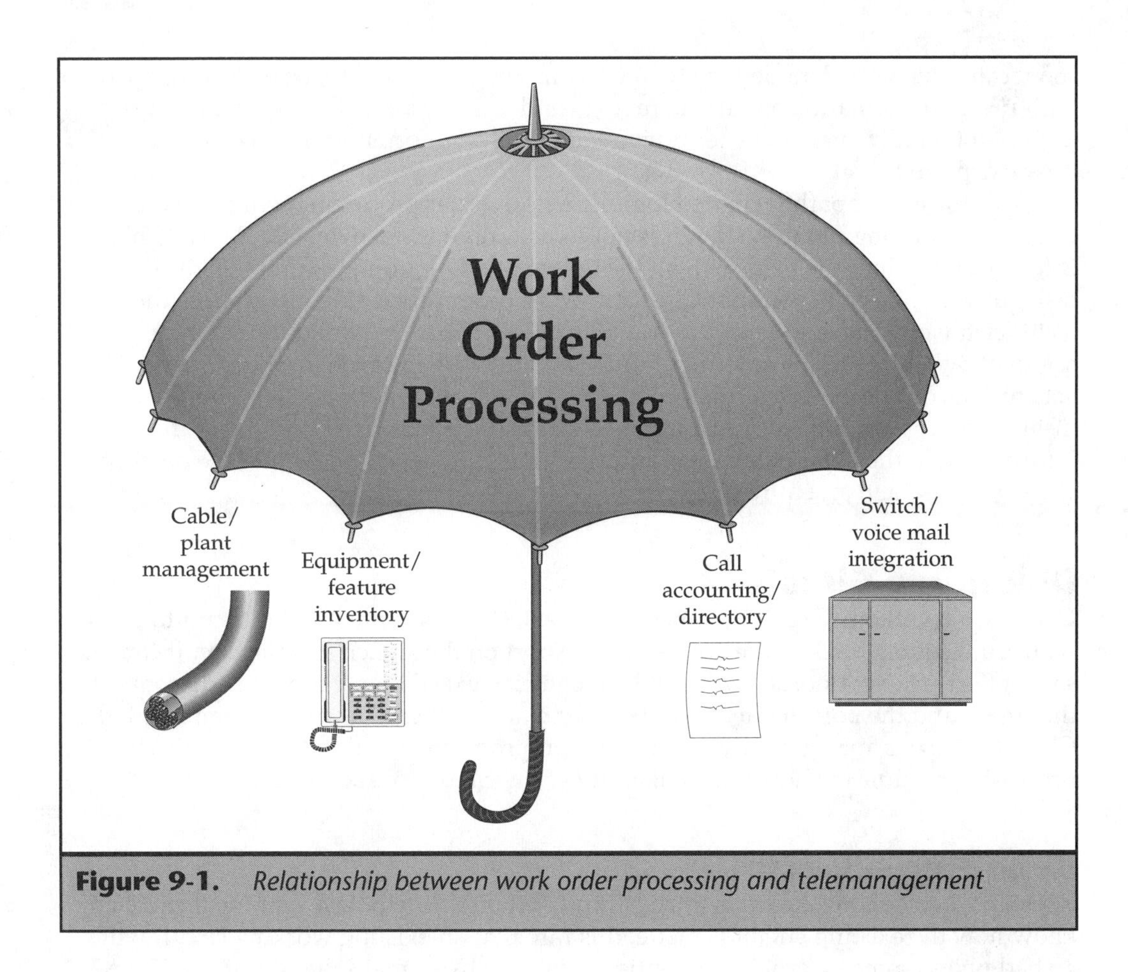

Figure 9-1. *Relationship between work order processing and telemanagement*

Integration with Cable/Plant Management (CPM)

As new circuits are created, old ones are removed, and existing ones are repaired, work orders are generated to perform these functions. The work order processing module draws information from the CPM database to provide information to the work order entry person (typically a help desk) and the technical person who will be configuring or working on the actual circuit.

Adding a Circuit

Adding a new telephony circuit can be a simple task, or it can be extremely cumbersome. The CPM module provides valuable information that will drastically increase the odds that the task will be a simple one. The technician looking at the initial work order can determine where the telephony circuit will end. The cabling information about that specific location is also available. A path from the terminating location back to a switch room can usually be identified easily. The technician going out into the field is supplied with precise information about where to go and which pair of wire to use. If a direct path is not readily available after reviewing the CPM information, alternate routes can be explored.

Once the circuit path is created logically, work order processing will place the circuit in a pending mode, so that it is not in use and it is not available. The technician is then sent out into the field with all the information required to complete the CPM task quickly and easily. Without CPM and work order processing integration, the technician would have to hunt for available pairs of wire and build the circuit, segment by segment. The amount of time and steps saved can be tremendous. Technicians can be up to ten times more efficient on certain difficult telephony installations. When the work is completed and the work order is closed out, the information in the CPM database is automatically updated to show that the pairs of wire are in use and an entire CPM circuit is created.

Deleting a Circuit

Deleting a telephony circuit becomes a breeze. Deletions are not that difficult to begin with, but keeping track of the now-available port on the switch and the circuit can take some effort. The work order will tell the technician exactly where to go to disconnect the circuit and the port that needs to be deactivated. When the work is completed, the CPM database is automatically updated by programs included in the work order processing module to show the availability of the circuit for future use.

Repairing a Circuit

Repairing a telephony circuit is much simpler when the help desk and the technician know how the existing circuit is made. It is rare that an existing working circuit will go bad unless some external intervention, typically by humans, has occurred. If the original diagnostic procedure from the trouble ticket portion of the work order does not reveal a solution, then the remaining variables are provided by CPM through the work order module. The port information about the switch is provided and can be easily checked. The actual cable circuit can also be reviewed for any problems that might readily be seen.

For instance, say the broken telephony circuit happens to use an outside feeder cable that was accidentally cut by a contractor. The technician can inform the user that the problem is already being addressed. If a wire has come loose, the circuit path is known and can easily be checked. If a new circuit must be created, the system can supply the information to create the new circuit. When the work is completed and the work order is closed out, the work order processing module can automatically update the CPM database to show the in-use and/or bad pairs of wire.

Integration with Equipment/Feature Inventory (EFI)

A work order is often required to pool information from the EFI database and supply information to the EFI module for billing purposes. Depending on the work being performed, a number of functions might occur.

Adding Equipment and Features

If you are adding a new telephone or other telephony-related device that the EFI database is tracking, you will need to check the inventory system for availability. The work order processing system will automatically take the inventory out of stock (logically) and place it in a pending mode. Any and all features needed by the work order as defined are also reserved. The technician is told exactly where to go to get the necessary equipment needed to complete the work order and informed of any special programming that might need to be performed. Keysheet definitions and programming may also be input and provided to the technician. When the work order is completed, the EFI in-use record is created, the keysheet is created, the item information is updated, the warehouse information reflects the change in stock, and the billing information is automatically generated.

Deleting Equipment and Features

The value of having integration between EFI and work order processing is readily seen when you are deleting equipment and features. More times than anyone would ever admit, a technician is sent out into the field to remove a piece of telephony equipment or deactivate a feature, and the work is performed. The only problem is that the party that used to be responsible for that equipment and/or feature is still billed just as though the work had never been performed.

Telephone companies are notorious for continuing to bill for equipment that has been removed or reallocated to some other area. Why does this happen? The phone company does not have integration between their billing system (they don't even have the added functionality that a complete EFI module provides) and their work order processing system. The transactions are typically entered by hand from the technician's notes on the work order. Companies are overbilled millions of dollars every month and they don't even know it.

If you are running your own turnkey internal telemanagement system, the integration between EFI and work order processing will make your accounting and billing practice as accurate as humanly possible. A final billing item will be generated to reflect the prorated amount for the time the equipment/feature was in use for the period, if applicable. In addition to the billing function, the work order processing module will prompt you to tell it where to place the equipment back in stock and then make that equipment available for use. The features that are made available through the work order are also updated by the module, if necessary.

Repairing Equipment and Features

Repairing equipment and features requires that the technician know precisely what equipment and features are in use. A work order can include all the EFI information necessary to make the work order process go as quickly as possible. If the item actually needs to be physically repaired, the item stock inventory can be accessed and the item taken out of the appropriate warehouse. The broken item can also be placed in the appropriate repair warehouse for accounting and later for allocation back into stock once the equipment is repaired by the appropriate vendor. If there are any billable items that need to be added to EFI as a result of the work order, the work order processing integration with the EFI module automatically takes care of it for you.

Integration with Call Accounting

The most important function of telemanagement is call accounting. More times than not, call accounting is implemented separately from facilities management. The result of this low level of integration is unidentified extensions and authorization codes making telephone calls. Call accounting integration with work order processing prevents this problem from happening.

How does a telephone get activated or an authorization code get turned on? Typically, this work is performed in a work order. As the paperwork shifts between areas, the information is eventually entered into the call accounting system. If communication between the technicians and the people responsible for the call accounting database is slow or breaks down, then call records end up in an undefined section of the accounting reports. If the extension is shifting from one department to another, the billing information will not be 100 percent accurate.

Work order processing integration into the call accounting database files provides a speedy and accurate method for updating the call accounting module of telemanagement. The moment a work order is closed, the line goes active and the call accounting system is updated.

Integration with Switch and Voice-Mail

Some facilities management packages provide a degree of integration with switch and voice-mail platforms. As time progresses, switch and voice-mail integration with the work order processing module will become the final phase in implementing the concept of single point of entry. Since switch and voice-mail systems are also computer-based, it is possible for a telemanagement system, through the work order module, to program the switch and voice-mail systems in an automated fashion with information gathered in a work order.

Ideally what should happen is as follows: A work order is created. The work order is approved, and a technician configures all necessary labor and assets needed to complete the work order. All work needed to complete the work order is done. When the work order is closed, the TMS then opens up a communications link to the switch and another link to the voice-mail system if necessary. The final step in making the entire work order complete is to have the TMS issue the commands to the switch and the voice-mail to activate the telephone, program the buttons, set the hunt group, assign the class of service, create the voice-mailbox, and any other functions that may need to be carried out.

For the most part, complete switch and voice-mail integration to a work order processing system does not exist yet. Some vendors have implemented switch and voice-mail integration for specific switches, but the ideal facilities management system with complete integration to the work order processing module has not yet been invented. If you find a facilities management package that does have integration with a switch you have or with a switch you are considering purchasing, weigh this functionality greatly in favor of the two systems. Accuracy and efficiency is increased along with complete single point of entry.

With all of the different switch vendors, versions of switches, revisions of software, and types of voice-mail platforms, you will find that complete integration is still a ways off. TAPI and TSAPI (discussed in Chapter 18) may help pave the way to a smoother integration with the more modern switches. These standards can potentially be used by telemanagement firms to create new, more advanced links between the computer and the switch, but the jury is still out on how much these standards will help with work order processing integration to switching platforms.

Work Order Tasks

Work order tasks are the individual functions that must be performed in order to complete an entire work order. Defining the tasks that you perform within your organization is a critical step in creating a smooth-running telecommunications department. Tasks in telephony include the functions that must be performed in the real world and the telemanagement functions themselves. Each module of telemanagement discussed in Chapters 6 through 9 must be maintainable in part through a work order that is made up of tasks.

An example of a typical task performed in a work order might be to define the cable path that will be used by a telephony circuit. This is a CPM task. When you are defining tasks, you must also define who will perform the task. In this example, the task probably should be performed by a technician, but if your CPM database is detailed and complete, the system may be able to select the cable path for you. A technician will also be required to perform the physical work, but this may be defined as a separate task.

Another task might be to delete a cable path from use. Although a technician may be required to actually remove the circuit from service, the CPM database portion of the work order can be performed by an administrative person instead of a more costly technician. Examples of classifications of work that may be performed in a work order are: management, help desk, user, accounting, technician, switch, and administrative. You must also define an hourly rate charged for each of the work classifications defined.

Some of the major telephony tasks that may be defined in the work order processing system are:

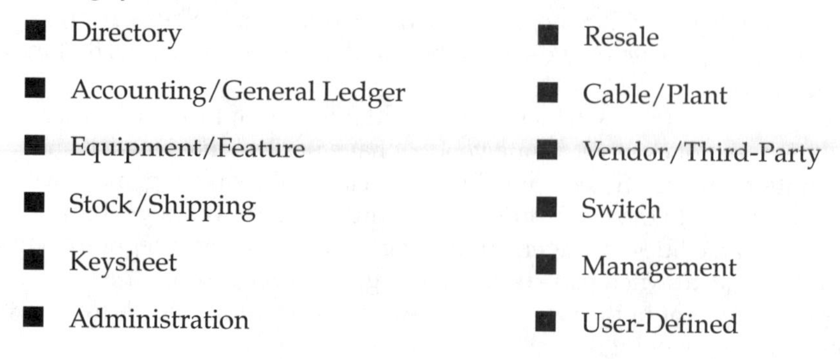

- Directory
- Accounting/General Ledger
- Equipment/Feature
- Stock/Shipping
- Keysheet
- Administration
- Resale
- Cable/Plant
- Vendor/Third-Party
- Switch
- Management
- User-Defined

A task defines who must perform the work and what part of the telemanagement operation is being affected, but it also allows the user to define which database elements are pertinent to the completion of that specific task. Case in point: A telephone station is being added to the system. You might call this a call accounting task. An operator entering the database information into the work order might be required to input all call accounting, directory, and general ledger information pertinent to the addition of the telephone. Another call accounting task might be to delete a telephone. The computer must initiate a final billing cycle for the extension, but the operator does not have to enter in any additional information other than the telephone extension number that is being removed. The actual work order, through the definition of tasks, can be instructed to only ask questions that are pertinent to the specific task that is being performed. When there are potentially hundreds of database fields that must be entered by multiple people, you will find that reducing the number of database questions asked by the work order processing system to only those that are pertinent to the specific work being performed is more efficient and user-friendly.

Tasks should also allow for flexible free-form questionnaire generation. You should be able to define a task that clearly states what should be done when that task is included in a work order. This free-form task allows you to define unique telephony- and nontelephony-related functions that you would like to be able to include in a work order. This functionality gives the work order processing system the flexibility to be used for applications other than telephony. CPM and EFI have already demonstrated this flexibility in Chapters 7 and 8, respectively. This makes facilities management a truly universal tool.

One final element of a task definition is the estimated amount of time that a specific task will take. This time estimate will later be used by the work order processing system to determine the labor time needed to perform the task. This time figure is also used to help calculate the labor cost of a specific work order.

You may begin to see that when you define a task's function within a work order, the work order processing system becomes more and more efficient as more and more specific tasks are defined. You will be able to more clearly define who needs to do what. You will also be able to streamline the work order process itself by defining what information in the telemanagement database needs to be addressed by the task being defined and used within the work order. Without task definitions, the overall function of telemanagement consolidated down to a work order processing system might be overwhelming.

Work Order Actions

Work order actions are entire operations that you want to perform within a work order: add, move, or repair a telephone; install a facsimile machine; etc. By the time your work order processing system is fully operational, your actions should be more specific: install a D12 digital phone with the executive feature package, create a new operator services station, repair a T16 digital phone with a broken base unit, etc.

In order to complete an action, several work order tasks must usually be performed. When defining work order actions, the tasks that are required to complete the action are defined. In addition to assigning the tasks to the work order, a set of dependent tasks may also be defined. For example, the tasks "configure the cable circuit path" and "assign the equipment and features" must be performed before the task that sends the technician out in the field. When the action code is assigned in a work order, all assigned tasks and the associated data elements are included in the actual work order. In addition to the action definition and task assignment, many work order processing systems will also allow you to assign a default lead time needed to complete the action(s) being requested. The information will be used by the system later to determine an anticipated completion date.

You can readily see the differences between the two levels of action detail definition. As you progress through the evolution of your work order system, you will want to create more detailed actions, which will be comprised of more detailed tasks. Addressing specific operations performed in the telephony arena instead of generalized operations is the key to a streamlined work order processing system.

When evaluating work order processing systems, pay close attention to how work orders are defined. Work order processing should be flexible to accommodate your environment, your equipment, and your way of doing business. As stated earlier, facilities management does not restrict the use of the TMS to telephony. The way you implement your computer system is up to you. By creating new actions and tasks to define your organization's needs, you can make the work order processing module an extremely powerful tool.

Trouble Tickets

Trouble tickets are a unique type of work order. When a person contacts your help desk or your department about a problem with telephony equipment, features, or services that your department supplies, a trouble ticket is the first course of action you will take to attempt to rectify the problem. A trouble ticket might be a form of a work order or just a progression of questions and answers to help solve a specific "trouble." A well-developed trouble ticket database can be an incredible timesaver. Trouble tickets have been known to reduce the actual number of work orders generated due to troubles and repairs by over 50 percent.

What goes into the creation of a trouble ticket definition? A trouble ticket definition should be developed for every type of trouble that is reported to your help desk or other individuals within your department. You may begin with a list of troubles that you, your help desk personnel, and your technicians can come up with. Once you have compiled a starting list of troubles to define, your staff, particularly your technicians, should try to mentally go through the steps that they would take to try to repair the problem. An example of a trouble ticket definition is provided in Figure 9-2. Trouble codes, descriptions, a default action code (in the event that a work order is generated), and a set of questions and possible actions are defined.

Work Order Processing
Trouble Code Definition

Trouble Code: STATIC-D12
Description: Static reported on a D12 Telephone Set
Default Action Code: REPLACE-D12

Trouble Inquiry

1. Does the problem occur: all of the time, only when you move the handset cord, only when you move the phone, or other?
   If answer to question 1 is "all of the time" or "other," go to question 2.
   If answer to question 1 is "when you move the handset cord" go to question 3. If answer to question 1 is "only when you move the phone," go to question 4.
2. Try the telephone in another telephone jack and test.
   If they have and the problem goes away, issue work order with CHK-LIN-ST action code.
   If the problem remains, issue default action code.
3. Try to borrow someone else's handset cord and test.
   If the problem goes away, issue work order with SHIP-LC action code.
   If problem remains, issue default action code.
4. Try to borrow someone else's line cord and test.
   If the problem goes away, issue work order with SHIP-LC action code.
   If the problem remains, issue default action code.

Figure 9-2. *Sample trouble ticket definition*

A trouble ticket can provide a low-level user with extremely high-level information to help solve complex problems. By stepping through the preliminary technical diagnostic issues with the user and the help desk operator, the solution to the problem, quite often, can be determined before a technician is ever put on the job. After each trouble ticket question, the work order processing system will allow for responses to be documented. Sometimes the results of the trouble ticket questionnaire solve the problem immediately. Other times the solution to the problem is found by the questionnaire, and the necessary part or service can be provided quickly, again eliminating the need for a technician to trouble shoot the problem directly. When the result of the questionnaire does require a technician to solve the problem, that technician does not have to check the items that have already been checked on the questionnaire, and many times the questionnaire tells the technician precisely what needs to be done to correct the problem.

Trouble tickets and trouble ticket definitions should be an ongoing process. As new forms of trouble are reported or new questions are thought of, new trouble tickets should be created or old ones should be updated. As the trouble ticket definitions evolve, the help desk becomes more valuable. Many troubles can be resolved over the phone. The trouble is tracked by the work order processing system, along with the resolutions documented in the trouble ticket process. As trouble tickets are issued, you will find that an increasing number of them do not result in a work order being generated, and the net result is greater productivity.

Labor Tracking

Labor tracking allows you to allocate your personnel resources efficiently. As discussed earlier, different tasks must be performed by select individuals. If you are running a large staff of individuals, scheduling their time is an important part of good telephony management.

Labor tracking begins with defining all of your personnel. You need to identify the employee, define the type of labor he or she is qualified to perform, and specify the number of hours that person is available per day or week for allocation to work order tasks. Once this information is input, the work order processing system will let you assign the tasks within specific work orders to individuals. As the system accumulates tasks under specific laborers, the amount of allocated and available time for that laborer is updated.

The work order processing system will be able to show you exactly what your work force is doing and how much of its time you have allocated directly through the work order process. When the work order is closed, the actual amount of time that it took the laborer to perform the task may also be updated. If this added step is performed, even periodically, you will be able to gauge the efficiency of your work force. You should be able to optimize your staff with the inclusion of this valuable feature of your work order processing system.

Headers and Footers

Work order *headers and footers* are customizable portions of a work order that you may define to meet your specific needs. In some systems, a header and footer may be defined for each action code defined; in others, the headers and footers can be assigned to the vendor or laborer that is assigned the primary functions of the work order. Headers and footers can be used to provide standardized information, manual field entry forms, questionnaires, and a host of other things. Having the ability to tailor a work order's appearance to your specific needs is a nice addition to a work order processing system.

Work Order Creation

You have been slaving over this darn telemanagement system for months. Let's assume that you did not bring your TMS online at the same time a new switch was being implemented, so all of the work you have put into the TMS was done in the individual modules, i.e., call accounting, cable/plant management, and equipment/feature inventory. Now you feel you are ready to create your first work order and get some major productivity out of this computer.

As you prepare to create your first work order, remember that many work order processing systems will allow you to perform work on multiple circuits within a single work order. This is very useful when you are moving a department or placing a group of contractors. You should also educate your users that performing multiple-move adds and changes to a specific department or group at one time costs less than processing each individual request separately. Budgets can be drastically cut if your users exercise a little planning in their telephony work order requests.

When a work order is created, the basic information about what needs to be accomplished by this work order is defined. At this point, the user tells you that the work that needs to be performed is trouble- related, or a phone (or some other piece of telephony equipment) needs to be added, or a new service is being added, etc. There are a number of ways to initiate a work order.

Using a Requisition Order

Requisition orders are still very popular in many organizations. The requisition order is a type of work order itself. It tells you the basics of what is to be performed, who the work is for, and who should be billed for the work. Sometimes the billing information comes in the form of a purchase order number, requisition number, billing ID, general ledger, project or fund number, but almost always there is some identifier assigned in the requisition order. The information is taken from the requisition order and input into the work order processing system, where it is assigned a work order number, the shell contents of the work order, and a due date.

Taking Orders by Phone

A work order can also be initiated by phone in many situations. Some organizations do not require their users to obtain written authorization for work to be performed. The purchasing and requisition department may issue the requisition number (in some form) over the phone, or many times a requisition number is not required and the work order including any and all charges can be assigned directly to the extension that the work will be performed on. The help desk operator initiating the work order can take all of the information necessary to get the ball rolling over the phone.

Almost all trouble tickets, on the other hand, are initiated by phone. When the call comes into the help desk, the operator is usually looking at a work order initiation screen. If the caller is reporting a trouble, the operator can look through a list of trouble codes and descriptions to try to find a trouble code definition that matches the one being reported. If none is found, the operator should make a note of the trouble so that a technician can develop a trouble code definition for later use. Let's say that the operator finds a trouble code that resembles the trouble being reported. The trouble code is selected and the operator then steps through the trouble ticket questionnaire with the user. Answers to the questions are noted and, if necessary, an action code is assigned. If the problem is resolved through the trouble ticket process, the ticket is closed and the results tracked by the system. If the trouble is not resolved, the appropriate action item is assigned and a work order is initiated.

Taking Orders Over the Network

One of the more powerful and productive tools you can implement for the organization is the power to have work orders initiated by your users. The work order initiation process is a very low-level function that almost anyone with some computer knowledge can operate. The real trick is to have integration between the TMS and the organization's LAN, WAN, or mainframe computer system. A gateway to the work order initiation form must also be established by your MIS department (for additional information on gateways, refer to Chapter 15).

Here is how it works. Once the network and gateway are established, there will be a command or menu choice that will connect any user to the work order processing initiation screen. A typical command might be **go telecom**. This is similar to using an online service to move about from one section of the system to another. The user is presented with a screen that asks questions about the trouble ticket or work order being initiated. The system prompts for all responses necessary in order to complete the trouble ticket or work order. Upon completion of the initiation process, a work order is created.

At the help desk, they monitor for any new work orders that are initiated and check them for completeness. If pertinent information is omitted by the user, the help desk contacts the user for further clarification before the work order is submitted for configuration. The help desk is also available to the user to assist with any questions or problems while the user is filling out the work order initiation form.

Work Order Configuration

Now that a work order has been initiated, it is the job of the work order administrator to configure the work order and/or delegate the various tasks within the work order to specific individuals. Based upon the labor type assigned to the task, the appropriate laborer must be assigned.

In the configuration process, each telemanagement function is detailed. If the task is a directory task, all of the directory information must be input. This task may be performed by your operator staff. If the task is to configure the cable circuit path, a technician may be required to fill out the information, and so on.

All parties involved with the work order process will have access to the work order lookup program. The work order lookup program will allow individuals to view work orders in progress to see their respective actions and due dates. The individuals who process the work orders must perform the tasks that are assigned to them before the work order can be closed. Some steps in a work order must be completed before the physical work order is even generated for a technician to perform the work. These mandatory steps are specified when the action codes are defined.

Finally, after all mandatory work order steps are completed, a work order is generated. A summary work order is depicted in Figure 9-3. Once all of the physical work is completed, the work order is returned to the administrator. Any changes made by the technician while in the field are noted in the appropriate space provided on the work order, and the work order configuration is updated. An example of a change might be that the circuit path chosen in the original configuration was bad, so the technician used another pair. The change is noted and the former pair is marked bad so the system will not try to use it again.

Closing a Work Order

Once all tasks have been completed within a work order, the final piece of magic performed by a work order processing system is done. Virtually all of the pertinent information for a specific station needed by the entire telemanagement system is included in the work order configuration. When the work order is closed, all affected databases are updated. The directory record is created, the financial information is generated, the cable record is defined, the equipment and features are put in use, inventories are updated, the keysheet is created—everything is updated automatically. The information existed before the physical work order was generated in the first place. If you are lucky enough to have switch and/or voice-mail integration, the modifications to those systems are also performed automatically; otherwise, they were probably one of the manual tasks that had to be performed within the work order. The best part about closing a work order is that all of the updates happen in a timely fashion. Since the labor-intensive part of entering the data into the system is performed in advance of the work actually being performed, your telemanagement

WORK ORDER REQUEST
XYZ COMPANY

WORK ORDER NUMBER: 00021 DUE DATE: 10/12/95

TENANT: 01 BUILDING: 100 SPECIAL SERVICES
AFFECTED EXT: 5432 FLOOR: 1 URGENCY LEVEL: M
BILLING ID: 5432 ROOM: 109 TRUNK:
DIV/DEPT: BUS BUSINESS ROUTE:
REQUESTOR NAME: DB Redd
CONTACT PHONE: 609-7 GL NUMBER: 56
ORDER DATE: 09/30/95 MASTER ACCT:
ORDER TIME: 10:47 TAKEN BY: AVP

COMMENT: GIVE HIM A SHORT CORD

TROUBLE CODE: KEYBK BROKEN KEY(S)

Q: 01. WHAT KEYS ARE BROKEN?
A: I A U V

Q: 02. IS THE PHONE STILL FUNCTIONAL?
A: BARELY

Q: 03. HOW DID THE KEYS BREAK?
A:

ACTION CODE: FIXKY FIX PHONE - BROKEN KEY(S)

TASKS REQUIRED

TASK NO.	TASK	COMMENT	TEC	SCHED DATE	EST TIME	ACT TIME	COMP DATE	C P	LAB TYP
01.	REPAIR	SEND REPAIRMAN		10/06/95	4.00	3.25	10/08/95	Y	DDD
02.	CABLE			10/08/95	1.25	1.00	10/11/95	Y	TEC
03.	DIR			10/10/95	.25	.50	10/15/95	Y	TEC
04.	EFI			10/12/95	1.00	.75	11/09/95	Y	TEC
05.	REVIEW	CHECK TIMES		10/12/95	.75	1.00	11/10/95		MGT
06.	PRINT			10/06/95	.25	.50	10/07/95	Y	
07.	UPDATE			10/14 95	1.00	1.50	11/11/95	Y	MGT
99.	RELSE	RELEASE W.O.		00/00/00	.00	.00	11/18/95		

CONTROL TRAILER

Figure 9-3. *Sample work order control*

databases remain as accurate as possible. There is no question that the information needed to perform the work order is present.

Conclusion

As the work order is closed, not only are all of the affected databases automatically updated, but vital information with regards to the operation of your company is created. History tracking of every operation performed is made available to you. Allocation of assets and resources can be seen at various levels of detail. You will also be able to identify wasted or repeated efforts by analyzing the trouble ticket and work order history information.

A work order processing system is very difficult to get off the ground. Work order processing is dependent on all of the other database modules of telemanagement to supply it with the necessary information to make it a productive tool. The best time to implement a work order processing system is when a new switch is being procured, but there is no bad time to begin managing telephony. Hundreds of millions of dollars are misappropriated, wasted, and improperly billed by phone companies and corporations every year. The savings that can be realized through the use of a work order processing system within telemanagement and throughout an organization can be immense. You will do yourself a tremendous favor if you take the time to build an accurate and fully functional work order processing system.

Before you do go out and implement a work order processing system, make sure you know what you are getting into and are prepared to follow through. Even with complete dedication to the vision of integrated telemanagement and work order processing, you are looking at several months of hard work that might turn into two or even three years of hard work. Work order processing is not for everyone. Your organization should have at least one thousand telephony connections to warrant a complete work order processing system, but if your organization is that large, the labors of work order processing will bear fruit.

PART THREE
Telephony
Connectivity

Chapter Ten

T-1s

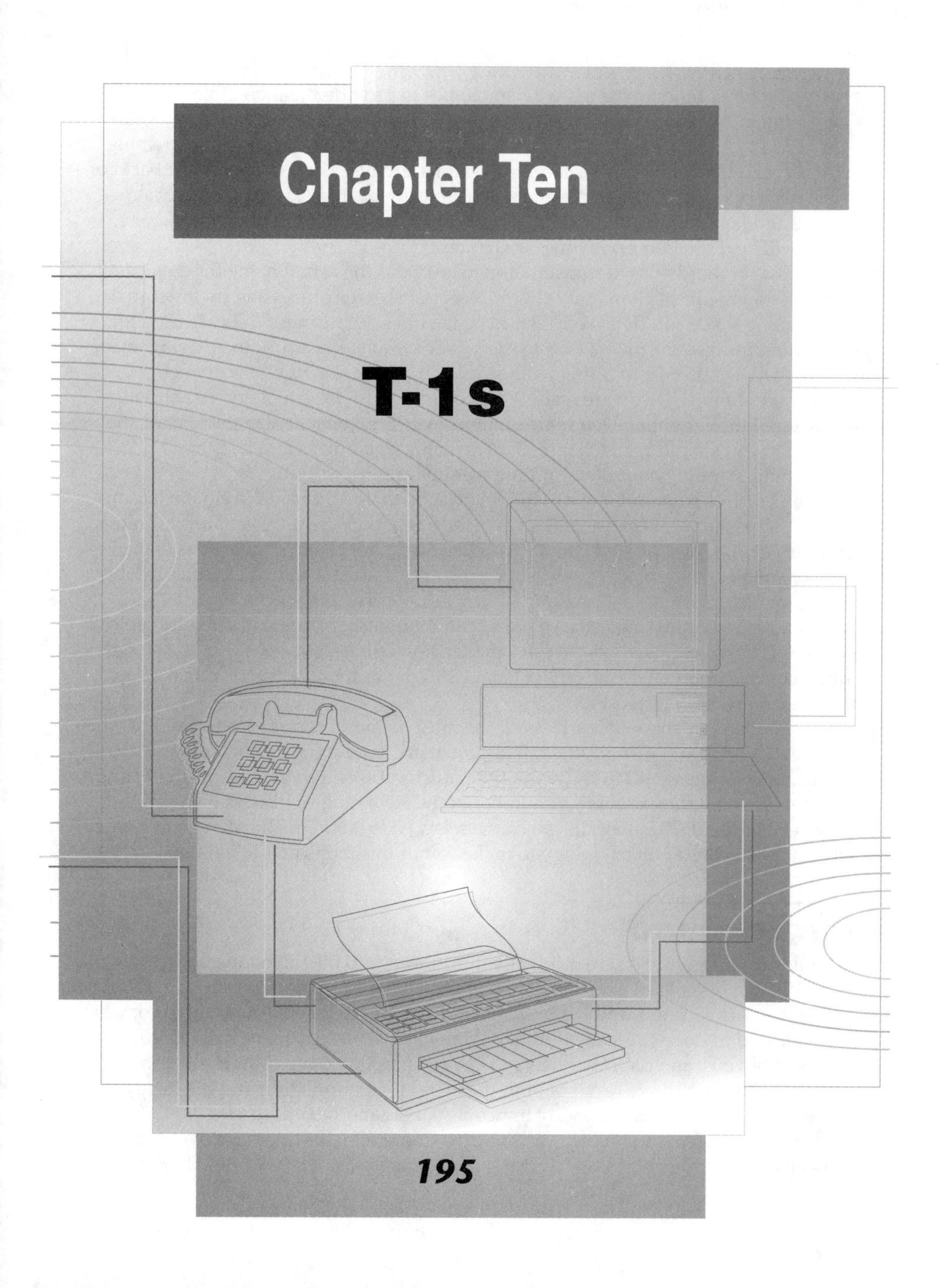

Telephony relies on so many technologies to provide the incredible communications that are available in the world today. The T-1 (also written as T1—both are acceptable) is the most widely used technology for creating switched and dedicated digital communications circuits. The T-1 is truly the backbone of digital service provided to the end user (typically business) in America today. Europe and Japan have similar standards that make up their telephony backbone, which will be discussed briefly in this chapter.

T-1 is a standards-based transmission format and the actual name for the transmission medium all in one—that is, a set of rules for processing digital signals representing voice and data over telephone facilities. In business, T-1s are used for everything from connecting a PBX to the public switched network to bridging and routing between LANs, to acting as short-haul transmission devices between microwave stations. If you are reading this chapter on T-1 technology, you should already be familiar with telecommunications in general or you may get lost. At the very least, you should read Chapter 3 on switching platforms and telephony to get a better understanding about telecommunications and networking.

Why are T-1s so popular and what other options do you have? Well, the reason T-1s are so popular is because of the circuit size that they create: 24 communications channels in one consolidated circuit. Another reason for their popularity is that all sorts of smaller *trunk groups*, such as long distance, local, tie-line, ISDN, etc., can be configured over a single T-1, making it very flexible. The other options to T-1 in America are: separate single *analog* lines brought in on separate pairs of wire typically on an RJ21 connector (see Chapter 3), smaller single-line service, or you have to go much larger—up to a T-3 (the equivalent of 28 T-1s) or more, which is much larger than most installations require.

In this chapter, you will learn about the most common forms of T-1s and how they are used. The T-1 and its peripheral equipment are among the most common connections that you will find used in the industry today due to their flexibility and the economy that they provide. A number of the technologies discussed in this book—including POTS lines, ISDN, frame relay, ATM, and videoconferencing—often rely on T-1 circuits and technology to make communications possible.

Background

Digital communications were originally introduced into the telephone network in the early 1960s and are used extensively today within the telecommunications networks in North America, Europe, and Japan. The reason was to reduce the amount of copper cable needed to carry the same number of telephone conversations, since large numbers of voice and data channels can be accommodated on a single physical circuit or facility. This is a direct result of both the increase in population and the increasing reliance and demand on telephony in people's lives and businesses.

One approach to the problem was to carry more than one message over the transmission facility, such as a T-1, using *frequency division multiplexing* (FDM) to divide the available bandwidth into segments that carry separate signals. For quality purposes, greater reliability, and faster speed, *digital* signaling would be used instead of analog. For example, the first call would be carried on band one (0 to 4,000 Hz), the second call would be carried on band two (5,000 to 9,000 Hz), and so forth. The 1,000 Hz left alone on each side of the band were known as "guard " channels, and they would prevent crossover of conversations into other bands. The biggest drawback to FDM was the rapid attenuation or degradation of the signal as the frequency of the signal became higher and higher. In order to combat the shortcomings of FDM, modern T-1 technology was introduced.

T-1 Technology

The spoken voice forms analog waves. The common telephone is an analog device in that it creates an electrical representation of the sound waves. As the pitch, or frequency, changes, either increasing or decreasing, the electrical current's frequency varies concurrently. In a similar fashion, the loudness or softness of the sound will increase or decrease the amplitude of the signal. Therefore, in order to produce a *digital* representation of these sound waves, some type of conversion or coding needs to take place (as shown in Figure 10-1).

To convert voice analog signals into a digital representation, samples are taken of these sound waves. Typical analog voice frequencies and levels are sampled 8,000

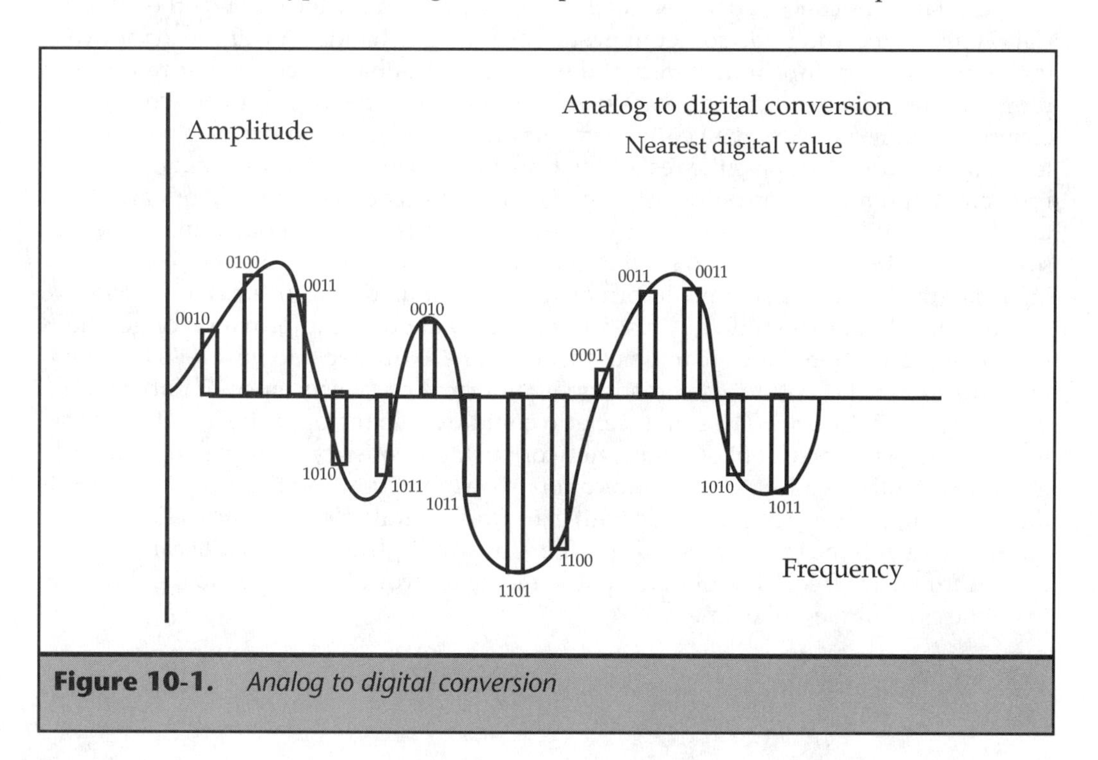

Figure 10-1. *Analog to digital conversion*

times per second. Their status is coded into an 8-bit word called a byte. The conversion is performed using a method called pulse code modulation (PCM) and an integrated circuit chip called a codec (coder-decoder), which performs the actual conversion from analog to digital and back again as shown in Figure 10-1. In this manner, $8 \times 8{,}000 = 64{,}000$ or 64 Kbps represents the voice conversation. These can then be multiplexed into a higher-speed transport system in which the conversation is provided one of several transmission slots (like seats on a train car). The T-1 transmission system, for example, has 24 slots (words), or 24 x 64 Kbps = 1.536 Mbps.

Multiplexing

Before you get lost in bits, bytes, frames, and synchronization, let's take a look at what multiplexing does and how it does it. Two types of multiplexing are typically used: frequency and time-division multiplexing.

Frequency Multiplexing

Frequency multiplexing is fairly easy to understand. Think of your FM radio. All of the radio stations in your area are transmitting their signal at the same time, but your radio picks just the signal from the radio station you want to hear out of the air and plays it for you out of your speakers. This is because each radio station broadcasts on a different frequency that you can tune in on and listen to a particular frequency signal.

Now, take this concept one step further and say that every ten seconds the radio station instructed you to do something, such as put your hands on your head, or drive in the right lane or stop, and the rest of the time they just have their regular radio programming going on. Obviously you wouldn't listen to them, but if you were a communications network, you could listen to the instructions and perform what they tell you to do and then pass the rest of the transmission along to its destination—the listener—but remove those silly instructions. The instructions to the network are usually things like: make a telephone connection, here is the telephone number, here is some information I want you to send to the other end, here is how big the piece of information is (just in case some information is lost along the way), etc. The listener doesn't care about that stuff, they just want to know about the information or get the voice communication directed at them. Just take one extra step and now let's assume that you are a radio station also broadcasting information (voice, video or data in the telephony environment). The same scenario could exist in the opposite direction, and thus you have basic concept of what a network using frequency multiplexing does. One end establishes a line of communication in one direction and the other end does the same thing. The same physical circuit can support multiple frequencies (just like the airwaves and the radio stations) and therefore multiple logical circuits can be created from each frequency of information by the network listening in to the different frequencies all at the same time.

Time-Division Multiplexing

Time-division multiplexing (TDM) is a little more difficult to understand, but it is also more similar to framing, which is the innate form of T-1 signaling. Time-division multiplexing is kind of like a guy with a remote control in his hand in front of a TV set. One moment, he is watching a baseball game, the next a football game, the next an infomercial, the next a sci-fi movie, and the next he is back to watching the baseball game. To the external observer, it looks like just a bunch of gibberish flashing on the TV screen, but our guy is actually watching all of the channels at the same time and probably feels that he is missing almost nothing of any of the programs he is watching. That is kind of what time-division multiplexing does.

A T-1 can carry 24 channels (kind of like TV channels). TDM is watching, and transmitting (which our zapper does not do) 24 channels at the same time. The incredible part about this is that TDM changes channels 192,000 times a second (24 channels × 8,000 samples per second). Now if you could change channels that fast and keep each channel separate in your mind, you could literally watch just about as many channels as you wanted and not miss anything (anything significant, that is). Believe it or not, that is what a T-1 and a frame are all about, being able to send and receive 24 different things on one circuit and being able to separate them out based upon when they are received, kind of in a cyclical fashion. Channels 1–24 are viewed in sequence and then 1–24 are viewed again and again and again. The framing part of a T-1's signalling is what keeps the channel switching in order, intact and accurate.

Frames

OK, now keep the multiplexing concepts in mind while you take a look at frames, because this is really how a T-1 is usually implemented over copper wire to a location. A single T-1 frame consists of 24 samples (8 bits each) from each of 24 channels (voice or data conversations). These are packaged together with a framing bit used for synchronization to complete a frame of 24 x 8 = 192 + 1 framing bit = 193 bits. Since each conversation is sampled 8,000 times per second, there are 8,000 frames sent each cycle. This means 8,000 frames × 193 bits per frame = 1.544 million bits per second that are sent to actually represent the 24 channels. Of the 1.544 Mbps, 1.536 Mbps is the actual amount of data; the single framing bit, sent 8,000 times a second, accounts for the other 8 Kbps.

Frames are clustered in groups of 12 to become superframes and then in groups of 24 to become extended superframes. Telecommunications providers use channel banks to achieve this analog/digital conversion and multiplexing, called D4 and D5. The channel banks sit between the standard voice-grade telephone lines and the digital (usually T-1) facility. They convert many telephone signals coming from the telephone lines into one high-speed digital transmission that can be broken apart by another channel bank or routed separately by the telephone network. Multiplexing is the combining and separating of multiple signals into one time-divisioned high-speed transmission.

Signal and control codes have been developed in the last framing bit so that the sixth and twelfth frame form special call-status messages. In fact, the last bit, sometimes called the F bit of the frame, can send repeated numbering patterns for synchronization of timing and provide several 4 Kbps signaling paths for an array of signaling information, out of band from the voice information. Thus, a hierarchy of transmission systems can transport a high number of multiplexed voice calls when changed to digital 64 Kbps channels.

The analog to 64 Kbps digital conversion is called pulse code modulation (PCM). Using different coding techniques called adaptive PCM, the voice could be represented at a subchannel rate of 32 Kbps. Other compression techniques existing today can rob bits (literally take bits out of the transmission), further reducing the voice representation to 24 Kbps; some even degrade the signal to 9.6 Kbps. However, the quality is degraded as more and more of the bits are removed. As a result, the industry has settled on 64 Kbps even though it is felt that this is twice the necessary sampling that the human ear can detect.

A single 64 Kbps channel is called a DS0, while the T-1 rate of 1.544 Mbps for 24 channels of 64 Kbps each is also referred to as a DS1. Similarly, T2, with 96 channels or 4 T-1s, is known as a DS2. Though T2 is used quite extensively in Japan, in actuality the U.S. industry skipped to T3 rates of 28 T-1s, commonly referred to as DS3 or FDS3 (fiber). Just think of a T3 as a really big T-1. It can handle up to 672 simultaneous, separate voice or data transmissions over a single strand of fiber optics—mind-boggling. Hence, T-1 and T3 transmission facilities became the backbone of the digital transmission telephone network of the telecommunications companies during the 1970s and 1980s. Table 10-1 shows the digital signaling hierarchy used in North America, Europe, and Japan for all of the digital standards.

Europe's Common European Postal and Telephone subscriber trunk dialing (STD—which is a fancy way to say "direct dial long distance") group has worked on a similar approach and has elected to use the E1 rate of 30 voice frequency channels and one channel for supervision/control signaling, as well as one channel for framing synchronization, as a $((30+1)+1) \times 8 = 256$ bit frame times $8,000 = 2.048$ Mbps, known as 30 B + D.

T-1 Lines

A T-1 system could be created by taking 24 analog voice circuits, converting each one into a digital signal, and then multiplexing those signals onto a single four-wire T-1 line (commonly referred to as a *span*). This method is often called *pair gain* or *pair reduction* since you are gaining additional pairs of wire or reducing the amount of wire pairs needed to support the transmission. One pair is designated for receiving voice or data, and the other pair is designated for transmitting voice or data.

Digital multiplexing signal level	Number of voice channels	North America in Kbps	Europe in Kbps	Japan in Kbps
0	1	64	64	64
1	24	1,544		1,544
	30		2,048	
	48*	3,152		3,152
2	96	6,312		6,312
	120		8,448	
3	480		34,638	32,064
	672	44,736		
	1344*	91,053		
	1440*			97,728
4	1920		139,264	
	4032	274,176		
	5760			397,200
5	7680		565,148	

*Intermediate multiplexing rates

Table 10-1. *Digital Hierarchy Used in North America, Europe, and Japan*

For distances longer than one mile, a repeater is placed every mile to regenerate the signal. Since the signal consists of ones and zeros, it's fairly easy to identify whether an "on" (a zero) or an "off" (a one) was sent and then re-create the signal and transmit it further along. This is the primary reason digital transmission provides a better quality signal: Any noise or distortion picked up along the way is eliminated when the signal is regenerated.

A new type of T-1, called a *pair-gain repeaterless T-1*, uses a different type of repeater and can send a T-1 signal up to five miles (versus one mile with the more common version of T-1). Look for this technology if you are planning for new T-1 installations. You will save in equipment costs without giving up any functionality.

T-1 Performance

With the high-speed data transmission capability provided by a T-1 also comes a great need for clarity. Garbled data or lost information is simply not acceptable in a T-1 environment. Because of the incredible speeds that are achieved over a T-1 (typically, the two-pair copper wire version, not a fiber-optic T-1), there are many factors that can interfere with the quality of the T-1 signal.

Possible Hazards

The growing need to change existing analog plants (installations) to support digital T-1 has always been hindered by *bridge taps*, parallel connecting nodes somewhere between the end points of a copper wire circuit, such as telephone jacks connected in a home that are not in use. They were attached anywhere along the circuit to provide new service at some point in time. The balance of the circuit (line) is not cut off but remains in place as a dangling open pair. These "taps" do not usually affect voice conversations, but their cross talk (noise they add on the line) frequencies cause unacceptable data bit error rates. These taps are hard to detect when changing analog voice-grade lines to digital, where they degrade T-1 and especially data.

Echo is another demon you must watch out for on a T-1 circuit. During a regular telephone conversation, a little echo is expected, even desired. The echo is how you hear yourself in the speaker of your receiver. But too much echo is not desirable, especially on data calls. If you hear yourself and then you hear yourself again a moment later, it becomes a bit annoying. If you have ever made a telephone call using a satellite (whereby the signal takes significant time to get to its destination), you know what echo is all about. Your voice gets to the other end and is picked up by the far end speaker and sent back to you—is that annoying, or what? In order to combat against echo, a device called an *echo suppressor* can be installed to block out the echo signal. Unfortunately, the full-duplex nature of a telephone call (the ability of two signals on a single circuit to travel in two directions at the same time) also loses some quality.

Load coils can also cause problems for digital T-1s. Load coils are used to boost the voltage, and the quality, of analog voice signals. If copper wire is being used for analog signaling, then load coils are desirable, but if T-1 signaling is to be carried over the copper wires, then the load coil must be removed to avoid enormous numbers of data errors.

These are just some of the hazards that can befall a T-1 circuit. Water seepage into the cable can cause all kinds of havoc. Signal crossover from other circuits is another pitfall to watch out for. And of course you don't want to overlook the normal noises that are often picked up on copper wire from time to time. These problems are usually tackled when the T-1 is originally installed, but you never know when you might start to have problems with your high-speed digital T-1 circuit. Keep your guard up.

Migration of Voice to Data

As data was integrated with "digitized voice" channels to share a common digital transmission facility (the T-1), there arose a major concern over the loss of *frame synchronization* due to a long pattern of zero bits. Frames could get out of synch, meaning that the time slot reserved for one channel on the receiving end got mixed up with the time slot of another channel from the transmitting end. This frame error would happen because zeros were sent when there was no data to be sent on voice lines. A sample lost here or there for voice communications is no big deal—you probably would not even notice it—but data must be 100 percent accurate. Given the short amount of time in a sample (1/8,000th of a second), lots of silence (zeros) is sent down the T-1, thus potentially causing the frames to go out of synch.

Since the eighth bit of each byte channel had been used for voice-status signaling, it was also available for data. Therefore, for every seven bits of data sent per channel, the office *channel units* (the devices used by the phone company to separate and combine the channels on a T-1) added one bit for data calls in the eighth position. When the channel was idle, they inserted a zero in the eighth position and all ones in positions two through seven, thereby preventing eight consecutive zeros. Therefore, seven bits of data inserted 8,000 turns per second becomes a 56 Kbps data link in a 64 Kbps channel. Thus, T-1 could send 24 channels of 56 Kbps data. This less efficient 56 Kbps is still very common in North America and causes the rest of the world to accommodate this obsolete version of coding.

Then there was a need to obtain a *clear-channel* 64 Kbps for data handling capabilities, so that the eighth bit could be used to represent information and the T-1 could transport 24 channels of 64 Kbps data, especially since the true Integrated Services Digital Network (ISDN) standard is 64 Kbps per channel (see Chapter 11 for information on ISDN). To achieve this, two techniques were introduced: B8ZS and ZBTSI.

B8ZS

Bipolar with Eight Zero Substitution (B8ZS) was one method introduced to free up that extra bit in each channel to get true 64 Kbps data transmission speeds. B8ZS provided bit stuffing and substitution changes, as shown in Figure 10-2. Notice that the fifth and seventh bits of the B8ZS substitution byte are of the same polarity. This violates the alternate mark inversion (AMI) rule, which states that every other "one" must be of opposite polarity. This is more commonly called bipolar violation (BPV). This can cause problems in many networks that are based on AMI. Hundreds of thousands of repeaters that are partially based on AMI still need to be upgraded to handle this clear channel solution, and thus B8ZS is more expensive to implement.

ZBTSI

Zero Byte Time Slot Interface (ZBTSI) requires overhead bits in the extended superframe format to convey information about the location of all zero bytes. Put simply, instead of stuffing a predetermined byte pattern into an empty slot, ZBTSI tells the circuit about the zeros (empty slots) by encoding the information in every sixth frame. The biggest advantage to this method is that it does not violate the AMI rule

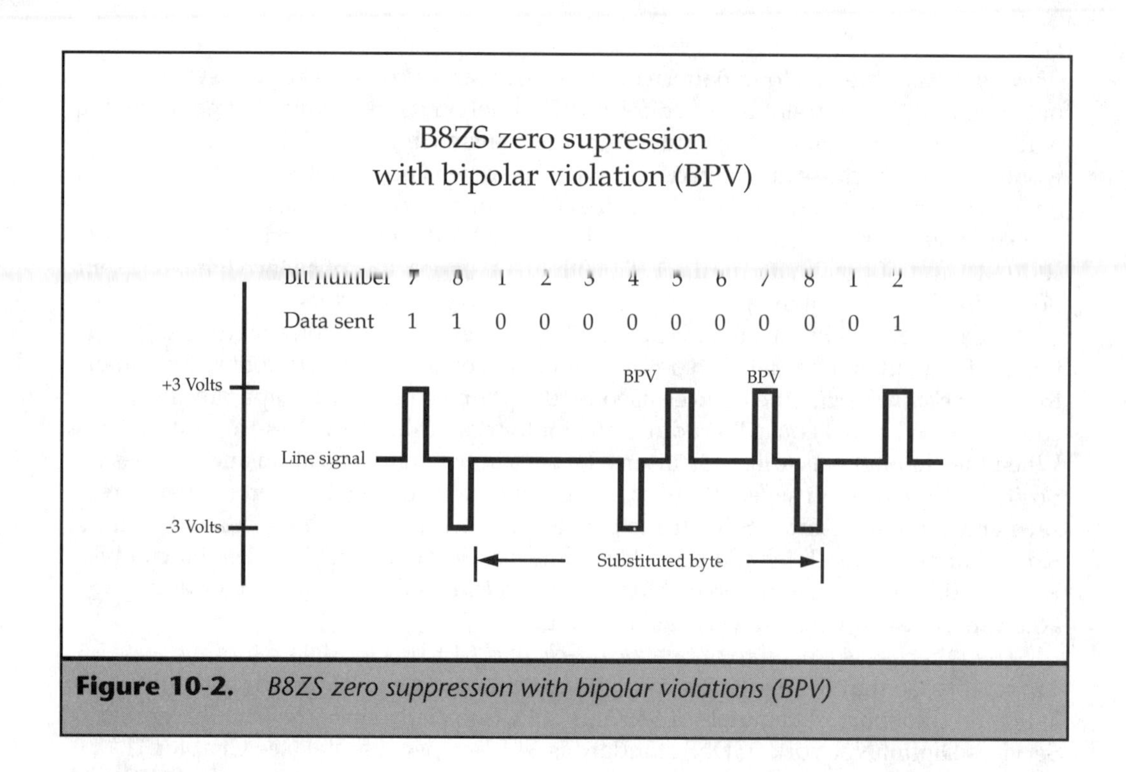

Figure 10-2. *B8ZS zero suppression with bipolar violations (BPV)*

and it works with virtually all existing T-1 networks. The main disadvantage is that there is just over a 500 microsecond delay added to the data—not that big of a deal.

T-1 Networking

The demand for faster and easier operations, the need to remove the costs of manually supplying special services, the desire of customers to have more control of their networks, and the dropping price of T-1 facilities led to the deployment of *digital cross-connects*, more commonly known as DACS (Direct Access and Cross-Connect Systems). Some of these DACS have the ability to perform DS1/DS0, DS3/DS1, or even DS3/DS1/DS0 switching to add or drop DS0s, DS1s, and DS3 channels as well as open up the access for testing.

Subsequent systems allowed customers to dial up and request a T-1 facility to control (reconfigure) their own network. This led to the need for advanced T-1 networking, which gives customers access to such features as:

- Compressed voice transfer: A slight reduction in overall voice quality in exchange for a greater amount of voice over data capability.

- Teleconferencing: The ability to connect more than two people together using telephony. The link can be voice only, one-way video and one-way voice, two-way video and voice, etc.
- Enhanced data services: The ability to control how you handle data without having the intervention of the phone company and potential tariffs to worry about.
- Compressed video: Video transmission capability that only uses between 56 Kbps and 3 Mbps instead of 45 to 90 Mbps. On certain switches this is known as bandwidth on demand. A very powerful multimedia tool.
- Automatic selected priority links: You have the ability to tell the network what information has the highest priority, rather that having the network determine this for you.
- Best-case routing in case of link failure: Also known as automatic route selection (ARS). This is what allows you to choose a WATS line for long-distance calls and a tie-line for calls over to your other office. It also lets you determine what telephone facility to use, instead of the phone company. See Chapter 3 for more information.
- End-to-end diagnostics: You have the ability to troubleshoot your own equipment and the phone companies if you choose. Again, you gain self-reliance.
- Circuit redundancy with automatic link restoration and increased mean time between failures: This is a basic function of T-1s and the CSU/DSU units that encapsulate the T-1 service.
- Aggregate trunk rates: You can bring more logical telephone trunks to your location for less money. You get an economy of scale that reduces your per-line cost dramatically.
- Passthrough: Gives you the ability to access your network through your own facility (the T-1).
- DS1 framing: This is the standard signaling convention used by T-1s.
- DCS (Distributed Communications System) compatibility: This gives PBX users the ability to distribute their T-1 capability throughout the organization through multiple locations instead of being stuck with one gigantic hub to handle all communications. This is analogous to a LAN in contrast to a mainframe computer system.
- Dynamic bandwidth allocation: As more bandwidth is needed, multiple channels within the T-1 can be combined to provide the service.

The T-1 network is an incredible piece of technology. It is capable of supporting so many specialized operations and functions that they can't all be listed, because new ones are being developed on a regular basis. Because of the bandwidth it supplies to the end user and the economical cost of the service, T-1s are guaranteed a place in the telecommunications networks of the future.

Fractional T-1

"Fractional T-1" is something of a misnomer in that, although the *cost* of the product is generally reduced in price, the service itself is a full T-1. In other words, the customer pays only for the bandwidth ordered and not the bandwidth delivered. Fractional T-1s are based on the same standards as regular (full) T-1s. T-1 multiplexors formulate various voice and data channels over the transmission line using D4 or D5 extended superframe format. DACS and network management systems enable multiple partitions and even subpartitions to coexist on the same backbone network, as large and small users are combined in the T-1 facility with all of the other users.

Equipment

A wide array of equipment can be connected via a T-1 to provide various functions and services. Here is a list of the more popular pieces of equipment that are connected to T-1 facilities.

- Multiplexors: Split the main channel or communications link bandwidth into apportioned, fixed, predetermined parts (time slots) among the various users.
- Private branch exchanges (PBXs): Private telephone exchanges, or the switch your company owns.
- Central office switching systems: Run by such companies as U.S. West, PacTel, and NYNEX (see Chapter 3).
- Toll and tandem switching systems: Run by such companies as AT&T, Sprint, and MCI.
- Channel banks: Connect multiple voice channels to high-speed links by performing voice digitization and time-division multiplexing (TDM).
- Transcoders: Allow compression of voice and voice-grade data from two to four T-1s onto a single T-1 facility.
- Echo cancelers: Eliminate the echo often heard on long-distance calls (see Figure 10-3).
- Channel service unit (CSU)/data service unit (DSU): Installed on customer premises at the interface to telecommunications company lines to terminate a T-1. They provide network protection and diagnostic capabilities. The DSU is

normally combined with a CSU to convert the user's data stream to bipolar format for transmission.

- DACS: A time-slot switch that allows T-1 or E1 lines to be remapped electronically at the DS0 level. Also called as DCS or DSX.
- Host computers: Mainframes or their equivalent that allow remote terminals to dial in directly or provide leased-line connectivity to a port or ports on the computer.
- Bridges: Interconnect LANs into a single logical network. They operate at the media access control (MAC) level of the OSI model (see Chapter 12).
- Routers: Connect separate networks into an internetwork. They operate at the network layer of the OSI model.
- Gateways: Connect networks running different protocols. They operate above the network layer of the OSI model, serving as protocol converters.
- Repeaters: Regenerate the network signal over long distances.
- Signal transfer points: High-speed packet switches designed to provide reliable connectivity, support intelligent networks, and route the control and signaling messages on a Signaling System 7 network used for frame relay and ATM (discussed in Chapter 12).

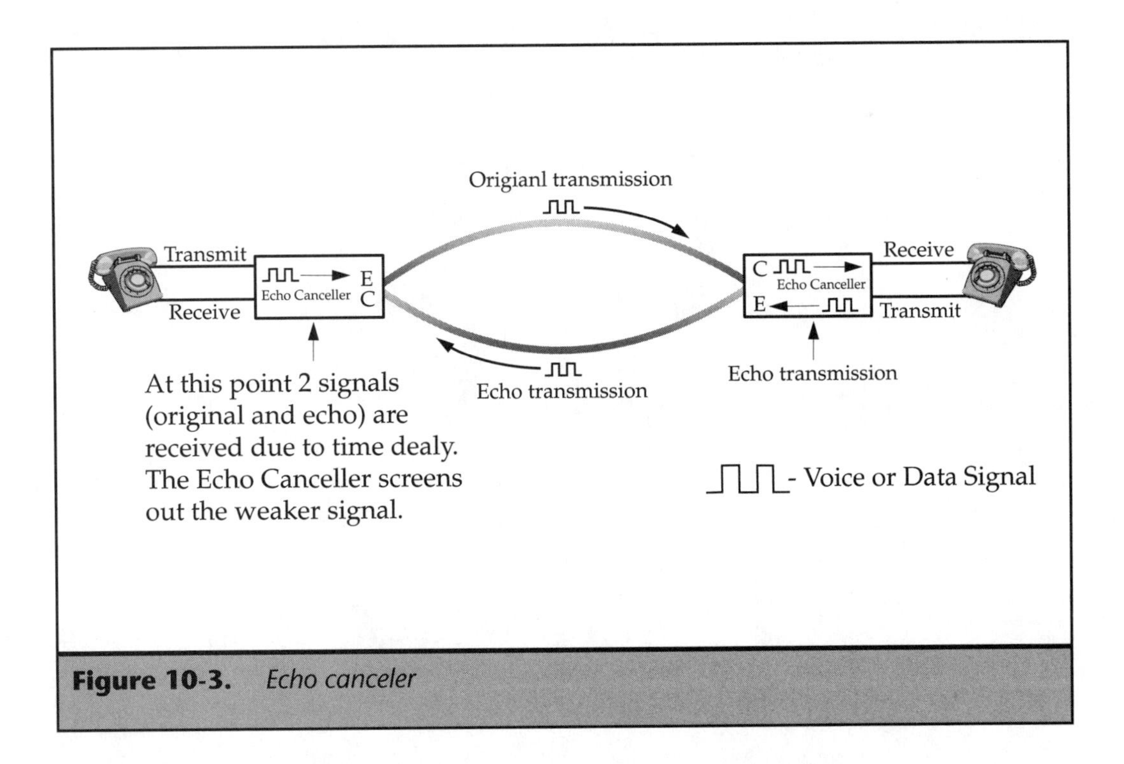

Figure 10-3. *Echo canceler*

Conclusion

T-1s are absolutely vital when it comes to modern telephony. They fill that middle ground of need for so many telephony applications. When a bunch of single telephone line circuits provided on a separate pair of wire for each circuit is just not enough, you make the jump to a four-wire T-1 circuit that can carry 24 channels (circuits). Or, if you can afford it, T-1s are also supplied on fiber optics. T-1s can be used to carry plain old analog voice signals; they can be used for both voice and data; and they can be configured for ISDN service. They can even be used to carry video signals. You will also find that more and more T-1s are appearing in smaller and smaller organizations, as well as individual dwellings, because of the bandwidth they can provide. The T-1 may eventually become obsolete, but for the foreseeable future it will be the predominant method of providing wide band telephony to private organizations.

Chapter Eleven

The Integrated Services Digital Network (ISDN)

The first "electronic" device widely available to both business and homes was the telephone. It is remarkable, given the rapid changes in technology, that the familiar plain old telephone service (POTS) analog line has remained a viable communications vehicle for so long. For the last twenty years, the equipment of choice for wide area network (WAN) communications has been a modem (Chapter 15) and an analog line. As the rest of our world (i.e., computers, document storage, networks, compact discs) was changing to high-speed digital technology, it became unlikely that our telephone interface could remain analog for long.

In the 1980s, the Consultative Committee on International Telegraph and Telephone (CCITT)—now the International Telecommunications Union - Telecommunications (ITU-T) Standardization Sector—began to establish the standards for an *all-digital* public switched telecommunications network, the Integrated Services Digital Network (ISDN). Because of the length of this standardization process and the need for international acceptance and implementation, ISDN will remain the most widely available digital public switched network interface available to individual and corporate users, both in the U.S. and internationally, for at least the next decade.

Overview

ISDN telephone lines use digital, in place of analog, communications protocols. Instead of using analog modems to send and receive data at less than 40 Kbps, ISDN allows digital connections at up to 128 Kbps over the same twisted-pair telephone line. Voice conversations, which were previously digitized in the network, are now digitized inside the ISDN telephone set, allowing voice and data to communicate over a common interface. While analog telephone lines are limited to dialing one call at a time to a single destination, ISDN can establish connections to several remote locations. In addition, ISDN devices support a broad set of signaling-based interactions with the network, allowing sophisticated messaging and feature interaction between the user and the network.

A line in an ISDN environment isn't necessarily analogous to an analog line. You cannot assume, as you could in the analog environment, that a line means one pair of wires to a single set, one telephone number, and one voice or data call. In ISDN, all these elements are controlled by software programs running in the telephone company's central office equipment. It used to be fashionable to state that an ISDN line was 1.x times as efficient as analog. In fact, the relative effectiveness of ISDN depends on the application. For some applications it may be no better than analog, in some data applications ISDN may be five to ten times more effective, and some ISDN applications can't be successfully implemented on analog lines at all.

ISDN Reference Model

In order to standardize ISDN equipment and networks, the ITU-T has defined the ISDN reference model, shown in Figure 11-1, which specifies ISDN equipment and network components and the standard interfaces at which service providers, equipment manufacturers, and users can connect and interoperate. In order to understand the functions of ISDN, we will start at the telephone company central office and work our way out the ISDN telephone line to the computer on your desk.

Local Exchange Carrier

The local exchange carrier (LEC) is usually your local telephone company's central office (CO) switch. By far the most common switch platforms used in the U.S. are the AT&T Network Systems 5ESS ® Switch and the Nortel (Northern Telecom) DMS-100™ with some companies implementing switches from other global suppliers such as Alcatel, Ericsson and Siemens. The CO switch is the source of digital ISDN "dial tone" as well as the older analog lines, and it provides the direct connection from your voice and data equipment to the public switched telephone network (PSTN).

U Interface

The ISDN U interface is the two-wire digital "telephone line" that runs from your phone company's (i.e., local exchange carrier's central office switch, under the street

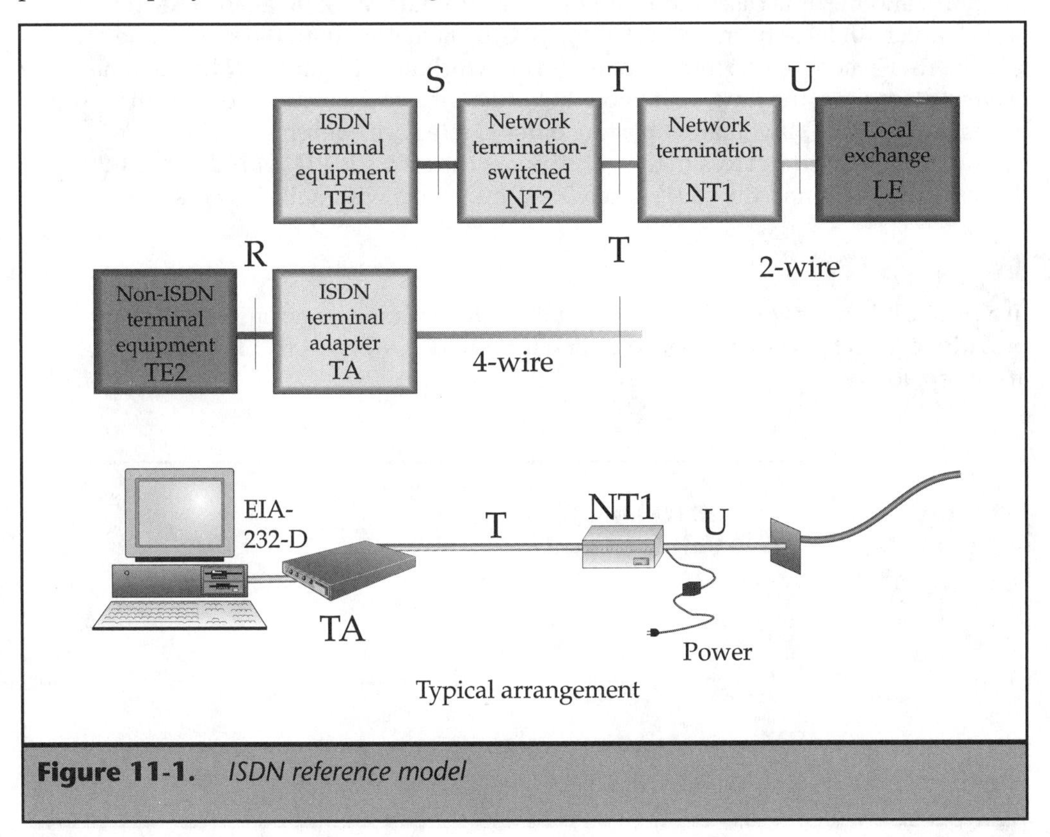

Figure 11-1. *ISDN reference model*

or on telephone poles, and into your home or business. Since ISDN was designed from the start to work over the existing twisted-pair telephone network, the U interface has been implemented as a single-pair connection that can reach over the relatively long distances between the central office and your telephone. The U interface, unlike many other portions of ISDN, is left open to national standardization, and the implementation varies among Europe, Japan, and the U.S.

The American National Standards InstITU-Tte (ANSI) T1.206 defines the standard American ISDN Basic Rate U Interface as using a 2B1Q line coding scheme. 2B1Q means that the coding method contains two binary (2B...) information elements expressed in a single quaternary (..1Q), or four-level signal, shown in Table 11-1. All ISDN voice and data communications are coded into a single stream of zeros and ones, which are paired up and expressed as one of these four voltage levels. This scheme allows the ISDN U interface to perform well over the typically long distances between American homes and the telephone company, often up to 18,000 feet of twisted-pair cable. The 2B1Q coding allows an ISDN device to "listen" for half as many voltage changes (bauds—see Chapter 15) as there are bits in the communications protocol, permitting a more accurate interpretation of weak signals.

Since 1992, almost all new ISDN installations in the U.S. have used the 2B1Q standard, and in some cases older pre-existing alternate mark inversion (AMI) lines (see Chapter 10) have been converted to 2B1Q by changing both the switch line card and the user's network termination. If you are working with an ISDN line that has been in place for some time, remember to verify with your service provider that your line is, in fact, 2B1Q if you are replacing older U interface equipment. The most common AMI-based devices in the network are the AT&T NT1U or NT1P-100 (-200 devices are 2B1Q), and the 6504U and 6508U ISDN sets with built-in NT1s.

NT1

In a public ISDN network, the two-wire ISDN U telephone line from a service provider connects to your ISDN Network Termination—Type 1 (NT1 or NT-1 are acceptable).

Binary Data Pair	2B1Q Line Voltage Level
10	+3
11	+1
01	-1
00	-3

Table 11-1. *ANSI T1.206 ISDN Line Encoding*

The NT1, while totally passive to your voice and data communications, performs important network performance and integrity checks and enables network-controlled loopback testing, which verifies that your digital line is connected and working properly. In many countries, the NT1 is considered part of the network, and it is provided and installed by the telephone company. In the U.S., however, the NT1 is considered customer premises equipment (CPE) and is furnished by the user, either as a separate package or as an integrated function within an ISDN device. NT1 prices have ranged from below $100 to over $300, but seem to be stabilizing in the $75 to $150 range.

The NT1's most important role is to convert the national standard, two-wire ISDN U interface (e.g., ANSI-2B1Q) into an ITU-T-standard (I.430) four-wire digital telephone interface. In many cases, the NT1 will be a separate device, which will not only provide the two-wire/four-wire conversion but will also give you status lights for the condition of your ISDN line connection and your ISDN equipment connection and a power-on indicator. If you plan on working with ISDN lines, an NT1 which has previously been verified as working ("gold" NT1) is the most basic ISDN line tester available and is usually the first item changed when trying to diagnose a dead connection.

The NT1 adds cost (due to the hardware) and complexity (due to the additional cords and connections) to an ISDN installation. To minimize these effects, some manufacturers provide an option for including the NT1 functionality as part of an ISDN data or voice device. While reducing the confusion of your ISDN installation, some of these arrangements limit your flexibility in adding services to your ISDN line. An integral NT1 that allows access to the four-wire S/T interface allows additional ISDN equipment to be installed on the same line, while an NT1 that directly terminates only the U interface to the ISDN device will need to be replaced if new or different capabilities are desired.

Power

ISDN devices, including the NT1, are microprocessor-controlled devices, and like computers, they require a power source. Unlike an analog line, the U interface does not provide central office battery power to your set, and ISDN equipment must be powered locally, usually from the commercial power in your building or home. This means that after fire, flood, hurricane, or earthquake, or even during a simple power blackout, your ISDN equipment *will not* work without power. Strategies to allow emergency 911 access or other critical communications can be as simple as leaving at least one central office-powered analog line in place at your home or business, up to a full uninterruptible power supply capability for an ISDN installation serving an entire enterprise.

ISDN power units generally are provided in one of two configurations: a stand-alone power block with either an AC cord or a plug, or a bulk power unit serving many ISDN lines and devices with optional battery backup. The ISDN standard requires 40V DC, which is fed to the NT1 on an additional separate power pair, resulting in a 2+2-wire cabling scheme.

Equipment Location

There are a number of options in locating NT1s and the associated power units. For a single line, the most common solution is to provide both a power block and NT1 underneath your desk. As soon as your location needs several lines in one location, however, the use of stand-alone equipment quickly results in a rat's nest of tangled cords, and a centralized arrangement will be more satisfactory. Any installation with six or more ISDN lines in a single wiring closet, or with mission-critical applications requiring UPS capability, should consider bulk or rack-mounted NT1s and a large capacity power unit. A downside to this centralized arrangement is that your telephone distribution wiring will need to allow for both the four-wire S/T interface and (usually) an optional Power-2 pair for powering common types of ISDN equipment.

NT2

The Network Termination—Type 2 (NT2) is an intelligent device that provides additional services such as switching or data multiplexing at the user's location. Typical devices include ISDN voice/data-capable PBX and key systems, ISDN data multiplexers, and ISDN-LAN hubs or gateways. In the ISDN PBX arena, the AT&T Definity G3, Nortel Meridian, and Siemens-ROLM 9751, as well as similar products from Fujitsu, Ericsson, NEC and Mitel, all support ISDN NT2 functionality. In an ISDN PBX, network access is usually provided via the Primary Rate Interface (PRI, described below). The line-side (station) interface may implement the ISDN interface with proprietary extensions, usually along with other proprietary digital or analog-hybrid sets.

For large data applications, an ISDN LAN bridge or router connected using the primary rate interface (PRI) is the most commonly available NT2 device. Direct ISDN to ethernet LAN access devices are available from a wide variety of LAN and WAN equipment manufacturers, including Ascend, Cisco, Combinet, Teleos, and Network Express. (See the "ISDN Applications" section.)

S/T

The most global and accessible ISDN standard is the four-wire ISDN S/T interface. While the S and T interfaces are actually specified separately, in practical application they are one interface, and common usage refers to T or S/T interchangeably. Broad international implementation of the S/T interface at the physical and low-level software levels allows equipment providers to build a wide variety of global ISDN products, based on widely available chip interfaces, with software customization for national or vendor-specific implementations. In public-network applications, the T interface originates from your NT1, while in PBX, key system, or private network applications, an S or S/T interface may originate directly from an NT2 line card.

Wiring Limitations

The S/T interface has two characteristics that affect your planning for an ISDN installation: wiring distance limitations and the multipoint passive bus.

The S/T interface is a two-pair (transmit/receive) balanced connection providing a very high-speed (160 Kbps total, 144 Kbps per user) digital interface and is intended for use with inside wiring over relatively short distances. The *theoretical* maximum distance between an ISDN device and the NT1 or NT2 is 1,000 *meters*, but in the field it rarely achieves 1,000 *feet*. This distance is reduced even further by the use of the multipoint capability.

Multipoint Passive Bus

An ISDN multipoint passive bus allows up to eight separate ISDN devices to be addressed on a single digital line behind an NT1 or NT2 (see Figure 11-2).

The ISDN line needs no additional equipment to implement the passive bus capability, only passive Y-splitters (i.e., 8-pin RJ-45), strapped 66-blocks, or 110-connectors in the wiring closet, wherever the ISDN T interface is to be branched. There are two multipoint implementations: the short passive bus, using ISDN devices all *within* several hundred feet of the NT1/NT2, and the extended passive bus, with ISDN devices located hundreds of feet beyond an NT1/NT2. Use of any complex multipoint configuration (meaning that you intend to split devices further than adjacent desks or offices) should be reviewed by a qualified wiring specialist familiar with both ISDN and your existing building wiring.

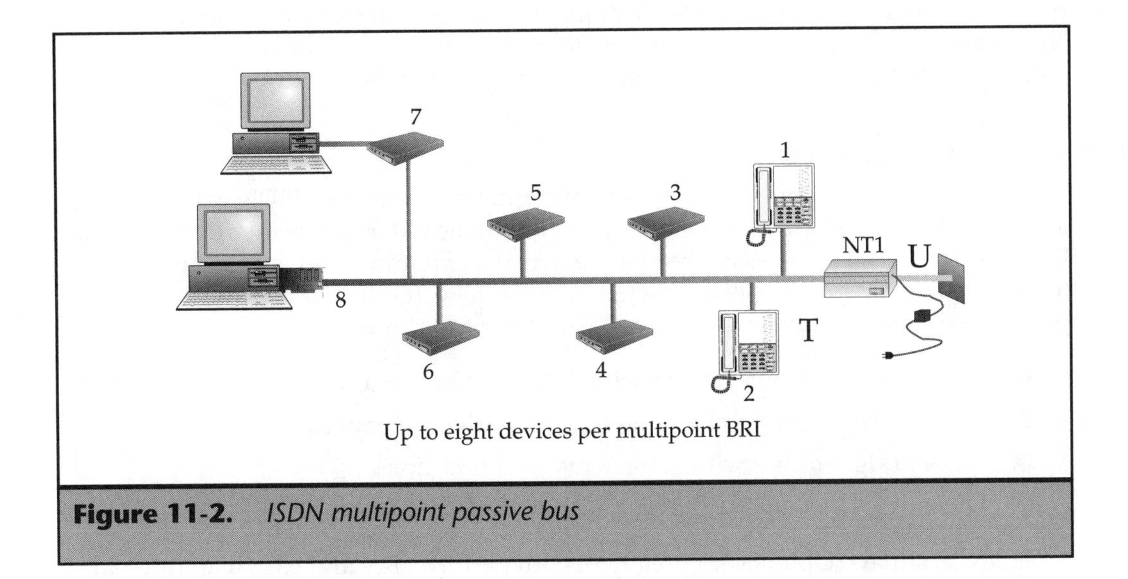

Figure 11-2. *ISDN multipoint passive bus*

Terminating Resistors

Terminating resistors are usually required at the ends of long S/T interface wire runs to eliminate unwanted signal reflections on the ISDN line, and they may be located at the cross-connect field, in a cord adapter, or in your ISDN equipment. The availability of a cord-type terminating resistor (such as the AT&T 440A4 Adapter) in your parts kit lets you transform a marginal or troublesome ISDN connection into a working one simply by plugging it inline. If this happens, the configuration of all possible terminating resistor options—at the NT1, at the cross-connect field, and in the ISDN equipment—should be reviewed according to the manufacturer's wiring guidelines.

TE1/TE2

Terminal equipment (TE) is your active voice or data device, and Terminal Equipment Type 1 (TE1) uses a direct (i.e., T interface) connection to an ISDN line. Examples of TE1 devices are ISDN digital voice sets, integrated voice/data terminals, ISDN Group IV fax machines, and ISDN videophones. Your existing analog voice set, fax machine, modem, or V.35 WAN hub are all considered Terminal Equipment Type 2 (TE2), or non-ISDN equipment.

R Interface

Any of the non-ISDN interfaces that you find on the back of your computer or communications equipment, such as the RJ-11 voice telephone jack, or RS-232, V.35, or 10base-T connectors for data, are considered to be R interfaces and are not covered by ISDN standards. Because of the desire of manufacturers to attain compatibility with international standards, essentially every commonly used interface (and some proprietary ones) has equipment available to adapt to the ISDN network.

Terminal Adapter

The ISDN terminal adapter (TA) is the most common type of user ISDN equipment in the network today. The TA's function is to translate the older R type interfaces into the ISDN protocols and services offered by the network. For most users, the TA replaces the analog modem previously connected to the serial communications port on your personal computer. The TA will generally be one of four types:

- Internal PC/computer board with data or voice/data connections
- External TA with a combination of voice and data connections
- ISDN bridge/router with LAN network connections
- Digital ISDN telephone set with a data interface

ISDN terminal adapters vary greatly in both performance and cost, with internal PC-board solutions ranging from $250 for single-purpose devices, to the $500–$1000 range for co-processors equipped intelligent communications interfaces. Similar functionality in external packages will be slightly more costly, with prices starting in the $350 range and extending up to $1500.

ISDN LAN adapters have experienced a dramatic reduction in price as their use increases, with functionality previously costing $1500 now available for $650 to $900. ISDN telephone set prices range from voice-only sets at less than $200, to executive voice-data display sets approaching $1000.

At a data center or information services gateway, the ISDN TA devices can become extremely large and complex, offering support for hundreds of connections, including both digital and voice-grade modem calls. ISDN hub or gateway devices typically handle multiple ISDN interfaces and can effectively reduce data center clutter by consolidating all WAN connections into a single architecture with an all-digital network interface.

ISDN Standards

ISDN standardization covers not only the physical connections to the telephone network but also the digital services and capabilities that are provided by various software-controlled configurations of ISDN lines. A user's ISDN connection is made up of a series of channels, each of which has a range of voice and/or data communication capabilities. These services are set up in the ISDN switch through *provisioning*, or software configuration processes, and are accessed on a call-by-call basis by your ISDN equipment.

Channel Structure

ISDN connections consist of communications *channels*, which are information pipelines from user to user or from the user to the network. Technically, the channel is a digital time slot within the ISDN communications protocol, assigned the ability to carry a specific communication type or bearer capability. The D, B, and H channel types currently deployed in the ISDN network are shown in Table 11-2.

D-Channel

The D-channel provides a signaling path between you and the network and provides a way to control the use of all the channels available in the ISDN interface. Using the D-channel, you indicate what action the network should take, on which channel, and using which type of service (e.g., "Place a call to (700) 555-1212 on channel B2 using circuit-switched data at 64 Kbps"). The network uses the D-channel to provide detailed information about incoming calls and network status, as well as to provide user interface and control of network features such as voice call forwarding or call transfer. In addition to signaling, the ISDN basic rate D-channel can provide X.25 (X.31) packet data communications, allowing effective transport of bursty data communications such as e-mail or telemetry.

Channel	Bearer Capability	Speed	Basic Rate	Primary Rate
D	- Signaling - X.25 Packet	16 Kbps	1	
D	- Signaling	64 Kbps		1 or 2
B	- Circuit Data - Packet Data - Voice	64 Kbps	2	23 or 24 (U.S.) 29 or 30 (for International)
H_0	- Circuit Data	384 Kbps (~6B)		1 to 4
H_{10}	- Circuit Data	1.472 Mbps (~23B)		1 (U.S.)
H_{11}	- Circuit Data	1.536 Mbps (~24B)		1
H_{12}	- Circuit Data	1.920 Mbps (~30B)		1 (International)

Table 11-2. *ISDN Channel Types and Quantity*

The major strength of ISDN is the standardization of the ISDN D-channel and the ITU-T Q.931 signaling protocol carried within it, Q.931 signaling allows dynamic control of channels, bandwidth, and services in a multi-vendor , multi-network, multi-service provider environment. All ISDN calls require a D-channel connection to control them, and despite the wishful thinking of some equipment vendors, devices that cannot interact with Q.931 signaling are, by definition, *not* ISDN.

B-Channel

The 64 Kbps B-channel is the workhorse of ISDN, and it provides on-demand *bearer* services, including circuit-switched data, packet-switched data, and voice. B-channel calls make up the majority of all ISDN traffic because of their flexibility in placing and accepting a variety of call types, and more importantly, because of their ability to interwork with the existing non-ISDN public switched telephone network.

B-channel calls are set up, dialed, and administered using the signaling control obtained through the D-channel. The D-channel dynamically assigns bearer capabilities to a B-channel, and you can simultaneously use multiple B-channels for similar or different services. Your ISDN connection can be configured to allow any combination of voice or data call types, incoming or outgoing, with each channel connected to the remote destination of your choice.

H-Channel

If your communications needs include high-speed dial-up connections, you can use H-channels, provided over an ISDN-capable T-1 (1.544 Mbps) or E1 (international

2.048 Mbps) facility, to implement applications such as LAN-to-LAN bridging, data center backup, or videoconferencing. The H_0-channel provides clear 384 Kbps bandwidth (quarter-T-1) under the control of a single D-channel call setup message. Obtaining the same bandwidth using multiple 64 Kbps B-channels (i.e., 6x64) requires the user to coordinate and control six separate call setups and requires compatible inverse multiplexing hardware at each end of the call.

H_{10} and H_{11} services (and H_{12} internationally) provide dialable connections with the full bandwidth of a T-1 (or E1) facility. Once again, while fully defined by the ITU-T standards, network-wide implementation of H-channel dialing capability is limited by carrier service offerings, interswitch and internetwork facility restrictions, and switching hardware constraints.

Bearer Capabilities

Bearer capabilities define the types of information streams carried within the various channels of an ISDN interface. The ISDN standards specify over a dozen bearer capabilities, with more under consideration, but they fall into the general categories of voice, circuit data, and packet data.

Voice

We take for granted the idea that by using a B-channel, a voice call can be originated from our ISDN digital telephone set, and the network will automatically perform the conversions required to allow that call to be received by a powder-blue rotary analog phone. The existing voice-grade telephone network is built on a foundation of DS-0 (64 Kbps) connections, providing a 56 Kbps talking path, and 8 Kbps of in-band signaling. The ISDN switched network can convert these connections into a 56 Kbps payload inside a 64 Kbps B-channel, with the in-band signaling translated into ISDNQ.931 protocol and running on the D-channel.

To convert speech into a Pulse Code Modulation (PCM) encoded digital information stream, your ISDN voice device includes codec (coder-decoder) capability. The only analog voice connection in an ISDN voice set is from the handset to the codec chip inside the set. In an ISDN device with an RJ-11 or other analog interface capability, a codec is provided within the TA or is emulated using digital signal processor (DSP) technology.

Since modems use analog tones to represent zeros and ones to the network, the entire set of data communications applications that use modem technologies are considered voice bearer capabilities to an ISDN network. One of the most difficult tasks in ISDN is connecting a modem using voice bearer services to an ISDN TA using any digital data service. Since the network has no built-in capability for translating the various modem standards to one of the digital formats, you cannot connect analog and digital directly, and your ISDN equipment must make the conversion (using a codec or DSP) or you must dial through an external ISDN/modem resource pool.

Circuit Data

Circuit-switched data gives your data application full-time access to the bandwidth of a B- or H-channel. If digital information were liquid, the ISDN channels would be "pipes" and circuit data would be a steady flow of information through the pipe.

The ISDN B-channel operates as a 64 Kbps synchronous data stream, and every ISDN access connection between the user and the switch is capable of this speed. However, since large portions of the network remain interconnected at 56 Kbps (plus 8 Kbps in-band signaling), and since the most common computer data connections are asynchronous serial ports, running at speeds such as 9.6 Kbps, 19.2 Kbps, or 38.4 Kbps, two forms of B-channel rate adaption are required: switched 56 Kbps (64 Kbps-restricted) rate adaption, and asynchronous rate adaption, such as the V.120 standard.

Packet Data

Packet-switched data allows multiple small data transmissions to share the bandwidth of the 16 Kbps D-channel (basic rate interface only—see the next section), or a 64 Kbps B-channel. Using our liquid information analogy, your packet data would travel the ISDN pipe in colored buckets, and at the end of the pipe, all buckets of the same color would empty into a single pool. In reality, the data is contained within X.25 (X.31) logical channels, which permit up to 15 independent data sessions on an ISDN D-channel and up to 256 logical channels on a B-channel. B-channel X.25 packet may be requested either as a permanent or as an on-demand service.

A major distinction between ISDN BRI D-channel packet implementations and the B-channel configuration is that the D-channel service uses an integral X.25 packet assembler/disassembler (PAD) within the ISDN equipment, while B-channel devices implement a functionally external PAD. The D-channel PAD is provided for the use of a sixteenth logical channel, in which the Q.931 signaling protocol is carried, and the spare PAD capacity is available for user data applications.

The X.25 capability, while fully supported by the ITU-T standards, is not as fully implemented (especially in the United States) as the circuit-switched data network. Some service providers lack full X.75 internetwork gateway capability (See "ISDN Applications"), and some "ISDN" PBXs do not provide X.25 packet handlers to implement the D-channel packet capability.

ISDN Access Interfaces

The rationalization of ISDN into "a limited number of standard user interfaces" is the key to providing the widest possible range of equipment options, all interconnected to the access network of your choice. Standardization ensures that the S/T interface chip is available to manufacturers from multiple suppliers at under $10 per part, and that ISDN products can be built using these chips, with only localized software changes required to adapt to the network requirements of various national implementations.

Basic Rate Interface

Your digital ISDN line connection is provided through a basic rate interface (BRI), which provides two 64 Kbps B-channels and a 16 Kbps D-channel. The 2B+D ISDN line shown in Figure 11-3 gives you up to 144 Kbps of dialable bandwidth and the means of commanding the resources of the ISDN network via the signaling channel.

2B+D ISDN Line

The 2B+D basic rate ISDN line offers unequaled flexibility in handling multiple voice, data, and image information streams, combined with a complex (and sometimes bewildering) array of user-interface and human-factor options. The international standardization of ISDN protocols allows a wide variety of BRI devices to be connected behind different service provider networks and switch implementations with, in most cases, changes only to user-configured software drivers and options. The BRI channels can be used independently, or the ISDN equipment can dial two B-channel data calls to the same destination and sum the capacity, resulting in up to 128 Kbps of pure digital communications.

An ISDN service provider may choose to offer you a BRI configuration without enabling use of all of the 2B+D channel capabilities. Many ISDN voice or data BRI lines are configured without any X.25 packet capability, either to simplify the options during setup or because the ISDN switch lacks packet handler or gateway capability. So-called 1B+D and 0B+D line terminology is sometimes used, but these describe limited services, since the ISDN BRI always carries the capability of a full 2B+D interface.

The tight software interdependence between your ISDN equipment and your ISDN service connection means that obtaining and sharing detailed configuration information *at the time you place an order* is the critical path element in a successful ISDN turnup. When you call and order an ISDN line, one of the first questions you

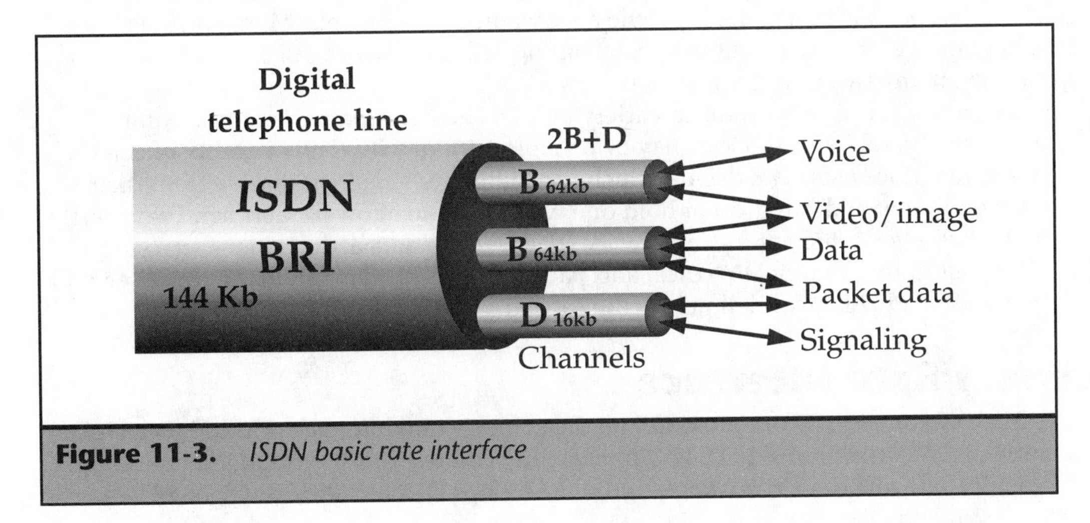

Figure 11-3. *ISDN basic rate interface*

will be asked is "What type of equipment are you going to use?" You will need to pass on your equipment manufacturer recommended ISDN configuration, along with your preferences on optional services, in order to obtain a correctly translated ISDN line. In return, your service provider *must* provide you with information describing their switch type, software version, and service configuration, along with your new telephone number and (if required) a service profile identifier (SPID).

Service Profile Identifiers

The service profile identifier (SPID) is a unique identification code that you enter into your ISDN device, similar to the PIN number on a bank card. The SPID allows the switch to associate multiple ISDN bearer services with your ISDN devices. Failure to obtain correct SPIDs from your service provider is the surest (and most common) reason for being unable to get an ISDN line into service. AT&T 5ESS switch point-to-point (single ISDN device) configurations do not require a SPID, but all National ISDN, all Northern Telecom (DMS-100), and all multipoint passive bus lines require at least one, and usually two, SPIDs.

Multipoint Service

Once you've decided to implement BRI service and to use the T interface passive bus capability to put up to eight devices on the line, you will realize that without some enforcement of good design rules, you will quickly run out of capacity on the 2B+D interface. Since any device could *theoretically* access any or all of the B-channels or logical D-channels on the line, the resulting free-for-all (and inability to establish connections) is not likely to result in satisfactory service.

The major source of service contention involves the B-channel, since the packet D-channel has 15 logical channels to allocate (one or several at a time) to the various ISDN devices. A few criteria will help to clarify this picture: Voice stations require a B-channel capable of voice bearer service, so two voice stations on a 2B+D ISDN line are the reasonable limit if "real" people are assigned to these sets. More voice sets can be in place, but they will get busy on incoming calls and "voice call blocked" on outgoing if any two others are in use.

For a single user who wants a variety of ISDN access capabilities, multipoint service to several ISDN devices may be a viable option, with the possibility of channel contention, since users can decide which capabilities to enable or disable from their own equipment and can drop or hold one type of call to allow for another. Two users can realize a very efficient full-time voice/data+voice/data arrangement on a single BRI, by allocating channel B_1(voice) and packet D-logical $channel_1$ to the first user, and channel B_2 and D-logical $channel_2$ to the second.

Primary Rate Interface

Bulk ISDN connections are provided on the primary rate interface (PRI), Figure 11-4, implemented on digital T-1 (1.544 Mbps) facility connections providing (for the U.S.) either 23 of the 64 Kbps B-channels and one 64 Kbps D-channel (23B+D), or 24

B-channels (24B). In Europe, PRI connections will be provided as 30B+D service on a 2.048 Mbps E1 digital facility.

The PRI differs from the BRI in more ways than in the number of channels provided. The PRI does not provide software and signaling support for an individual user interface, and it expects to terminate on an intelligent NT2 device, such as a PBX switch, a LAN/WAN hub, or a data multiplexor. In addition, because of differences in implementation and service offerings, it is not safe to assume that a 23B+D PRI is equivalent (or even similar) to 11 ½ BRI 2B+D lines.

The first PRI T-1 facility is always provided as a 23B+D configuration, to provide the initial D-channel signaling capability required to establish control over all the B-channels. The major advantage of an ISDN PRI over a traditional T-1 digital facility is that any call type—voice, data, inbound, outbound, WATS, etc.—can be carried over this single facility on a call-by-call basis. The dynamic boundaries created by

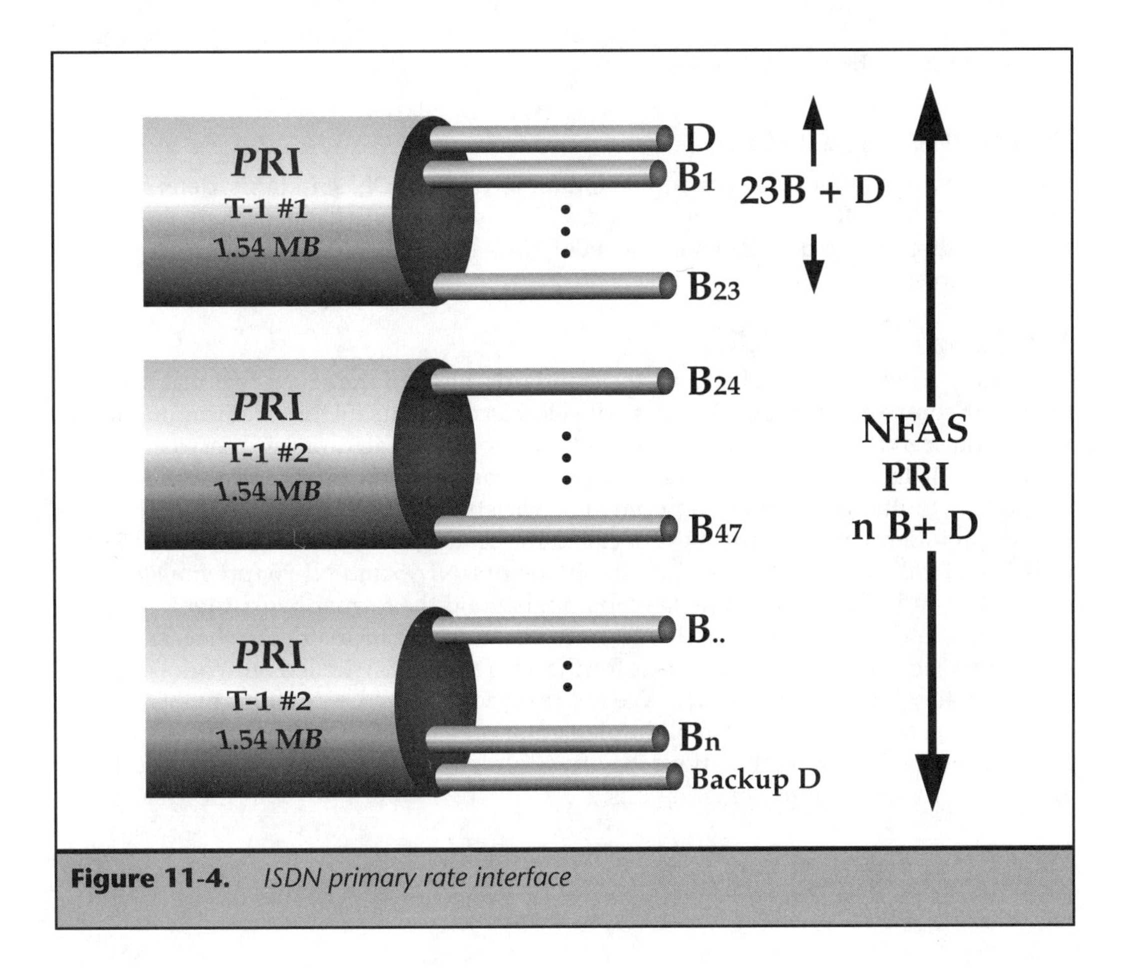

Figure 11-4. *ISDN primary rate interface*

call-by-call service selection eliminate the breakage caused by having to engineer single-purpose facilities for peak traffic loads.

Non-Facility-Associated Signaling

The second through *n*th T-1 may be added using 24 B-channels in each facility for calls controlled by the D-channel in the first PRI facility. This arrangement is called non-facility-associated signaling because the D-channel is not physically associated with the bearer channels.

Backup D-Channel

With potentially hundreds of B-channels spread over many T-1s, network reliability becomes an issue, and with all the eggs riding on the reliability of a single D-channel, a redundant signaling capability may be desired. While the Q.931 protocol contained in the D-channel itself is usually highly reliable, the hardware and connection path supporting the T-1 facility may be less so, and a backup D-channel, on a different facility than the first, is recommended to reduce the chance for failure, especially for arrangements of three or more T-1s.

ISDN Implementations

Service providers offer different implementations of the ISDN standards, determined primarily by the software capabilities of the ISDN switch they have implemented. In the U.S., these tend to be either some level of National ISDN or an earlier vendor-specific implementation.

National ISDN (U.S.)

Beginning in the late 1980s, Bellcore, the standards agent of the seven regional bell operating companies (RBOC), began to develop a consensus ISDN specification that would transcend the individual implementations of manufacturers and service providers. By late 1992, National ISDN-1 (NI-1) was ready for public demonstration, supporting multi-vendor ISDN equipment, a variety of switch platforms, and multiple service provider networks, in an event called the Transcontinental ISDN Project 1992 (TRIP '92).By the mid-1990s, NI-1 and its successors, NI-95 and NI-96 (previously called NI-2 and NI-3), have gained acceptance beyond the former Bell System companies, and they are well on their way to becoming de facto (although least common denominator) U.S. standards for the entire public switched network. Unfortunately, NI standards alone have not realized expected reductions in ISDN equipment cost, because uneven NI implementation has forced manufacturers to support several NI implementations in software, along with the pre-existing AT&T and Northern Telecom implementations.

AT&T

AT&T builds three ISDN switches: the 5ESS® Digital Switch, the 4 ESS™ Switch, and the Definity® PBX. The 5ESS is used by LECs, other service providers, and a few very large private networks to provide both BRI and PRI access, using both National ISDN and AT&T Custom ISDN software. The 4ESS Switch is used by IECs, international service providers, and several LECs, and it supports AT&T custom software on PRI access. The Definity G3 PBX supports AT&T custom BRI and PRI service in customer installations.

Northern Telecom

Northern Telecom's DMS-100 (and -200/-300) family of central office switches, used by LEC, IEC, and alternate access providers, and the related Meridian PBXs support both BRI and PRI access. A decreasing number of lines are being provisioned with Northern's pre-NI software, and NI-1 is available, with NI-2(-95) and NI-3(-96) to follow.

Euro-ISDN

The European Community, through the European Telecommunications Standards InstITU-Tte, has established a multinational ISDN implementation, EURO-2, which is being aggressively deployed throughout the continent. Equipment vendors based in Europe or doing business there can be expected to support the EURO-2 standard in both switches and ISDN devices. Since European telephone authorities provide the NT-1 (per their standard) as part of the network infrastructure, essentially all ISDN devices are built to the S/T interface. *Don't* try to use your ISDN/LAN adapter with an 2B1Q U interface for your office in Europe. A major networking difference between Euro-ISDN and the U.S. version is that the European network supports the E1 digital facility, which uses 2.048 Mbps to provide a 30B+D PRI interface, instead of the U.S. (and Japanese) 1.544 Mbps T-1 facility with 23B+D.

Networking and Compatibility

The global telecommunications infrastructure consists of the three distinct and separate components shown in Figure 11-5: the Signaling System 7 network, the X.25/X.75 packet network, and the 56/64 Kbps circuit-switched DDD voice/data network.

Signaling System 7

Signaling System 7 (SS7) is a required part of an end-to-end ISDN network implementation, because it supports the ability to provide a 64 Kbps connection on the trunks between central office switches and between networks. SS7 removes signaling information (i.e., the called and calling telephone numbers, the call type, and network status information) from the 8 Kbps in-band signaling channel of the voice and Switched-56 network and provides out-of-band signaling through a dedicated packet-mode network. Signaling information passes between the CO switch and a local SS7 signal transfer point (STP) over high-speed A-links. The implementation of SS7 also removed the opportunity for blue-box fraud (simulation of the in-band tones generated by an operator, in order to gain free access to long-distance toll calls), and let service providers decrease call setup time between switches and across networks.

Switched-56 and 64-Clear

Although SS7 has been "in-process" since the mid-1980s and is universal in Europe and Japan, the process of upgrading the network to 64 Kbps clear channel (64CC) is still not complete within the U.S. In addition to adding software, hardware elements, and SS7 links to each central office switch, service providers face the monumental task

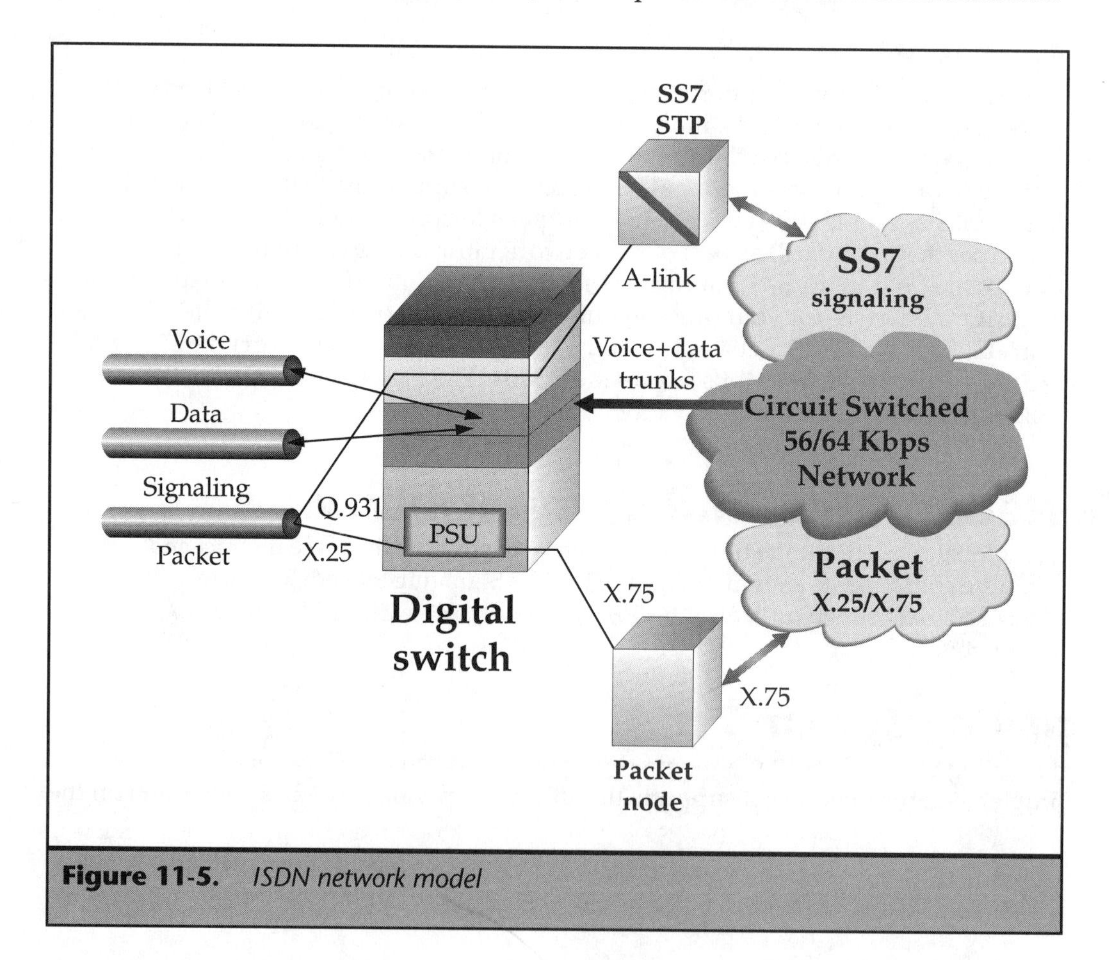

Figure 11-5. *ISDN network model*

of modifying millions of interoffice digital transport components to remove ones-density error-checking capability, which conflicted with user access to the upper 8 Kbps of the circuit. In the meantime, if you wish to make a data call that transits any portion of the network that is not 64-clear, or that terminates on a non-ISDN Switched-56 data line, your ISDN hardware must select the correct 64-restricted (i.e., 56 Kbps) B-channel bearer capability. Some international-style equipment, not expecting to see a 56 Kbps network, will not permit this type of rate adaption.

SS7/Q.931 Interworking

Signaling information originating in Q.931 format from an ISDN D-channel must pass into the network to place and receive calls, identify callers, and provide network status information. The Q.931 signaling interworks directly through the ISDN switch into an SS7 signaling link. Incoming calling line identification (ICLID or caller ID) is an example of a signaling service made available to an ISDN device by a Q.931 D-channel message, obtained via information passed from a distant switch using the SS7 network.

X.25/X.75 Packet

Packet network connections in ISDN are defined by the X.31 standard, which covers the use of X.25 packet-mode equipment with ISDN. Packet services may be provided as an integral part of the ISDN switch network, or they may be obtained as an adjunct service from a public packet switching network. In either case, communications to other networks are provided using X.75 (X.75 prime) intralata connections or full X.75 packet network gateways.

E.164/X.121 Addressing

The address, or telephone number, for an ISDN call uses the same E.164 numbering format (country code+area code+office code+number) standardized as the North American Numbering Plan. While today's numbers are limited to 12 digits, E.164 can handle up to 15 digits beginning in 1997. Packet-mode calls may be addressed in the 14-digit X.121 format (4-digit data network identification code-DNIC + 10-digit terminal number).

ISDN Applications

ISDN would be of little use if it were simply a digital re-creation of the existing analog telephone line. The increased speed of ISDN, up to 128 Kbps on a single line, combined with flexible access to many service types, makes ISDN an ideal match for today's need for remote multimedia access. By far the most widely implemented ISDN applications are those providing remote access to corporate LANs or to the Internet. A promising set of new technologies blends the shrinking cost of desktop videoconferencing with data networking capabilities to provide a new set of multi-user multimedia conferencing tools.

ISDN LAN Access

Access to corporate LANs or to the Internet with sufficient speed to allow transparent user interaction is a goal that has eluded us on the existing analog network. ISDN lines finally give us enough bandwidth on a digital connection to effectively communicate from the home office or from a remote work location. Figure 11-6 shows a data center remote LAN access application, with a large ISDN-connected gateway router at the central LAN site, capable of serving both ISDN and analog modem users. The LAN gateway connects to an ISDN PRI serving several hundred users (on a multiple-T-1 NFAS PRI) and reads the incoming ISDN bearer capabilities on a call-by-call basis. If the B-channel is 56/64 Kbps data, then the distant end is ISDN (or Switched-56) digital, and a transparent router or bridged data connection is made to the LAN. If the bearer capability is voice, the gateway knows that the incoming call originated on an analog modem, and it uses internal DSP or codec/modem resources to convert the call from analog tone format back to a digital LAN protocol. Security functions such as authentication or automatic dial-back may also be provided within the network gateway to protect instITU-Ttional LANs from unauthorized access (hacking).

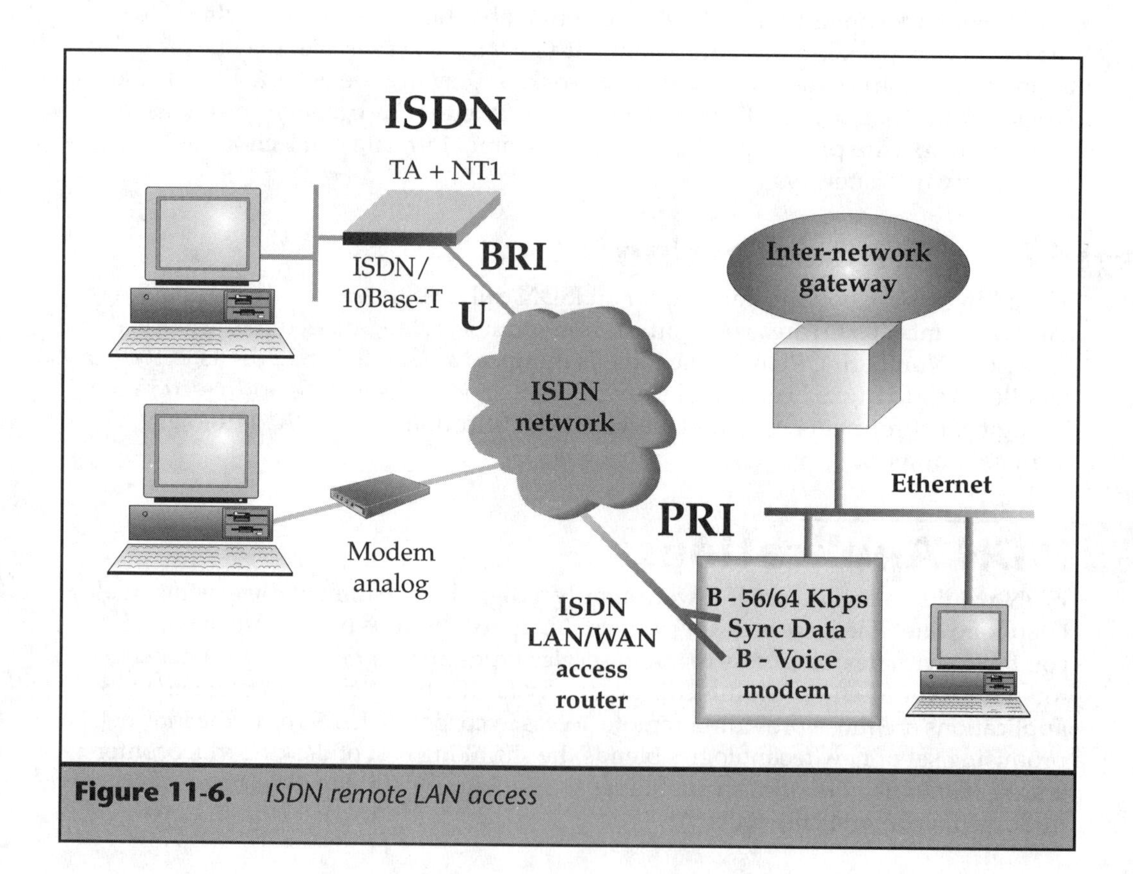

Figure 11-6. *ISDN remote LAN access*

Since a single ISDN channel at 56 or 64 Kbps is *good*, then multiple channels must be *better*, so two de facto standards, ML-PPP and BONDING, have emerged, which allow two or more B-channels to be combined or stacked by the ISDN LAN access device to create a connection with greater bandwidth.

ML-PPP

Multilink point-to-point protocol (ML-PPP or MP) has been defined by the Network Working Group of the Internet Engineering Task Force (IETF) as an extension to the TCP/IP LAN protocol set, and it is specified in Request for Comments 1717 (RFC-1717). ML-PPP makes a TCP/IP application channel-aware, and it allows the software protocol stack to originate and synchronize multiple ISDN B-channel calls. IP packets are distributed across the available B-channels, and the ML-PPP protocol assures correct reassembly at the distant end. As with most communications protocols, symmetry is required, and both ends of the ISDN call must be ML-PPP-compatible, so extensive testing has been conducted to ensure interworking between equipment from different manufacturers.

BONDING

The Bandwidth on Demand Internetworking Group (BONDING) standard provides a means for delivering multiple B-channel operation to software applications that are not ISDN channel-aware. The BONDING standard provides an inverse multiplexing (Imux) function that accepts a single high-speed data stream (i.e., 128, 336, 384, or 768 Kbps) from an application and splits it into multiple 56/64 Kbps B-channels. This capability is an adjunct to the user's data application and may either be provided in a separate ISDN TA or integrated as part of a more sophisticated LAN or video device.

ISDN Internet Access

Internet access, as shown in Figure 11-7, is a special case of the remote LAN access capability shown in Figure 11-6, and in fact the equipment configurations are often identical. ISDN's speed advantage allows graphical browsers such as Mosaic or Netscape to deliver graphic-rich World Wide Web (WWW) pages without the coffee-break delays associated with modem downloads.

The LAN being accessed by the user is at the Internet service provider's (ISP) location, which has high-speed TCP/IP access to an Internet point-of-presence (POP). ISDN access accounts are becoming widely available from both direct ISPs and commercial on-line services, and they are generally available at a modest premium over slower analog modem connections. Some service providers bundle (or resell) the ISDN access capability from the local telephone company, and they may be able to provide an Internet Centrex capability in your service area, which results in no calling charges for your access to the Internet. When setting up ISDN access with an ISP, make sure you negotiate a compatible set of channel capabilities (ML-PPP or BONDING) and protocol (PPP, ML-PPP, or serial line interface protocol—SLIP) connections.

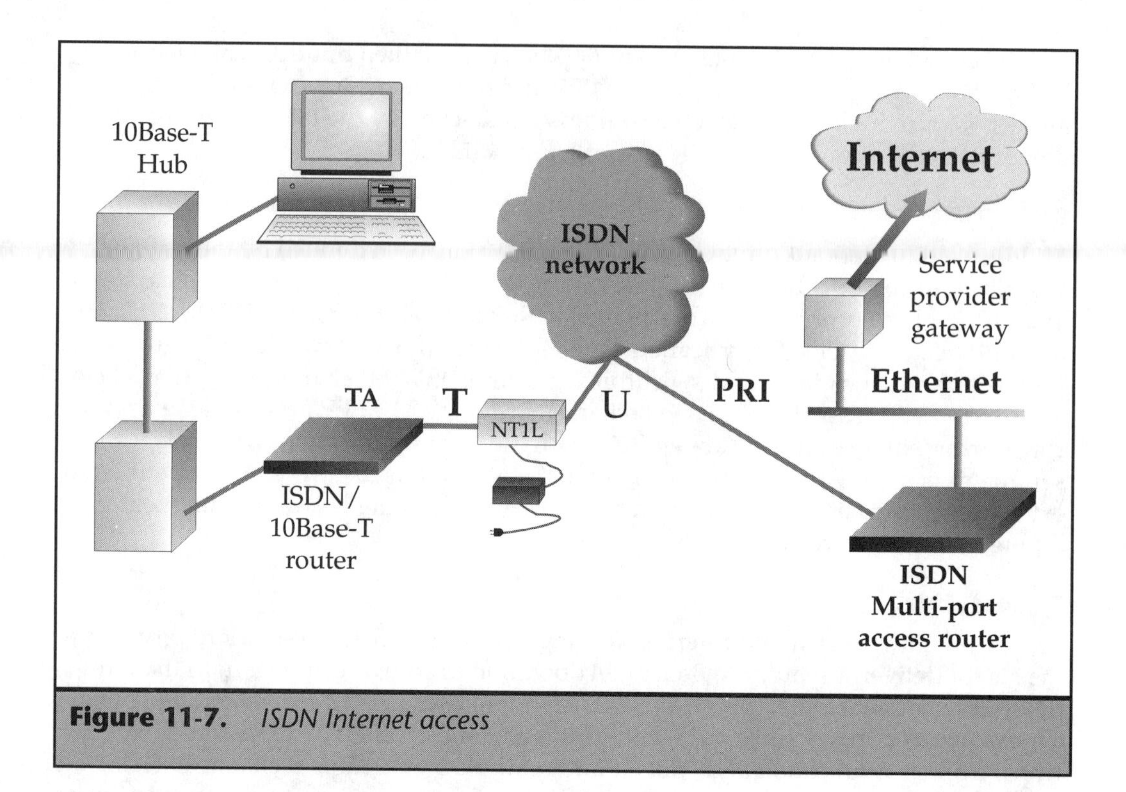

Figure 11-7. *ISDN Internet access*

ISDN Videoconferencing

The high-speed communications capability that ISDN provides for LAN access has been combined with a new generation of hardware and software video codecs to provide standards-based videoconferencing in the personal computer on your desktop. The cost of videoconferencing today is *one-tenth* of the cost as recently as five years ago, and while the price continues to drop, new capabilities such as multimedia bridging, Figure 11-8, are being added to the network.

Previous generations of videoconferencing equipment were often room-sized installations, set up for executive conference calls, and they used proprietary video codec technology and dedicated network connections. The current trend is to move away from large fixed installations and towards small group (roll-around) or individual desktop systems (see Chapter 17). These systems, while they may still process proprietary algorithms, almost always conform to the ITU-T standard H.320.

H.320/H.261—The Px64 Standard

The ITU-T standards for videoconferencing, which allow videoconferencing between different manufacturers' equipment, are a very large set of interrelated and coordinated documents. The H.320 standards umbrella covers the H.261 compression standard, generically referred to as P times 64 (Px64), which provides two picture

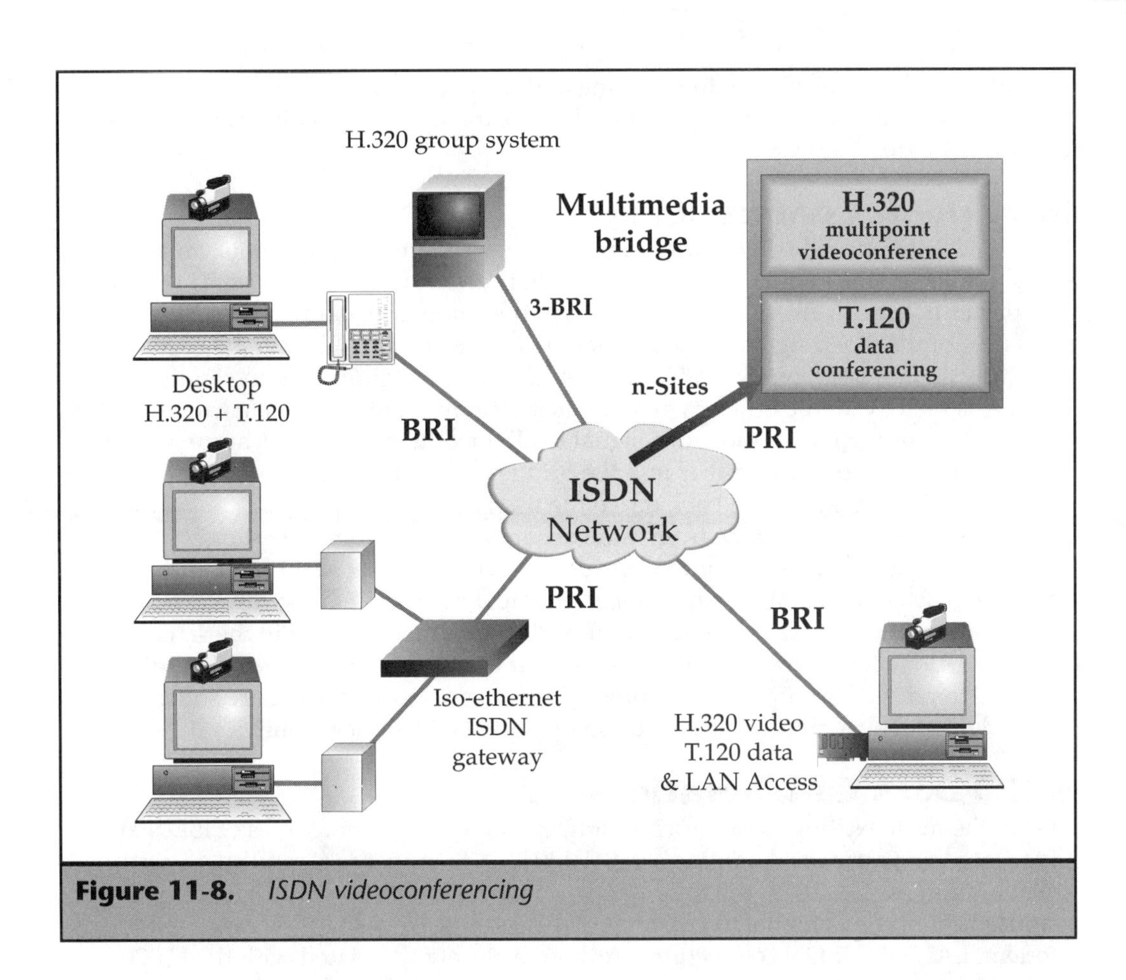

Figure 11-8. *ISDN videoconferencing*

resolutions: the common intermediate format (CIF) of 288 lines by 352 pixels, also known as Full-CIF (FCIF), and a one-quarter-sized picture, Quarter-CIF (QCIF), with 144 lines by 176 pixels. In addition, at least three audio compression algorithms are provided, G.711, G.728, and G.722, and call setup is controlled by H.221.

While H.320 standardized how videoconferencing is exchanged, it does not specify how video is processed by videoconferencing equipment, and a wide variety of choices in functionality and price exists. High-end systems will use dedicated codec engines and complex audio/visual management capabilities to provide extended standards support at speeds from 384 Kbps up to 2.048 Mbps on ISDN PRI or multiple BRI lines. Midrange systems, either roll-around or desktop, tend to be based on PC processor platforms, with hardware video codec/capture boards and ISDN interfaces supporting one to three BRI lines, with speeds from 56 Kbps to 384 Kbps. The newest type of desktop videoconferencing uses the built-in power of Pentium™ or PowerPC™ CPUs in a multimedia-capable personal computer to provide all or part of the codec functionality. The software codec functionality might provide H.320 decoding only, or with more power available, both encoding and decoding. Limitations in CPU power may limit low-end systems to a subset of the H.320 standard, usually QCIF video and one audio mode.

Multipoint Conferencing

In a videoconferencing call, in order to involve more than two end points, an external bridge, called a multipoint conference unit (MCU), must be used. The MCU is used to mix the audio signals from all of the parties in the conference, and it controls switching of the video signals. Several methods of video switching can be used for a multiway conference: voice switching, in which the user dominating the audio channel is seen by all the other parties; explicit conference control or chairman mode, in which one user controls the video signal to all the others; and quadrature, which allows four video streams coming into the bridge to be combined into a single four-panel video signal going back out to the users.

MCU services are available from interexchange carriers, local telephone companies, or independent service bureaus, or they are operated within large corporate, instITU-Ttional, or educational private networks. In most cases, every videoconferencing device on a multipoint bridge *must* be operating in the same standards mode for both video and audio and at the same speed. External capabilities required to provide audio or speed conversions or video transcoding are sometimes available, but they almost always result in degraded conference quality.

T.120 Data Collaboration

One of the most exciting capabilities arising from the core technologies of ISDN and videoconferencing is the emerging set of T.120 ITU-T standards for multiway data collaboration. While still in the final stages of international standards balloting, these standards offer the potential to create shared electronic workspaces across analog modem, LAN, and ISDN connections. Although closely associated with the H.320 videoconferencing standards and capable of supporting full multimedia voice, video, and data conferencing, much of the T.120 usage is likely to be in the data-only or voice/data modes. Once again, the speed of ISDN will allow sharing of graphical software and images as easily as low bandwidth text and spreadsheet applications.

ISDN Data Networking

Not all ISDN data communications are complex LAN access or videoconferencing applications, and one of the secrets of ISDN is that it can be used with your existing communications software and applications, just like a modem, but much faster. Figure 11-9 shows a variety of asynchronous data connections running through PC serial COM ports at speeds of up to 57.6 Kbps.

Since most asynchronous ISDN adapters handle the AT modem command set controls, usually all that users need to change in order to connect through the ISDN device is the speed setting on their communications software. Depending on the ISDN device and service provider options, the user may access the 16 Kbps packet D-channel or the 64 Kbps B-channel on a call-by-call basis. The asynchronous data speeds provided by most typical RS-232 (or EIA-232-D) communications ports are not directly compatible with ISDN B-channels, which provide synchronous communications at 56 or 64 Kbps and require rate-adaption to pass through the network.

Asynchronous Rate Adaption

The method used to permit asynchronous data traffic to conform to the synchronous ISDN network is B-channel rate adaption. Rate adaption is provided in the user's ISDN device, and it provides bit-stuffing to bring asynchronous port speeds (i.e., 9.6, 19.2, 38.4 and 57.6 Kbps) up to the ISDN channel capacity of 56 or 64 Kbps. Much as today's analog modems have several standards such as V.32, and V.34 (V.fast), which must match end-to-end for successful communications, ISDN has V.120, along with three older methods. V.120 is the newest ITU-T standard for rate adaption, and it is built into almost all present-generation asynchronous ISDN communications equipment. It is gradually replacing the older V.110 standard in international use and

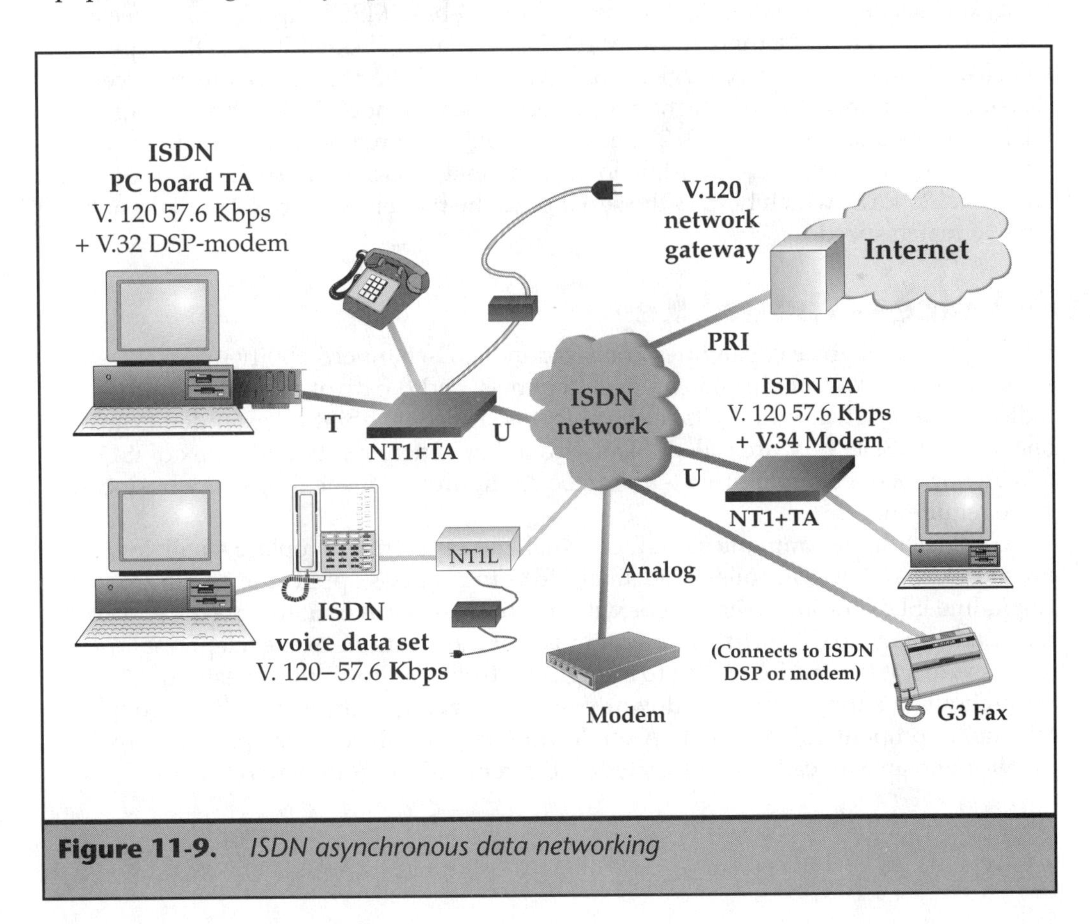

Figure 11-9. *ISDN asynchronous data networking*

has almost completely replaced vendor-specific rate adaption schemes such as AT&T's Digital Multiplexed Interface (DMI) and Nortel (Northern Telecom) T-link.

NOTE: It is possible to run a 57.6 Kbps serial connection through a 56 Kbps network, because you measure the 57.6 Kbps speed at the data terminal equipment (DTE) interface, including start and stop bits in the serial protocol, which are not passed through the B-channel.

Asynchronous data calls on ISDN data B-channels cannot talk to modems or conventional fax machines, because the data bearer capability *will not* connect to the voice-only capability of an analog line. However, several manufacturers have implemented DSP or codec/modem capabilities in their ISDN designs, and these access a *voice* B-channel bearer capability, which will provide the connection to a distant analog line.

ISDN X.25 Packet

D-channel packet capabilities are almost free in CO-based ISDN applications, because the D-channel is already there to support ISDN signaling. Depending on the type of switch and the type of ISDN service you have, you may be able to use one or several X.25 logical channels for low or moderate speed data connections, such as e-mail. ISDN devices that support the X.25 packet capability often have two or more serial port connections to provide multiple logical channel access. Because the ISDN devices have an X.25 PAD, which buffers the serial port, the two ends of an X.25 call do not have to match speed.

ISDN Voice—Virtual Key

ISDN's powerful voice capabilities are based in the D-channel's ability to simultaneously control the switch, the B-channels, and the display, buttons, and indicators associated with digital electronic sets. These capabilities, extended from one to many ISDN sets, are called ISDN virtual key functionality, and a mix of ISDN and analog sets, seen in Figure 11-10, can be configured to meet almost any business voice requirement.

Local telephone companies use the virtual key capability to replace small, low function key system controllers and small PBXs in business applications with single-line ISDN business services or with multiline services usually called Centrex. A major advantage to the virtual key system is that the locations of the telephone stations are not limited by wiring to a single controller or PBX but can extend throughout an entire campus or downtown area, serving multiple buildings from the telephone company ISDN switch. A single attendant or call coverage position can monitor and answer calls for associates in different buildings or across town.

Virtual key serves a flexible mix of analog and ISDN sets, which can be grown (or shrunk) on a line-by-line basis, and the ISDN voice sets can be provided either as a single end point (one set) or multipoint (two voice sets) configuration on an ISDN BRI line.

Few computers are located very far away from a telephone, and most power users are also power communicators for whom voice communications is as important as data. Whether at the corporate workplace or in a telecommuting or work-at-home environment, ISDN voice services allow state-of-the-art sophistication on a single BRI line connection. As computer operating systems become ISDN-aware, tools such as Microsoft Windows™ telephone application programming interface (TAPI) and MacOS™ Apple ISDN Telephone Manager allow computer telephony integration (CTI) applications (see Chapter 18), which integrate the power of the desktop computer and screen-based graphical user interfaces (GUI), with the flexible, software-controlled network interface and control capabilities offered only by ISDN.

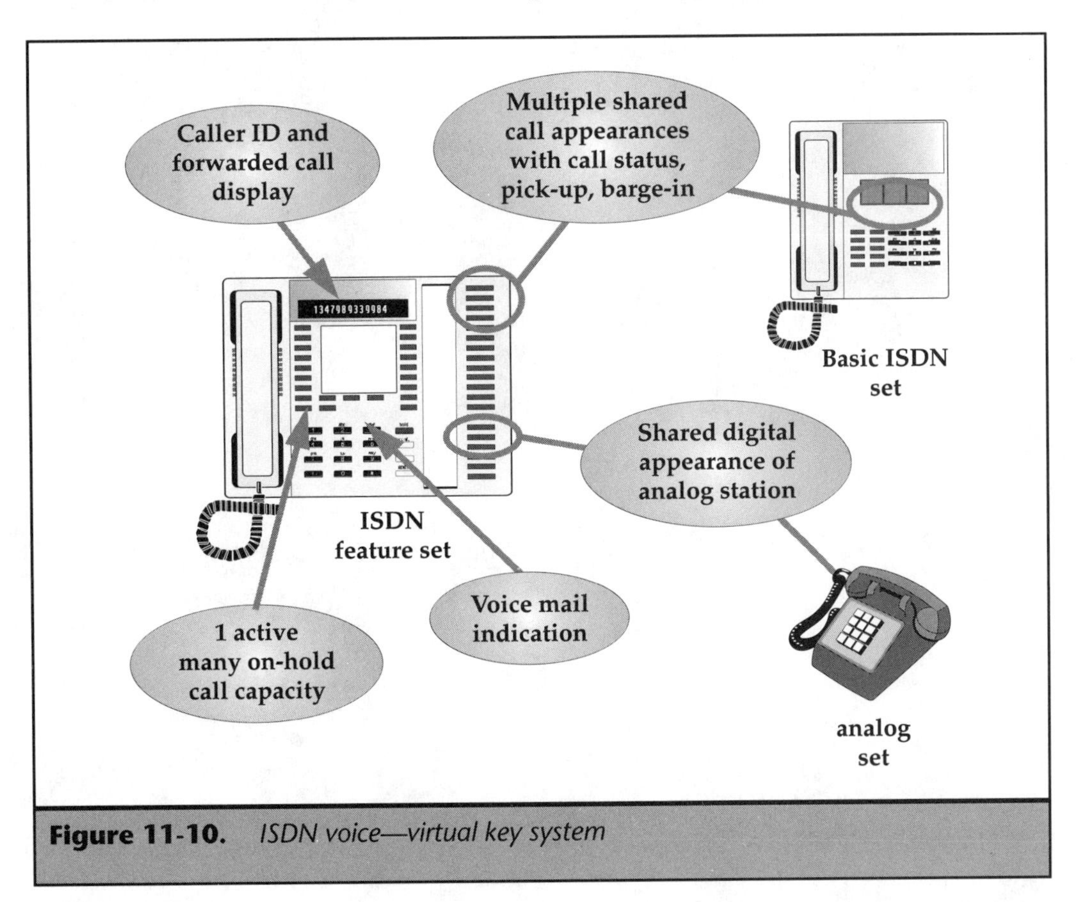

Figure 11-10. *ISDN voice—virtual key system*

Conclusion

ISDN is not just another set of standards, it provides a fundamental re-design of our telecommunications network into a digital medium. In "going digital,"we will be able to extend our communications and information technologies to the far corners of the world. Having weathered this first small step into a digital future, at first portions of and ultimately the entire public network can evolve gracefully into higher speed digital services, and a broadband networking future.

ISDN is already available over a large portion of the U.S., and is widely supported internationally, a position unlikely to be achieved by any other peer-to-peer digital communications protocol in the next decade. For the rest of this century, ISDN will be increasing its presence as the digital telephony access method of choice, and as the hardware and service cost continue to decline, the rate of acceptance will continue to accelerate.

Chapter Twelve

Frame Relay and Asynchronous Transfer Mode

During the last two decades, we have seen a dramatic transformation in the computer industry. We have moved from a centralized model, in which a large CPU had a few I/O devices (mainframe model), to a distributed model, in which all computers connected through a network have their own CPUs (microcomputers). With this distributed architecture came communication among nodes and a new standard for information transmission, or *packet switching*. (A *packet* is a group of fixed length binary digits, including the data and call control signals.)

X.25 is a form of packet switching. It is a technology originally devised to handle network traffic generated by the new philosophy of computer networking. It was defined as an ITU-T (then CCITT) standard in 1976 and has subsequently become very successful, almost ubiquitous. However, with the advent of Integrated Service Digital Network (ISDN), there was a need for a better packet switching technology (ISDN is discussed in detail in Chapter 11). Furthermore, ISDN, which has been in production since 1986, requires the transmission data to be in a digital rather than an analog format. As digital transmission is much more reliable and error-free, a lot of complex processing overhead required for the management of analog data, for example error recovery, was no longer required.

Frame relay was introduced to satisfy the need for fast packet switching in a digital data format. Probably the most important technical innovation brought about by the standardization of ISDN, frame relay is unlike the other new technologies incorporated in the ISDN standards in that it does not require any installation of new equipment or hardware; in fact, most applications require only a minor software upgrade to implement frame relay.

Asynchronous Transfer Mode (ATM) is the new packet switching technology designated for broadband ISDN. Multimedia applications require that information be delivered in a timely fashion. To meet the changing needs of advanced communications users, ATM was developed to get the job done. Wasted bandwidth on fixed circuits is another major concern undertaken by ATM. Equipment upgrades to accommodate the new technology are still under way, so the technology has not yet taken hold.

Frame Relay Technology

To comprehend the underlying principle behind frame relay, let's compare it to X.25. X.25 is an environment characterized by analog lines between the network and the customer premises equipment (CPE). The X.25 network was susceptible to high error rates. Moreover, the older CPE was capable of only very basic functions. In the X.25 environment, the link nodes have to carry the bulk of the responsibility of ensuring faithful delivery of all user data from source to destination. This required an extensive amount of processing at the network level of the Open Systems Interconnection (OSI) reference model. OSI is the only standard internationally recognized for communication between different telephony systems made by different vendors. OSI was developed by the International Standards Organization (ISO) and is continually updated. OSI includes error control and retransmission at the network nodes.

The evolution of the network towards digital facilities has increased network reliability tremendously, while at the same time the unintelligent CPEs at the network's periphery are rapidly being replaced by much more intelligent devices. Given these two factors, a large amount of network processing is now redundant. Frame relay does away with this redundancy and offers a more streamlined approach to packet switching.

As you can see in the X.25 network model shown in Figure 12-1, the data link control protocol requires the data frames, sent by the source towards the destination and vice versa, to be acknowledged at every hop through the network. Furthermore, at each intermediate node, tables must be maintained to allow for the flow and error controls and call management of each virtual circuit. Frame relay, on the other hand, does not require such acknowledgment at every hop. As shown in the frame relay

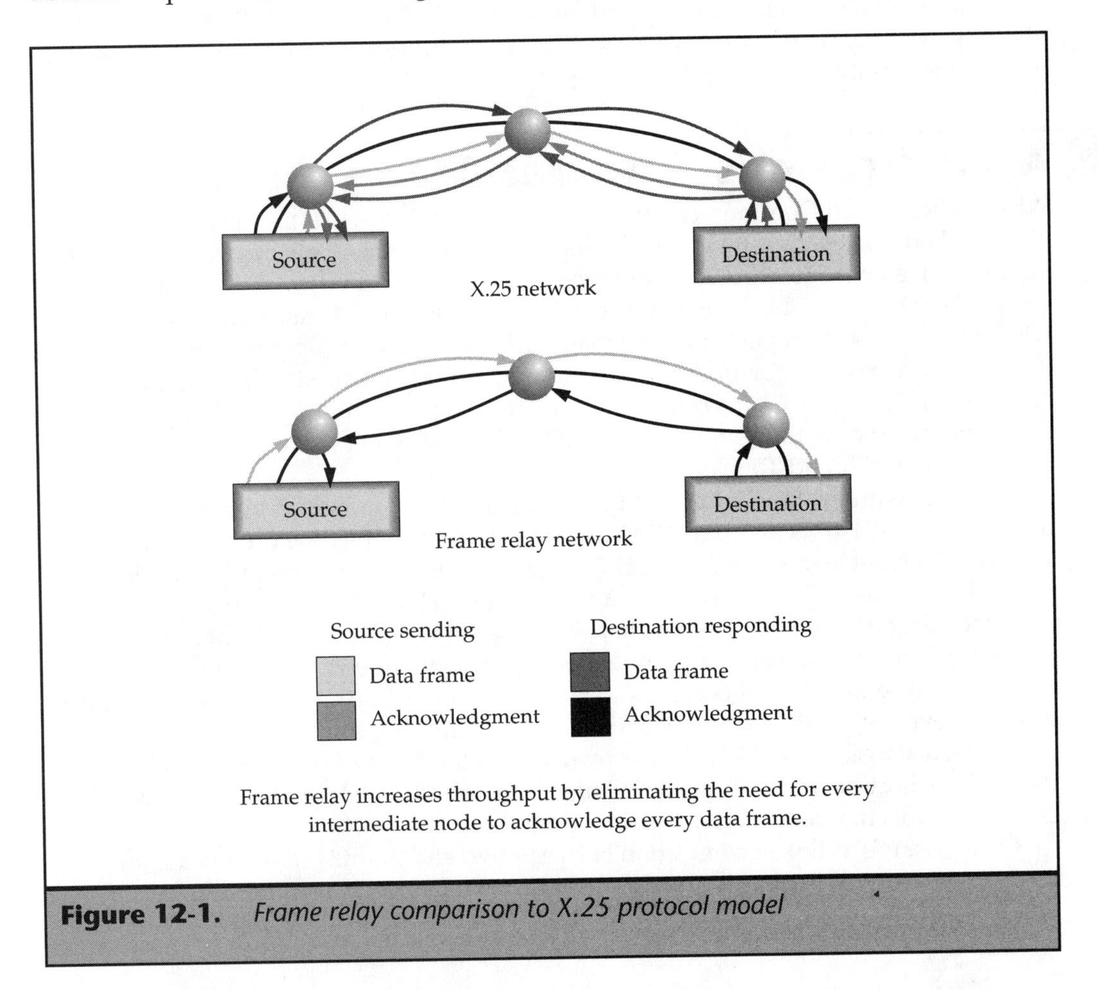

Figure 12-1. *Frame relay comparison to X.25 protocol model*

network in Figure 12-1, a data frame is sent from the source to the destination, and an acknowledgment frame generated at a higher layer is delivered back to the source. The limited amount of processing now performed by the network instantly translates into a significantly higher throughput (almost an order of magnitude) on the network.

While ISDN provides us with a comprehensive digital upgrade to the existing network with services like frame relay, new markets brought forth by recent advancements in computer products like multimedia and video-on-demand require a large amount of bandwidth—more than can be provided by frame relay. Networks will be required to carry in an integrated way both real-time traffic, such as voice and high-resolution video, as well as nonreal-time traffic, such as computer data and file transfer, which may tolerate some delay but will not tolerate data loss. Thus, the trend is to move towards gigabit/second Broadband ISDN (B-ISDN). Included in B-ISDN, which has been around since 1988, are services like asynchronous transfer mode (ATM) and switched multimegabit data service (SMDS), based on an optical fiber network called the synchronous optical network (SONET), which will provide a form of cell relay based on fixed-length cells. ATM's versatility makes it the most promising of all the technologies in B-ISDN at present.

Asynchronous Transfer Mode Theory

When Bellcore (Bell Communications Research) proposed ATM, they just took the idea of a better packet switching technology one step further. Bellcore was established just after divestiture to provide centralized services to the seven regional bell holding companies (shown in Chapter 6, Figure 6-3). ATM is analogous to fast packet switching with small fixed-length packets. The small packet allows for timely delivery of information across the circuit; multiplexing; and voice transmission having the same quality as STM (synchronous transfer mode) on switched circuit T-1 or T3 links used for telephone lines without the wasted bandwidth. (See Chapter 10 for more information on T-1 and T3 telephone circuits.)

In order to understand what ATM is all about, a brief introduction to synchronous transfer mode (STM) is in order. ATM was developed to overcome the shortcomings that were inherent in the STM model. STM is used by the telecommunications backbone networks to transfer packetized voice and data across long distances. It is a circuit-switched networking mechanism, where a connection is established between two end points before data transfer commences and torn down when the transmission is complete. The end points allocate and reserve the connection bandwidth for the entire duration, even when they may not actually be transmitting the data. The way data is transported across an STM network is by dividing the bandwidth of the STM links (familiar to most people as T-1 and T3 links) into a fundamental unit of transmission called timeslots that repeat after a certain period.

On a given STM link, a connection between two end points is assigned a fixed timeslot that the connection is always carried on. Therefore, if a connection has any data to transmit, it drops its data into its assigned timeslot, and if it doesn't, then that timeslot goes empty. Other connections waiting in line cannot use that empty timeslot. During peak hours, most timeslots may be carrying information across the entire bandwidth, but there will always be segments of waste. If there are a large number of timeslots going empty (which is most of the time), there is a significant wastage of bandwidth. STM also limits the number of connections that can be supported simultaneously. Furthermore, the number of connections can never exceed the total number of timeslots, 24 for T-1s, 672 for T-3.

As mentioned before, the gigabit/second B-ISDN cross-country and cross-oceanic links need to simultaneously carry both real-time and nonreal-time traffic. The problem with carrying the different kinds of traffic on the same medium in an integrated fashion is that the peak bandwidth requirement of these traffic sources may be quite high, while the duration for which the data is actually transmitted may be quite small. In other words, the data comes in bursts and must be transmitted at the peak rate of the burst, but the average arrival time between bursts may be quite large and randomly distributed.

For such connections, it would be a considerable waste of bandwidth to reserve them a timeslot at their peak bandwidth rate for all times, when on the average only one in ten timeslots may actually carry the data. It would be nice if that timeslot could be used by another pending connection during the down time. Therefore, using the STM mode of transfer becomes inefficient as the peak bandwidth of the link, peak transfer rate of the traffic, and overall burstiness of the traffic expressed as a ratio of peak/average, all go up. Research done by Bellcore shows that multimedia integration for voice, video, and data communications, which have a high demand for smooth, even transmission of their data, is definitely in high demand in the global economy for the next decade and unlevel amounts of information being transmitted is simply not acceptable. (For more information on full-motion video using telephony, refer to Chapter 18 on videoconferencing.)

Consequently, ATM was presented independently by Bellcore and several telecommunications companies in Europe. The critical point that was made is, instead of always identifying a connection by a timeslot, to put the connection identifier in the information packet (or cell as they classified it) being sent and keep the size of the packet small, so that if any one packet got dropped in passage due to congestion, not too much data would get lost, and in some cases it could easily be recovered. The fixed size of the cells arose out of the hidden motivation of the telecommunications companies to sustain the same transmitted voice quality in an ATM network as in STM networks. Thus, two end points in an ATM network are associated with each other via an identifier called the Virtual Channel Identifier (VCI) label instead of by a timeslot as in an STM network. The VCI is carried in the header. The best way to illustrate how this works is with an example.

Assume an ATM network exists with nodes in New York, Atlanta, Dallas, and San Francisco. Say that David, while vacationing in New York, decides to play NetTrek on an overcast day with his colleagues in San Francisco, who are still grinding away at work. Also assume that we have ATM cell interfaces at UNIs (User Network Interfaces) in both New York and San Francisco. This is what can happen:

David's portable laptop makes a connection request to the UNI in New York. After an exchange of connection parameters between his laptop and the UNI (such as destination, traffic type, peak and average bandwidth requirement, delay and cell loss requirement, how much money he has left to spend, etc.), the UNI forwards the request to the network. The software running on the network computes a route based on the cost function specified by David and figures out which links on each leg of the route can best support the requested quality of service and bandwidth. Then it sends a connection setup request to all the nodes in the path all the way to the destination node in San Francisco.

Let's say that the route selected was New York-Atlanta-Dallas-San Francisco. Each of the four nodes might pick an unused VCI label on its respective node and reserve it for the connection in the connection lookup tables inside its respective switch. Say New York picks VC1. It will send it to Atlanta. Atlanta in turn picks VC2, associates it with VC1 in its connection table, and forwards VC2 to Dallas. Dallas picks VC3, associates it with VC2 in its connection tables and forwards VC3 to San Francisco. San Francisco picks VC4 and associates it with VC3 in its connection tables, and pings the addressed UNI to see if it would accept this connection request. Fortunately, the UNI finds David's colleagues and returns affirmative. So San Francisco hands the UNI and David's friends VC4 as a connection identifier for this connection. San Francisco then acknowledges back to Dallas. Dallas acknowledges back to Atlanta and sends it VC3. Atlanta puts VC3 in its connection tables to identify the path going in the reverse direction, and acknowledges to New York sending it VC2. New York associates VC2 in its connection tables with VC1, and acknowledges the originating UNI with VC1. The UNI hands David's laptop VC1, and connection is established.

David identifies the connection with VCI label VC1, and his colleagues identify the connection with VCI label VC4. The labels get translated at each node to the next outgoing label like so:

	New York	Atlanta	Dallas	San Francisco	
David →	VC1 →	VC2 →	VC3 →	VC4 →	colleagues
David ←	VC1 ←	VC2 ←	VC3 ←	VC4 ←	colleagues

Other scenarios are also possible and would depend on a vendor's implementation of the ATM network.

When David has had enough of playing NetTrek and looks to see the sun out again, he is off to some serious scuba diving off the Atlantic coast. He logs off his computer, and the connection between himself and his friends on the West Coast is torn down. This task is accomplished by reassigning the network resources reserved for this connection to other connections, which means the VCI labels are then reused for other connections.

Frame Relay Architecture

Frame relay employs *Q.922*, which is an enhanced version of Link Access Procedure on the D-Channel (LAPD), or Q.921. When discussing frame relay, one must consider out of band control, which essentially means the separation of the control and the user planes. The *control* plane is responsible for establishment of the data connection, and the *user* plane is responsible for the transfer of data. As you can see in Figure 12-2, the user plane of frame relay is limited to only the core functions of Q.922.

These core functions constitute a sublayer of the data link layer, which enables the transfer of data from one user to another. Any additional data link or network functions can be selected as end-to-end functions, although these are not part of the frame relay service. The core functions also designate frame relay as a connection-oriented service. To gain a more thorough understanding of frame relay, one must look at the frame relay format.

To begin with, the format of the frame used in frame relay is very similar to the LAPD format. As you can see from Figure 12-3, the frames can be of variable length. By variable length here, we mean that the information field in the frame can vary in size. The other fields are fixed. The frames are also delimited by a flag field, which signifies the start and the end of the frame. A detailed view of the address field shows its various components.

To understand how frame relay operates, it is essential that you understand how a connection is set up. A connection is a conversation between two entities sending and receiving information. Rather than requiring separate physical facilities for each conversation, each frame contains a Data Link Connection Identifier (DLCI) that is

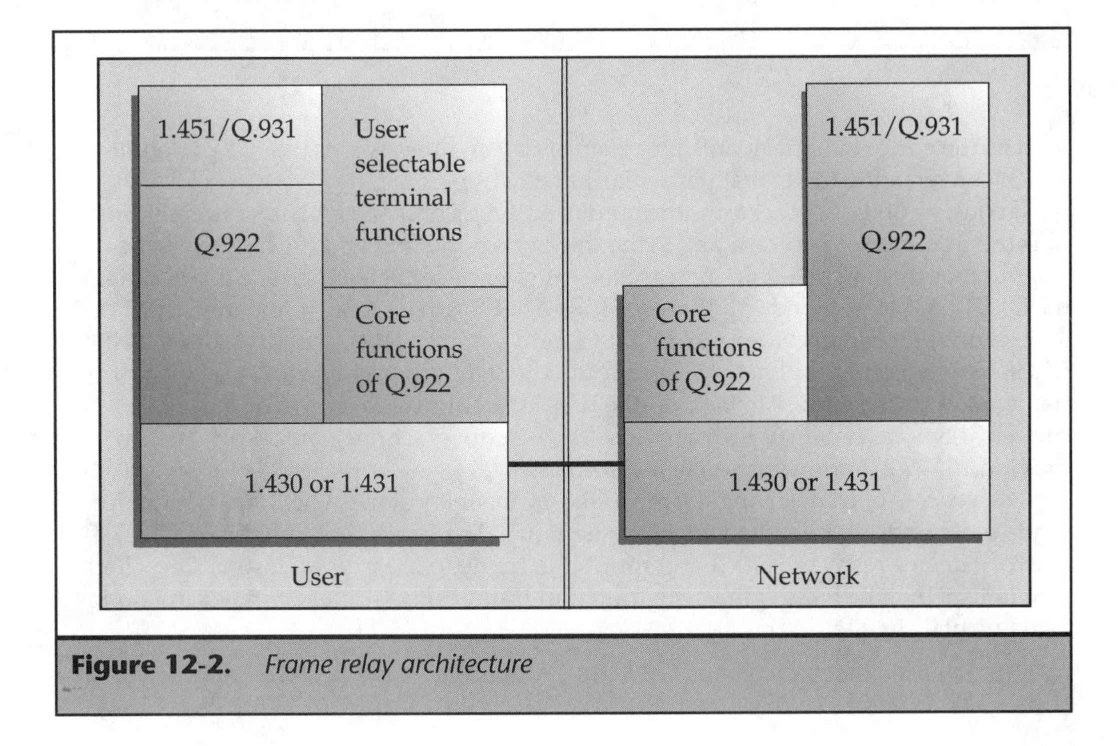

Figure 12-2. *Frame relay architecture*

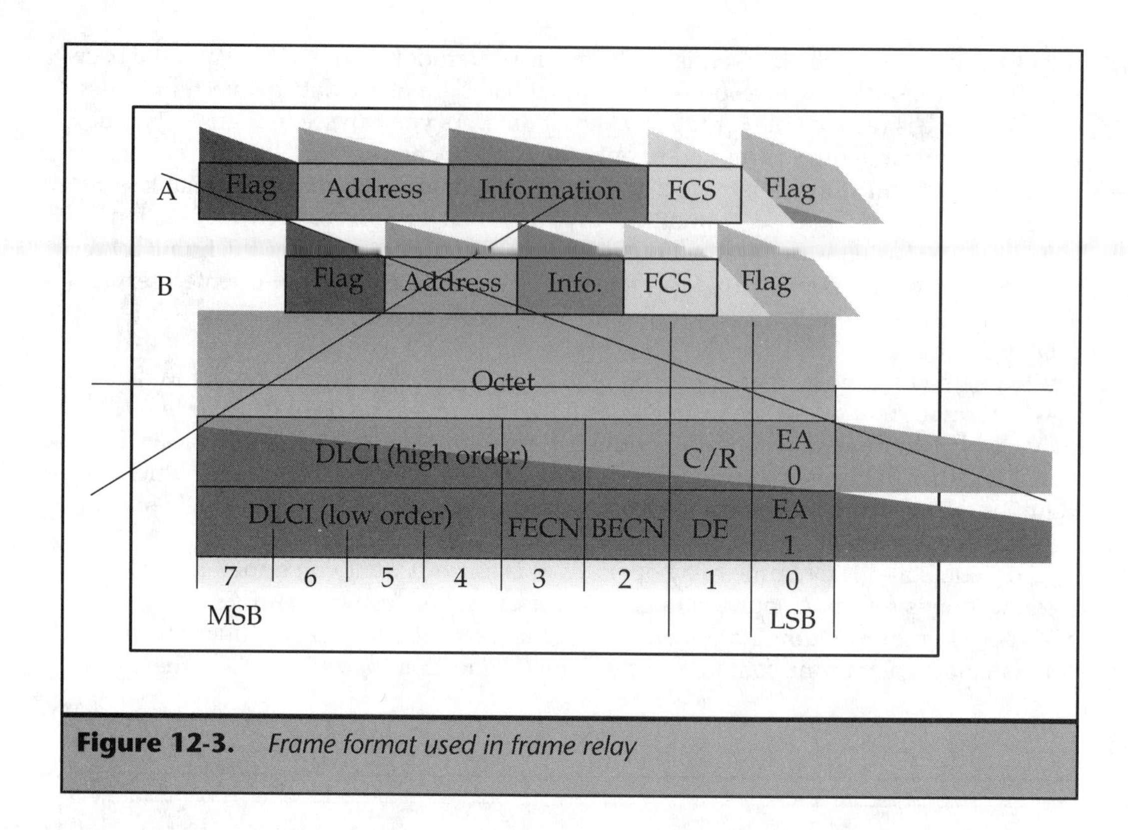

Figure 12-3. *Frame format used in frame relay*

used as the addressing mechanism of frame relay. In other words, the DLCI denotes which conversation owns that particular information.

It consists of the six most significant bits of the second octet (an *octet* is eight bits) plus the four most significant bits of the third octet (the first and the last octets are reserved for the flag field). To increase the number of conversations at a given time, the DLCI can be extended to a three- or four-octet format. As recommended by ITU-T, a three-octet format provides a 17 bit DLCI, and a four-octet format provides a 24 bit DLCI. Consequently, with a three- and four-octet DLCI, a user can increase the normal frame relay address space to beyond the present 10 bits (which provide 1,024 conversations) to 17 bits (which provide 131,072 conversations) and 24 bits (which provide 16,777,216 conversations), respectively.

The process of transferring data through frame relay starts when the user device sends the frames described above to the network. The frames contain the DLCI. The network device reads the DLCI and routes the frame to the proper destination. The mechanism by which the frames are routed in frame relay is called *virtual networking* using virtual circuits.

ATM Architecture

An ATM cell or packet as specified by ITU-T is 53 bytes. Figure 12-4 shows the layers of ATM. The 48 bytes of payload may optionally contain a 4 byte ATM adaptation layer and 44 bytes of actual data, or all 48 bytes may be data, based on a bit in the control field of the header. This enables fragmentation of cells and reassembly into larger packets at the source and destination, respectively. (Since the header definition may still be in flux, it is possible that presence or absence of the adaptation layer information may not be explicitly indicated with a bit in the header, but rather implicitly derived from the Virtual Channel Identifier label). The control field may also contain a bit to specify whether this is a flow control cell or an ordinary cell, for example, or an advisory bit to indicate whether this cell is droppable in the face of congestion in the network.

ATM UNI Cell Structure

ATM is a *connection-oriented protocol* inasmuch as there is a connection identifier in every cell header that explicitly associates a cell with a given virtual channel on a physical link. (In a connectionless protocol, there is no connection specifically set up; the data is usually broadcasted and expected to get to its destination without supervision. A good example of a connectionless service is a datagram.)

The currently defined structure of an ATM cell at the User Network Interface is shown in Figure 12-4. The connection identifier consists of two subfields, the Virtual Channel Identifier (VCI) and the Virtual Path Identifier (VPI). Together they are used in multiplexing, demultiplexing, and switching a cell through the network. VCIs and VPIs are not addresses; they are explicitly assigned at each segment (link between ATM nodes) of a connection when a connection is established and remain for the duration

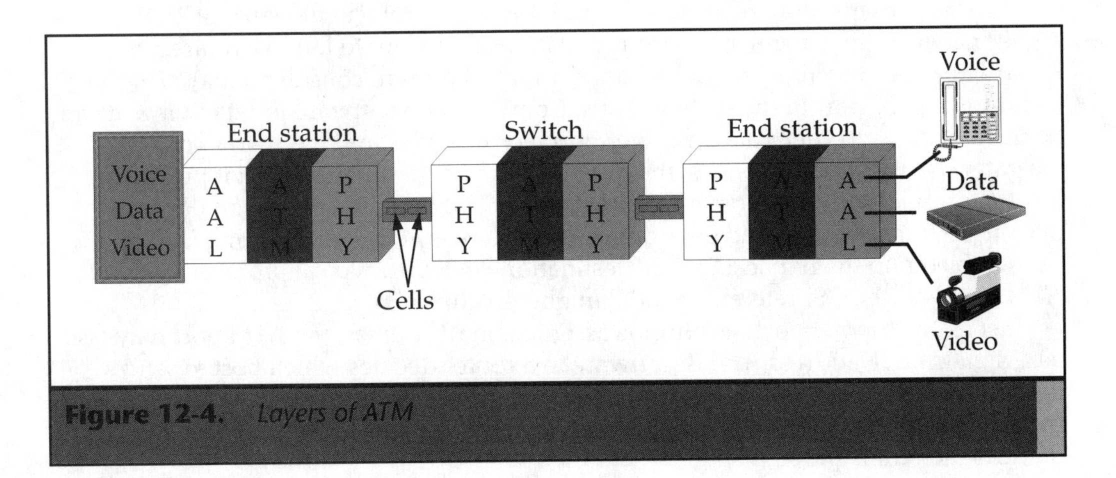

Figure 12-4. *Layers of ATM*

of the connection. Using the VCI/VPI, the ATM layer can asynchronously interleave (multiplex) cells from multiple connections.

The virtual path concept originated with concerns over the cost of controlling B-ISDN networks. The idea was to group connections sharing common paths through the network into identifiable units (the paths). Network management actions—such as call setup, routing, failure management, and bandwidth allocation—would then be applied to the smaller number of groups of connections (paths) instead of a larger number of individual connections (VCI). For example, use of virtual paths in an ATM network reduces the load on the control mechanisms because the functions needed to set up a path through the network are performed only once for all subsequent virtual channels using that path.

Now the basic operation of an ATM switch will be the same, no matter if it is handling a virtual path or virtual circuit. The switch must identify, on the basis of the incoming cell's VPI, VCI, or both, which output port to forward a cell received on a given input port to. It must also determine what the new values of the VPI/VCI are on this output link, substituting these new values in the cell.

ATM and the OSI Reference Model

As is probably evident by now, ATM is designed for switching short fixed-length packets of data over gigabit/second B-ISDN links across very large distances. Thus, its place in the OSI reference model is somewhere around the data link layer. The reason it does not cleanly fit into the abstract layered OSI reference model is because within the ATM network itself, end-to-end connection, flow control, and routing are all done at the ATM cell level. So there are a few aspects of traditional higher layer functions present in it. In the OSI reference model, it would be considered layer 2 (where layer 1 is the physical layer and layer 2 is the data link layer in the Internet protocol stack). But it is not very important to assign a clean layer name to ATM, so long as it is recognized that it is a hardware-implemented packet switched protocol using 53 byte fixed-length packets.

What is perhaps more relevant is how all this will interact with current TCP/IP and IP networks in general and with applications that want to talk ATM directly in particular. A convenient model for an ATM interface is to consider it another communications port in the system. Thus, from a system software point of view, it can be treated like any other data link layer port. For instance, in IP networks connected via gateways to ATM backbones, the model would be no different than it presently is for a virtual circuit connection carried over an STM link, except that an IP packet over an ATM network would get fragmented into cells at the transmitting UNI and reassembled into the IP packet at the destination UNI. So a typical protocol stack of ATM based on the OSI reference model might look like Figure 12-5.

Just as an ethernet port on a host is assigned an IP address, the ATM port may also be assigned an IP address. The IP software in a router decides which port to send a packet to based on the IP address and hands the packet to the port. The port then does the right thing with it. For an ethernet port, the ethernet header is tacked on and the frame is then transmitted.

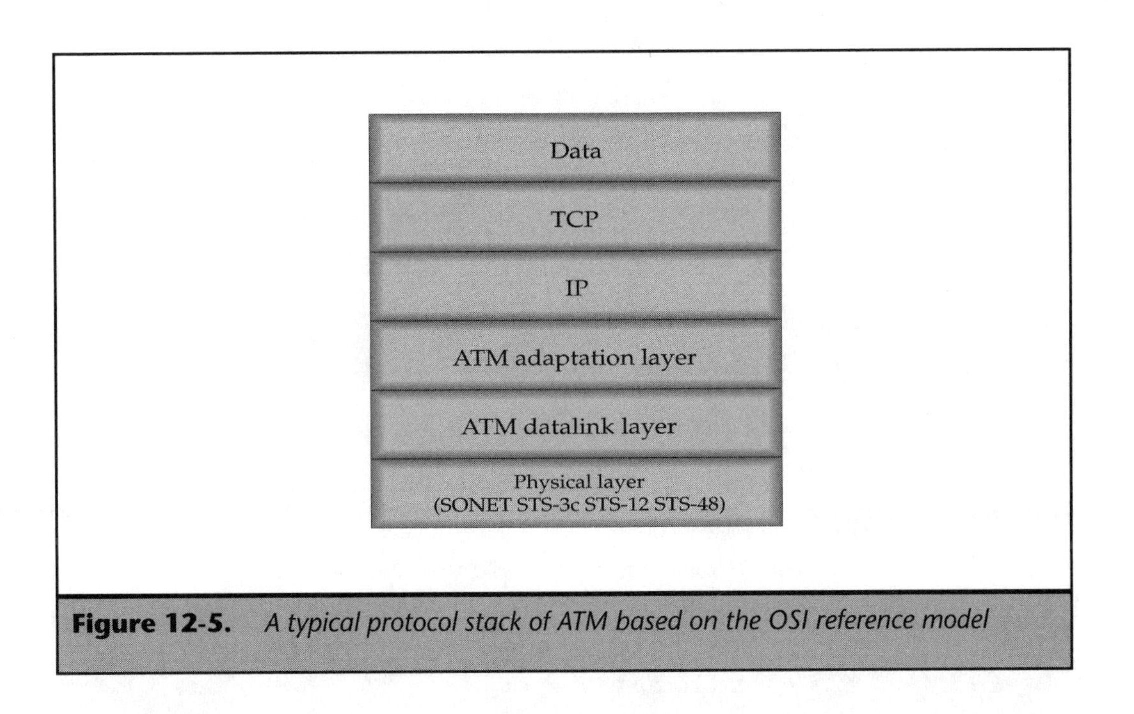

Figure 12-5. *A typical protocol stack of ATM based on the OSI reference model*

Similarly, for an ATM port, the IP datagram is fragmented into cells for which an ATM adaptation layer is specified in the standards. The fragmentation and reassembly is done in hardware on the sending and receiving sides. A VCI label acquired via an initial one-time connection establishment phase is placed in the header of each cell, and the cells are drained down the fat ATM data link layer pipe. On the receiving side, the cells are reassembled in hardware using the ATM adaptation layer, and the original IP packet is reformulated and handed to the receiving host on the UNI. The adaptation layer is not a separate header but is actually carried in the payload section of the ATM cell as discussed earlier. In a very similar manner, voice and video can also be encoded, transferred and decoded. This process is illustrated in Figure 12-6.

Frame Relay Versus ATM

Earlier, it was mentioned that frame relay has the advantage of being a packet switching technique with a high throughput and a low delay factor. In addition to these advantages, it provides for virtual networking and statistical multiplexing. By statistically multiplexing several connections on the same link based on their traffic characteristics, one can solve the unused slot problem of STM. In other words, if a large number of connections are very "bursty" (i.e., their peak/average ratio is 10:1 or higher), then all of them may be assigned to the same link in the hope that statistically they will not all burst at the same time, and if some of them do burst simultaneously, that there is sufficient elasticity that the burst can be buffered up and put in

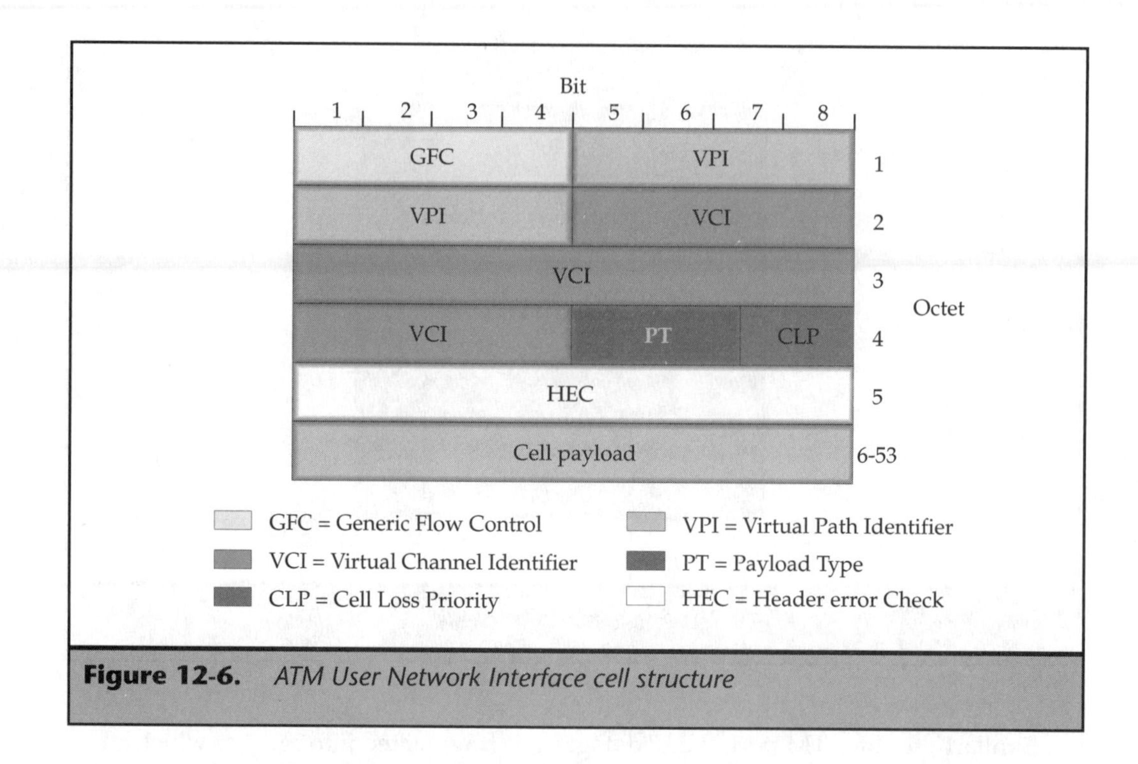

Figure 12-6. *ATM User Network Interface cell structure*

subsequently available free slots. This is called *statistical multiplexing,* and it allows the sum of the peak bandwidth requirement of all connections on a link to even exceed the aggregate available bandwidth of the link under certain conditions of discipline. Furthermore, frame relay can drastically simplify the design of routers and bridges for LANs through virtual networking, a concept that was explained earlier in the "Frame Relay Architecture" section. It also reduces management costs and the cost of the LAN itself, as it requires only N physical connections and not the N(N-1)/2 physical connections required for a full mesh of LANs, as explained previously.

People look for a combination of cost savings, ease of management, and high level of performance in modern networks. They don't want to have to build and manage complex mesh networks. Frame relay solves this problem for virtual networking. Manageability is a critical consideration. Users want to be able to hand off the management of their networks to carriers. Also, improvements in price and performance are definitely the measure by which users judge a technology. Frame relay's combination of price and performance is a primary reason for frame relay's growing success. In numerous case studies it has been found that frame relay users have the opportunity for tremendous savings. In addition to savings, users get the redundancy and performance of a meshed network that they would not otherwise be able to afford.

However, the frame relay service as it is defined by the current standard does not provide for link-by-link flow and error control, but that can be easily implemented at a higher level of the OSI reference model. It also does not provide for voice data integration, video services, and teleconferencing.

Users of frame relay equipment and services are vocal supporters of the technology. Network managers report measurable improvements in networking costs, performance, reliability, manageability, and ease of implementation. Corporations are pleased with the resulting improvements in productivity, customer satisfaction, and profits. The message is clear: frame relay provides a proven, cost-effective solution for data networking.

Although data networking at lower bandwidths may undeniably be a frame relay domain, ATM is essential for voice, data, and video integration. While frame relay and ATM networks may initially coexist as separate networks, eventually frame relay and other access technologies will feed traffic into the multimedia ATM networks. One can expect to see most new networks being ATM-based and supporting frame relay access as well. Over time, the predominance of LAN interconnections will be frame relay going to wide area ATM networks.

Why does frame relay work well as an access interface for ATM? There are remarkable similarities in the protocols. If one looks at the basic technology, the only real difference between frame relay and ATM is the variable length versus the fixed-length packets (or cells). That's where all the differences stem from in terms of bandwidth efficiency and ability to support very delay-sensitive applications like voice and video.

ATM's 53 byte fixed cell is optimized for handling multimedia traffic at high speeds. However, with ATM, approximately ten percent of each full cell is overhead. In very short transactions, where cells are not full, an even greater percentage of bandwidth is wasted. As a general rule, the lower the speed (e.g., T-1 and sub T-1), the more concerned you are with overhead for data applications. At levels of T3 and above, the benefits of ATM outweigh the efficiency considerations. At relatively lower speeds (T-1 and below), frame relay more efficiently transports data.

Frame relay's success as an access interface to ATM will be enhanced by an industry-wide agreement on the interworking between frame relay and ATM. The Frame Relay Forum and the ATM Forum (discussed in more detail at the end of this chapter) have been working on just such an agreement. These two industry organizations represent companies ranging from wide area networking service providers, to equipment vendors of CPE, switches, computers, and multiplexers, to major networking software vendors. With this formidable array of networking talent, the industry has actively taken charge of ensuring broad-based interoperability between these two technologies.

The goal of these interworking activities is to develop specifications for two different interworking scenarios. The first scenario allows frame relay transport from end-to-end across an ATM network. The frame relay emulation may take place in the ATM network itself, or in the customer premises equipment (CPE) at either end of the connection. The second scenario is to support frame relay at one end of the connection and ATM at the other end. This may be attractive to users who want to aggregate lower speed data over the network into one higher speed ATM interface to a mainframe or centralized server. Commonly known as *service interworking*, this scenario requires that a protocol conversion function take place between frame relay and ATM.

In early 1994, the Frame Relay Forum, in cooperation with the ATM Forum, released the Frame Relay/ATM Network Interworking Implementation Agreement for final approval. Both groups endorse the Agreement as the industry-accepted specification for frame relay-to-ATM network interworking. The Agreement is expected to be widely implemented by both frame relay and ATM equipment vendors and service providers.

How to Obtain Frame Relay and ATM Services

The realms of frame relay and ATM services are very different from each other. Frame relay is already an established technology. ATM, on the other hand, mostly exists in various testbeds across the country. However, ATM is rapidly gaining popularity and is definitely destined to be a mainstream WAN as well as LAN implementation in the future.

Public Frame Relay Services

Public service providers (carriers) offer frame relay services by deploying frame relay switching equipment. Both frame relay access equipment and private frame relay switching equipment may be connected to services provided by a carrier. The service provider maintains access to the network via the standard frame relay interface and charges for the use of the service.

Access to the frame relay service involves three elements: customer premises equipment (CPE), a transmission facility, and the network itself. The CPE may be any of the types of access equipment, such as a frame relay-compliant router, or even a private network switch with a frame relay-compliant interface. The access facility must be appropriate for the speed involved—generally a 56/64 Kbps or T-1/E1 link. (When fractional T-1/E1 is service is desired, a full T-1/E1 link is still generally used for access. However, depending on the carrier offering, the unused portion of the T-1/E1 may in some cases be used for transporting other traffic, such as voice.) A standard CSU/DSU is used in conjunction with the 56/64 kbps or T-1/E1 service.

At the network interface, the carrier will be responsible for terminating the circuit appropriately. The carrier is also responsible for transporting the information to the appropriate transmission facility at the other end of the virtual circuit. The transmission facilities at the two ends of the circuit, though, may be of different speeds. This allows the users to mix and match so the speed matches the actual aggregate traffic needs at each site.

Carriers generally offer several options for buying the services. One option is to buy a given amount of service as if the service were a dedicated facility. This has the advantage of fixed pricing (no surprises) and straightforward comparison with dedicated bandwidth alternatives. Buying service by the frame or megabyte transmitted is also an option. This gives you the advantage of only paying for actual information transmitted, but it is more difficult to predict the exact usage (and thus the cost).

Almost all of the frame relay services (FRS) are provided on T-1 lines. In 1991, there were various carriers offering FRS. The first carrier to offer the service was Williams Telecommunications Group, also known as WilTel, which started FRS in the U.S. in March, 1991 and based its tariffs on fixed rates. Other carriers following the lead were BT North America Inc. which in September, 1991, extended its service to France and London as well as the United States, and CompuServe with its Frame-Net service. CompuServe also added Europe and the Far East to its Frame-Net service. Cable and Wireless Communications plans to help integrate large global companies with its international Platform-Network (Planet), which includes FRS. LMEricsson Telephone Co. of Sweden is also planning to integrate FRS with its internationally available X.25 switch line. As far as common carriers in the U.S. are concerned, as of late 1991 only three carriers (NYNEX Corp., Pacific Bell Co., and MCI Communications Corp.) were offering FRS. However, other local carriers have also joined this group since then. As an example, the following is an estimate of the Pacific Bell frame relay charges.

Service	**Installation Fee**	**Monthly**
56 Kbps Frame Relay	$995	$125
128 Kbps Frame Relay	$1669	$313
384 Kbps Frame Relay	$1699	$563
1544 Kbps Frame Relay	$1699	$663

If frame relay is a data networking alternative that you are presently evaluating as a way to more effectively meet your current and/or future internetworking needs, then you might want to consider the public services from local exchange carriers (LECs) as part of the overall plan. To date, over 200 customers are realizing the benefits that these services offer. The industry sectors represented by this customer base include state, county, and city government, retail, banking, finance, insurance, manufacturing, health care, education, utilities, and engineering.

LECs began entering the public frame relay services market in April 1992. Since then, all regional bell operating companies (RBOCs) and GTE have introduced commercial service offerings. Plans for deployment in about 40 metropolitan areas across the country by year-end 1993 were publicly announced in May, 1992. However, in response to high market demand, these companies report that their initial plans have been exceeded, and service availability will continue to expand in 1994.

Customers subscribing to LEC frame relay service will typically access the network over dedicated (leased/owned) 56/64 kilobit per second (Kbps) digital lines or T-1 facilities. Each digital access line terminates on a switch in the frame relay network (often a switch is located at or near the customer's local serving office).

There are three basic LEC frame relay pricing components. The first component is the monthly charge for an access line to the frame relay network. This is usually a fixed charge; however, depending on the distance from the customer's local serving office to a frame relay switch, a mileage charge may also apply. The second pricing component is the monthly charge for establishing and maintaining a UNI connection on a frame relay switch. This is a fixed charge that depends on the line speed of the UNI (e.g., 56/64 Kbps, 384 Kbps or 1.5 Mbps). The third pricing component is the monthly charge for establishing and maintaining PVCs. Since the ability to establish and maintain multiple connections over a single UNI is a key advantage inherent to frame relay, the incremental per-PVC charges are consequently low.

Additionally, according to most carriers, frame relay should actually save money for most users. The reason for this is that frame relay pricing is not distance-sensitive. Users can expect a savings of 50 percent or more on circuits longer than about 125 interexchange carrier miles.

Public ATM Service

Five key areas define public ATM networks:

- Physical access
- Logical connections
- Class of service
- Information rate
- Network management

Public ATM services can be accessed, currently or in the near future, at line rates ranging from T-1 (1.5 Mbps) to OC-3 (155 Mbps). In addition, gateways will soon be implemented to connect ATM users to users of existing services like frame relay and the Internet.

T-1 ATM access provides several benefits to users, including a common protocol in their LAN/WAN networks and efficient integration of data, video, and voice on a single, relatively low-cost, transmission facility. However, T-1 ATM has a significant overhead compared to other data link layer protocols. The throughput of T-1 ATM using HEC-based mapping as defined by ITU G.804, with AAL Type 5, is 1.39 Mbps.

Initially, carriers will implement T-1 circuit emulation as a means to allow users with T-1 sites to access public ATM services without requiring upgrades to their existing time division multiplexing (TDM) equipment. Existing T-1 sites can interwork with locations that have converted to T3 ATM. Circuit emulation allows a smooth evolution because users can convert existing high-volume network sites to T3 ATM, interwork those sites with T-1 locations using existing TDM equipment, and then slowly migrate to end-to-end ATM as appropriate. In time, carriers may implement end-to-end private line services over an ATM backbone via T-1 circuit emulation specifications.

The n x T-1 access fills the bandwidth gap between T-1 and T3. It allows users to access public ATM services by combining multiple T-1s (typically two to eight) using an inverse multiplexer.

SONET OC-3c (155 Mbps) access will be available in the near future. SONET access will provide large amounts of bandwidth and dramatically improve restoral times in event of access failure. Soon, gateways connecting ATM to other services like frame relay and the Internet will provide interworking between users of different services. Additionally, they will allow the current installed base of frame relay and Internet protocol (IP) users to interwork with higher-speed ATM sites. In fact, many believe that frame relay will be the preferred access method for low-speed (i.e., T-1 and under) data-only traffic.

It should be noted that T3 ATM is the only access option widely available. The other options should be available by the time this book is in print.

Fujitsu makes a 4 x 4 switch element chip set (MB86680). Note that there *are* other ATM/AAL chip sets out there, besides the Adaptive design, now that the industry is rolling. Other vendors include Brooktree (Boulder, Colorado, (303) 494-4484) and Integrated Telecom Technology (IgT), which both have ATM UNI chips and other cool ATM chips.

Future Prospects

When B-ISDN is finally implemented, we will see a sort of frame to cell evolution, and therefore a means for frame-cell interworking might be required. A solution such as the one shown in Figure 12-7 can be implemented for this purpose.

In this case, the adaptation layer of ATM, AAL 3/4 will perform the conversion from LAPD frames to cells and vice versa. The adaptation layer will be in the FRS node at the Network-Network Interface (NNI). Thus it will be possible to provide a network with multiple access options, that is, one that offers both connection-oriented service (frame relay) and connectionless services (SMDS) to the users.

As far as the technology is concerned, there are many areas where further research might be needed, of which a few critical areas are worth mentioning. First, one of the most important features of ISDN is that it can provide for voice, data, and video service integration. However, in frame relay it still needs to be seen if voice-data integration is possible, because of the undeterministic latency produced by frames of variable length. In isolated environments, frame relay technology has been able to carry voice by adjusting for the variable delay. In any case, further research is required to ascertain if this can still be done in a real-world network environment.

Second, it still remains to be seen if cells are better than frames as forms of packet multiplexing. Frames are excellent devices for data transmission. Cells, on the other hand, as they are of fixed length, are very suitable for voice and video transmission. Furthermore, the processing requirements for cells are much lower. However, while

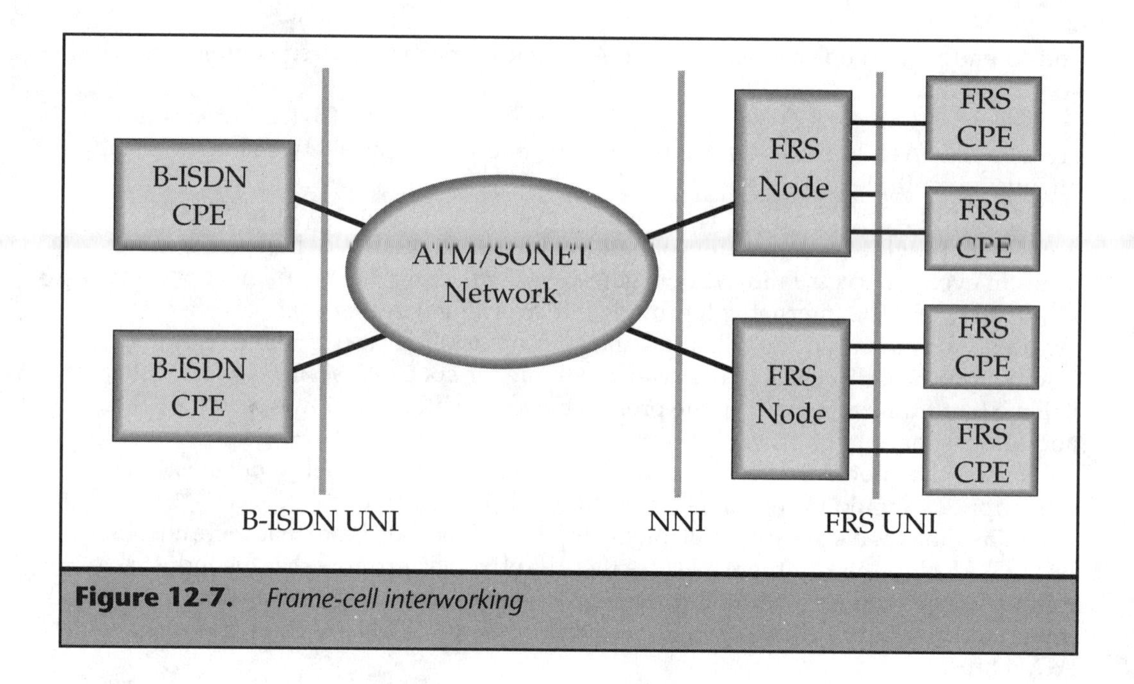

Figure 12-7. *Frame-cell interworking*

most people believe that cells become more appropriate as facility speeds increase, this is not the case. Research in this area might provide us with the correct answer. Third, frame relay is currently based on PVCs. Research needs to be done to find a marketable switched virtual circuits solution to replace the PVC, as SVCs might provide better performance.

Frame relay services are successful at the moment and will be even more successful in the future, because users get more for their money with this service. In addition to the service demand, the technology is readily available and a majority of vendors have agreed on a standard. As mentioned before, it has key advantages for end users in terms of performance as well as cost efficiency. In essence, it is a standardized interface that provides multiplexed access to bandwidth-on-demand backbone networks and delivers LAN-like performance over a wide area. Frame relay can enable reduction of direct and indirect costs and simplify network design and operation, and is applicable to LAN interconnection, host-to-host communications, and traditional terminal environments. Frame relay can provide a common interface method for many traditional and emerging data communication environments and is an effective complement to cell relay backbones.

Conclusion

Frame relay is a simplified form of packet-mode switching, optimized for transporting today's protocol-oriented data. The result of this simplification is that frame relay offers higher throughput, while still retaining the bandwidth and equipment efficiencies that come from having multiple virtual circuits share a single port and transmission facility. Thus, frame relay can:

- Reduce the cost of transmission facilities and equipment; provide increased performance, reliability, and application response time
- Increase interoperability through well-defined international standards, and be implemented as a private networking technology or as a carrier service

A major reason for the high level of interest regarding frame relay is that it is a technology that has been developed in response to a clear market need. With the proliferation of powerful end point devices (such as PCs and workstations) operating with intelligent protocols (such as TCP/IP, XNS and DECnet), users are seeking wide area network communication methods that offer higher throughput and more cost-effective use of digital transmission lines. With that need in mind, frame relay has been rapidly developed and standardized to have exactly the combination of characteristics needed by today's corporate networks.

On the other hand, ATM is a revolutionary technology which provides a form of fixed cell relay instead of a variable packet or frame relay like frame relay. This enables it to be faster and allows integration of voice and data in ATM networks. ATM has several key benefits that include:

- **One network:** ATM will provide a single network for all traffic types—voice, data, and video. ATM allows for the integration of networks, improving efficiency and manageability.
- **Enabling of new applications:** Due to its high speed and the integration of traffic types, ATM will enable the creation and expansion of new applications such as multimedia to the desktop.
- **Compatibility:** Because ATM is not based on a specific type of physical transport, it is compatible with currently deployed physical networks. ATM can be transported over twisted pair, coax, and fiber optics.
- **Incremental migration:** Efforts within the standards organizations and the ATM Forum continue to assure that embedded networks will be able to gain the benefits of ATM by incrementally upgrading portions of the network based on new application requirements and business needs.
- **Simplified network management:** ATM is evolving into a standard technology for local, campus/backbone, and public and private wide area services. This uniformity is intended to simplify network management by using the same technology for all levels of the network.
- **Long architectural lifetime:** The information systems and telecommunications industries are focusing and standardizing ATM. ATM has been designed from the onset to be scalable and flexible in geographic distance, number of users, and access and trunk bandwidths (as of today, the speeds range from megabits to gigabits).

Finally, users can already use frame relay as an access interface to ATM. In fact, many ATM switches already support frame relay as an interface. It is very probable that computer-based training and document imaging applications may require ATM. For users who are just now upgrading to 56 Kbps or fractional T-1 lines from 9.6 Kbps, T3 (and even T-1) access requirements are a long way off. You don't need a DS3 line to an office with ten people. At that office, even T-1 may still be too much. At these lower speeds, frame relay will be more efficient and cost-effective.

Frame relay is available today, is broadly implemented in hardware and service offerings, and is cost-effectively meeting existing user and applications needs. ATM networks will be available universally in the near future. Industry efforts are underway to make sure that frame relay and ATM interoperate, and important successes have already been achieved. Both technologies will continue to be available in the future, working in concert with each other and even with other technologies to support networking needs well into the 21st century.

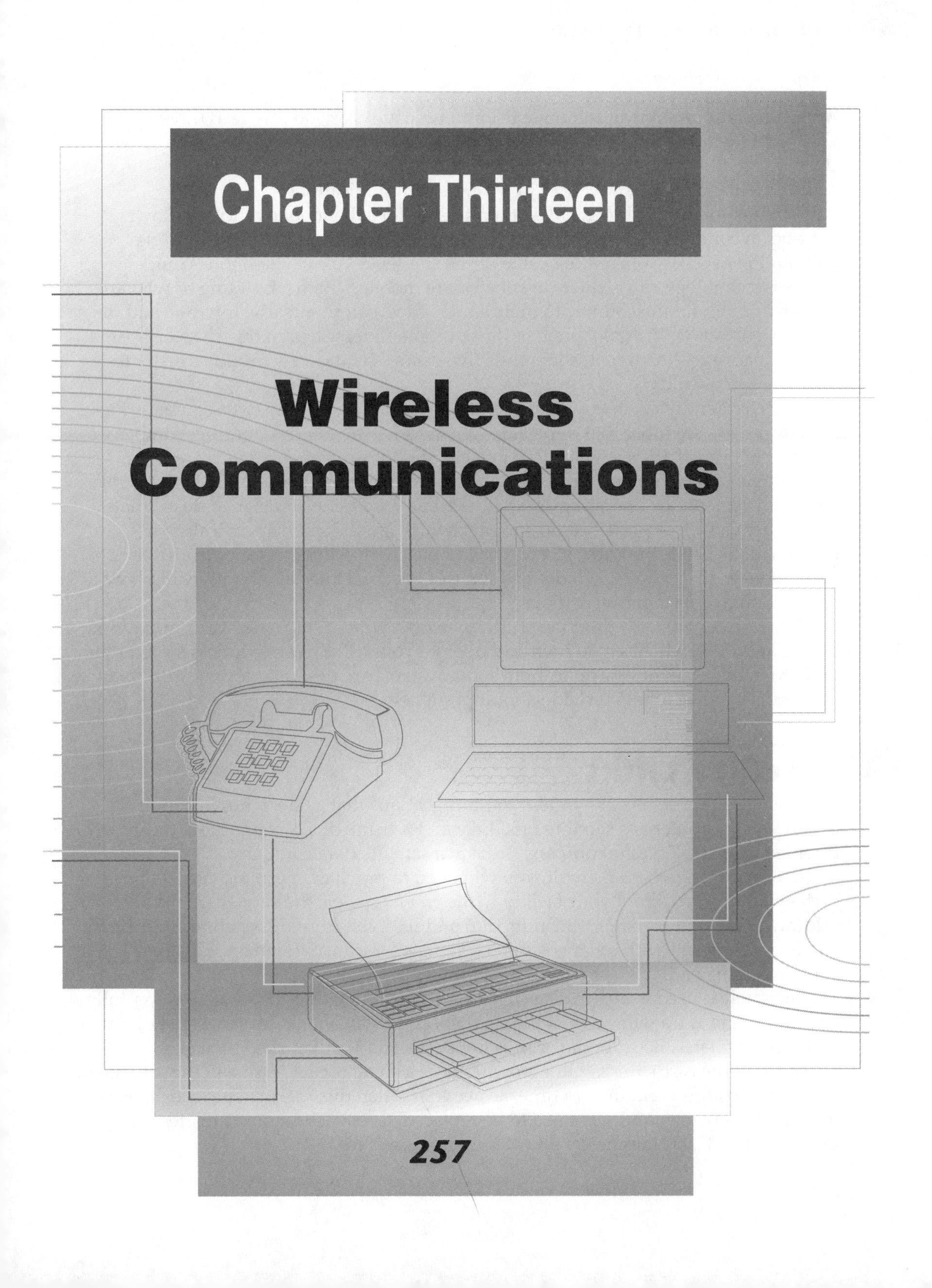

Chapter Thirteen

Wireless Communications

Portable computers and mobile telephones have changed low-tech traveling salesmen into high-tech "road warriors." Rather than needing to find a phone booth, the road warrior can be in contact with his office at all times. This link is made possible by recent advances in wireless communications.

Wireless communications have other uses besides the mobile office. Radio or light-beam communications can be used to make connections between buildings without the necessity of running cable or renting wires from a telephone company. Wireless technology can connect users who are mobile within a building or who are mobile throughout the country to their home office, computers, the Internet, and the list goes on and on. Virtually none of the equipment described in this chapter existed twenty years ago. Much of it is less than five years old, and new products are entering the market constantly.

This chapter divides the wireless communications world into two segments. *Public wireless systems* are those that depend on a network provider or operating company to provide an infrastructure, which you pay charges to use. These systems are comparable to the wired switched telephone lines or "private line" circuits that you rent from a telephone operating company. Public wireless systems include cellular telephones as well as packet radio systems, which are designed specifically for carrying data. While you may already be familiar with cellular phones from a user's point of view, this chapter will discuss some of the principles of operation and how to apply them to data communications.

This chapter will also describe *private wireless systems*, which, like your in-building wiring and PBX switches, you own and operate yourself. Private wireless systems are divided in this chapter by application: some systems are designed for point-to-point connections, while others provide a multipoint network.

Satellite Links

Perhaps the simplest use of wireless communications that a telecommunications expert will encounter is a satellite link. Like an FX trunk or T-1 connection, the circuit is rented from a telephone company or from a satellite data company.

To use a satellite-based circuit, you will need to purchase a ground-based line at each end of the link, from your facility to the *ground station*, the site where the antenna is located and your signal is transmitted to and received from the satellite. Your signal will be multiplexed with other customers' signals for transmission and demultiplexed at the receiving end.

The satellite used is in a *geosynchronous orbit*, meaning that the satellite is carefully positioned and is traveling at just the right speed to stay above the same spot on earth as the earth rotates. Because the orbit is not perfect and is disturbed by irregularities in the shape of the earth, the satellite's owner uses a small rocket engine in the satellite to correct its position from time to time, as needed. When this rocket's fuel is exhausted, the satellite will drift and will need to be replaced. Geosynchronous satellites are located about 35,000 kilometers above the earth's surface.

The wireless communications link between the ground station and the satellite uses a very high-frequency radio link, which travels in a straight line. Ground stations at both ends of your link must be able to aim at the satellite. You can see how this makes it impossible to have a single satellite link that connects opposite sides of the earth.

To the user of the circuit, there is very little difference between a circuit carried by cable and one carried by a satellite link. If the circuit is carrying voice traffic, there is a delay, which is noticeable when the other party begins to speak. The delay is caused by the long distance that the signal travels going up to the satellite and then down again. Even at the speed of light, the 70,000-kilometer round trip makes this delay noticeable. While users find the delay annoying, they are generally tolerant of it, especially when the link crosses an ocean.

If your satellite link is carrying data, the delay will be less of a concern. Some computer network software will need to be tuned to expect a slightly longer response time, but most will work without any adjustment on your part.

Cellular Telephones

Cellular telephones have become very common in the past few years. The widely deployed analog cellular telephone system is called *advanced mobile phone service* (AMPS). Since its introduction in 1983, the subscriber base for AMPS has grown to in excess of 8 million users in the U.S.

Cellular phones work by dividing the world, or at least the coverage area, into cells. Each cell is the geographic area served by a single base station, which contains a transceiver, antenna, and dedicated lines to a *mobile telephone switching office* (MTSO). Cell sizes range from about 20 miles across, in flat areas which have light traffic load, down to a few city blocks, in very densely trafficked areas with lots of buildings to block the radio waves. The MTSO manages the interconnection of the cellular network with the external wired telephone network. The MTSO also controls the base stations in the cells to manage the switching of calls between them, and it tracks which cell each mobile unit is in, to contact in case of an incoming call.

In the cellular system, adjoining cells use different radio frequencies to prevent interference between units. Often, the cells will be arranged in a hexagonal pattern with seven different groups of radio frequencies used. Figure 13-1 illustrates such a typical arrangement.

Obviously, the cell shapes aren't perfectly hexagonal. Rather, since the customer communicates with the base station with the strongest signal, the actual cell being used varies with radio *propagation*, the ability of the radio to travel through the intervening space or objects. Unlike AM broadcast radio, which can be received at long distances because it reflects from the atmosphere, cellular telephones use radio frequencies which travel by *line-of-sight*, in straight lines. Radio waves at these frequencies are limited by intervening buildings and hills. The operating company will vary the cell shapes and sizes according to several guidelines, including

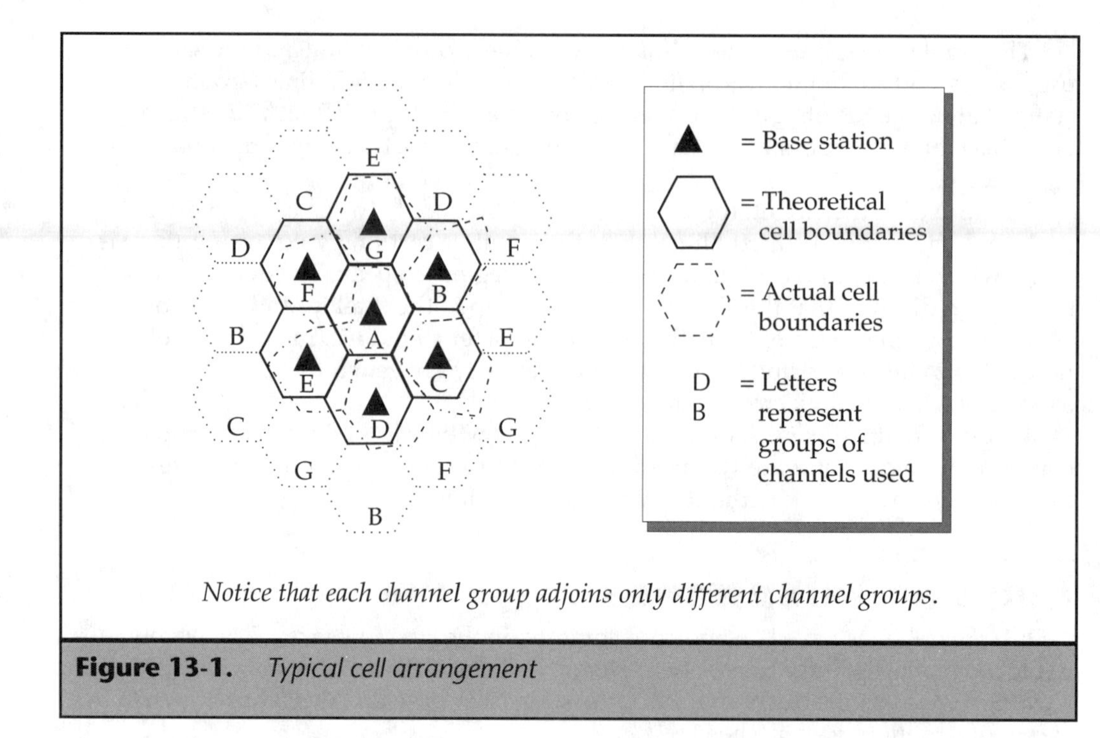

Figure 13-1. *Typical cell arrangement*

interference and the expected amount of traffic. The problem of finding a site for the base station can force the operating company to vary the cell shapes as well.

As you would expect, the cellular phone system works by passing responsibility for a particular customer from one cell to another as the customer moves. If a call is in progress, the MTSO connects it to a free channel on the new cell's base station and then directs the cell phone to *hand off* to the next cell by changing to a new channel. This happens automatically without action from the mobile customer.

AMPS Cellular Phones

AMPS cellular phones transmit on a frequency of 824 MHz to 849 MHz, and they receive the transmissions of base stations operating on 869 MHz to 894 MHz. A channel spacing of 30 KHz gives 832 full-duplex pairs of channels. These channels are divided into two groups called *A* and *B,* to allow two operating companies in each marketplace.

Within the A or B channel ranges are defined specific control channels. Each base station transmits control messages on a control channel. A newly powered-on mobile unit searches for the strongest control channel and communicates with that base. This process is called *registration.* As long as that mobile unit has power, it continues to report its presence to the base station and monitor the control channel. This *standby* operation does not consume much battery power, since the power-hungry transmitter in the mobile unit need not be turned on except for the very short digital control messages that maintain the unit's registration.

When there is a call for a mobile user, the base station sends a control message telling the phone to ring. If it is then answered, a reply is sent, and the base station connects the incoming call audio to one of its data channels and directs the mobile to connect to that channel. The mobile begins transmitting audio from the handset on the corresponding uplink channel, and the call is in progress.

When the mobile phone originates a call, the process is similar, but the dialing message is sent by the mobile to the base. The base then connects the call audio to a data channel and directs the mobile to connect to it. The mobile's transmitter remains on continuously for the duration of the call, transmitting the audio. Because the transmitter is on continuously, the *talk time* of a cellular phone is very much less than the standby time for the same battery. Typical talk times of one hour are obtained from the same unit that will operate for 12 or 15 hours on standby.

The base station constantly measures the signal strength of the mobile unit. If the mobile signal strength falls below a threshold, the MTSO will direct neighboring base stations to measure that same signal strength. If there is another base that can better receive the mobile, then the call will be handed off, with the mobile directed to change to a channel assigned to that base. The base inserts control messages directing the mobile to hand off into the audio channel. These are not passed to the microphone, but do cause momentary gaps in the audio.

Cellular phones are billed according to the call time. Both calls you originate and calls you receive are charged. Rates vary by area, but they typically range from 15¢ to 90¢ per minute during business hours. A monthly fee is also charged, typically starting at $15 to $30. Larger monthly fees charged for "calling plans" include a specified number of minutes, and the plan usually includes a better price for additional minutes. Minutes during nonbusiness hours are much cheaper, even free under some calling plans or new customer promotions.

Because the mobile unit must identify itself to the base station, it is possible for thieves to intercept that identification and modify their phone to have the same identification. There are two identifying numbers involved: the cellular phone number, and the *electronic serial number* (ESN). Equipment manufacturers are required to make the ESN difficult to change, requiring special test equipment or passwords, which are not given to the user. Unfortunately, this has not deterred thieves, who have reverse-engineered the manufacturers' safeguards. By monitoring traffic at a busy cell site, such as a freeway junction, thieves have found it fairly easy to get numbers to "clone." The operating companies attempt to detect this fraud by programming the MTSO to observe when two mobile units are operating simultaneously with the same ESN.

The call audio is transmitted by both the base station and mobile unit as FM radio, which can be received by widely available receivers. Although there are laws prohibiting the interception of cellular phone calls, it is unwise to believe that a cellular phone call is private. News reports involving the private calls of public figures have made this clear to many, but our habitual assumption that a telephone call is private is ingrained.

Digital Cellular Phones

Two standards for cellular phone systems, using digital communications instead of analog voice, are currently being installed. These systems use the same channel assignments as the AMPS system, and they are designed to coexist with it. Advantages of digital cellular include the following:

- More calls per cell, that is, more calls in the same number of channels, by time division multiplexing of multiple calls onto a single channel.
- Calls are free of radio interference, because the digitized audio is re-created at the receiver.
- Amateur interception is reduced, because the digitized signal is more complex to decode.

Because of the large expense of upgrading each cell's base station equipment, operating companies are only deploying the digital cellular equipment in cells where the traffic volume is high. The mobile units sold feature dual modes of operation, so they can operate in a fully analog cell or work on a digitally modulated channel.

Personal Communications Service

The federal government is currently auctioning licenses to use an additional band of frequencies for mobile communications. These frequencies, designated for *personal communications service* (PCS), are likely to compete with the current cellular system. The name PCS has been chosen because of the assumption that these frequencies will be used to provide lower-cost products that would be more applicable for personal use than the present cellular frequencies. You have undoubtedly noticed that both business and personal users are using the existing cellular network, and the same will occur with the new PCS services as they are introduced.

The new PCS frequencies are at about 1800 MHz. At this frequency, the radio waves will not travel as well as the 900 MHz of the present cellular system. To make use of these frequencies, the operating companies will need to install more base stations using smaller cells. While this will make the infrastructure more expensive to install, it will allow the individual mobile units to be lower power, which translates into lower cost.

There are several competing proposals for the protocol to be used by the PCS mobile units. It is likely, however, that the PCS network will use digital encoding. The expected deployment for PCS is not before 1998.

Data Communications: Connecting a Modem

As with a conventional, or *landline*, telephone, you will need to use a modem to do data communications over a cellular phone link. Even the digital cellular systems present an audio interface to the caller, so a modem must be used. Because the cellular phone can be switched between base stations while the call is in progress, momentarily interrupting the voice path, you should make certain that you select modems with error detection and correction features, as described in Chapter 15. Line quality on a cellular data call is very dependent on conditions, so you will often find that the modem falls back to lower data rates than the maximum. If you normally experience some line noise on your cellular calls, you may find that the fastest modems, using V.34 protocol at 28,800 baud, do not achieve full performance. Most cellular users continue to use V.32bis modems at this time, believing that the additional cost for the V.34 modem is not justified, since the modem generally falls back to V.32bis or lower speeds anyway.

As a cellular modem user, you have an additional problem when you attempt to attach a conventional modem to your cellular phone. Because the cellular phone sends the dialing information to the cellular base station in a control message rather than as dialing tones, the tone dialing unit included in a conventional modem is not usable for dialing the cellular phone. Worse, cellular phone manufacturers have not standardized the interface between a modem and the cellular phone, if they provide one at all. Generally, you will need to order a package that includes a specific cable and downloadable software for your modem and cellular phone. These packages are generally available for modems designed for laptop computers, especially modems in PCMCIA format (also known as PCCard, a trademark of the PCMCIA, which stands for Personal Computer Memory Card International Association), the standard interface for adding small peripheral cards to portable computers.

CAUTION: Before buying a cellular phone and a modem for use on the road, make sure that they can be connected. Some major cellular telephone manufacturers' units are only compatible with modems manufactured by the same company. For example, Motorola cellular phones, which are compact, popular, and widely available, only work with Motorola modem cards.

Because each cellular phone must have a unique ESN, a combination cell-phone/modem would be a separate, special-purpose phone with a separate registration cost. At this point, manufacturers have assumed that it is more economical to carry a single, general-purpose cellular phone and attach the computer modem to it when desired.

Air-to-Ground Telephones

Public air-to-ground telephones have been installed in commercial airliners in North America for about the last ten years. At this point, there are three licensed operators for these systems: GTE AirPhone, AirOne, and AT&T Wireless (formerly Claircom). Which provider you use depends on the aircraft you are flying, since the airline company will make an agreement with a provider to install the air-to-ground link. Each company operates a network that is very similar to the ground-based cellular system, but it uses very large cells because there is good line-of-sight to an aircraft for a long distance.

While there may be a handset installed for every row of seats, typical installations have from two to eight voice channels from aircraft to ground. The equipment carried on the plane arbitrates access to the voice channels, giving a "Please wait" message if all the channels are busy. The rates charged, from two to five dollars per minute, generally keep contention from being a problem. Callers arrange payment by swiping a credit card through a reader on the handset before making any calls.

Until recently, calling was originated in the air only, and the aerial handset did not have an identifying number. Recently, one provider, AirPhone, introduced a system that allows regular users to obtain a credit card from them, which can be used to inform their network at which station a customer was sitting; a small fee is charged for this (about one dollar). Once registered with the network by swiping this card, the user can receive pager-style messages on a display located on the handset or even receive return calls. This registration card also acts as a charge card for making calls, and the user earns a quantity discount for monthly usage.

The signal quality of the air-to-ground links can't match that of a conventional telephone or even a cellular telephone. Early units, which used analog communications between the air and ground stations, were prone to static-like interruptions and fading. Later units encoded the voice channel digitally to reduce noise, but, unfortunately, also reduced the bandwidth per channel to allow more channels in the same frequency space. These encoding schemes are proprietary to the providers, and you will find that some providers have substantially better signal quality than others.

CAUTION: FCC regulations do not permit you to use conventional cellular telephones from the air, even from a private plane. The reason for this rule is that the frequencies used by cellular phones travel by line-of-sight. In the air, your phone would interfere with communications in several cells that would be blocked from you if you were closer to the ground.

Some newer air-to-ground systems provide a jack, which is intended to connect modems in travelers' laptop computers. These jacks attempt to mimic the RJ11 jack used by most standard telephones. To use your modem with one of these jacks, you will need to follow the operating instructions, including swiping your credit card, and then your modem will be able to dial. Hardware in the onboard electronics translates the dialing tones into the internal messages used by the air-to-ground system so that your call can be completed.

You will want to use a modem that includes error correction and compression if you must make a data call from the air. Because the links are very noisy, actual performance will probably be only 2400 baud. In particular, AirPhone's newer digital units (which can be distinguished by the display on the back of the handset) have very poor link quality. Since the call will cost several dollars per minute, you will want to keep the transfer short. Also, don't forget to allow for the rule that you cannot use electrical devices for several minutes after takeoff and several minutes before landing, or you could be forced to shut down mid-transfer.

Cellular Packet Data

While the cellular telephone network is designed to handle voice calls and can support data only using modems, there are also cellular systems that are designed specifically for data. Two of these systems, Ardis and Mobitex, operate on their own frequencies, while the third, Cellular Digital Packet Data (CDPD), coexists on the same channels used by conventional cellular phones. (CDPD was developed by a consortium of cellular operators, and it is not really operating yet. The tradeoffs between the other providers and CDPD are discussed later.) Table 13-1 summarizes the providers of the three systems as well as some of the system characteristics.

Network Operation

These systems are called *packet data* systems because, rather than establishing a call that remains connected, they transport data between users in units of a few hundred bytes called *packets*. Each packet can be a complete transaction between two users, or the users can transfer a larger amount of data by sending multiple packets. The mobile

System	North American Providers	Frequency	Data Rate	Coverage Claimed
Mobitex	RAM Mobile Data Cantel Mobile Data	896-902 MHz Transmit 935-941 MHz Receive	8,000 baud	90 percent of metropolitan areas
Ardis	Ardis Corp.	806-824 MHz Transmit 851-869 MHz Receive	19,200 baud	60 percent
CDPD	Cellular operators	824-849 MHz Transmit 869-894 MHz Receive (same as AMPS cellular)	19,200 baud	Deploying now

Table 13-1. *Cellular Data Network Systems*

user carries a unit that combines the modem, packet message protocol logic, and radio transceiver functions in a single unit. This unit is called the *packet radio modem*.

It is important to understand that the data rate at which the system communicates (quoted in Table 13-1) is not the rate at which user data is transferred. There is substantial overhead in the communications channel of the packet data system. Also, each channel is shared among multiple users. As in cellular systems, the base station directs the mobile unit as to which channel it should tune to. Packet modems that have a packet to send to the base station contend for the channel and then retry if a *collision* (two or more users transmitting at the same time) causes the messages to be garbled. A checksum is used to validate the packet. The base station acknowledges each valid packet, so the packet modem can retry if no acknowledgment is received.

Even on an unloaded network (a network with no other users), a packet modem that is fed continuous messages to transmit can only use about one-third of the channel. The remaining time is used for acknowledgments and free time to allow for the possibility of some other station transmission. You can see that the maximum throughput of a packet modem depends on the number of channels available to the local base station, as well as the amount of local traffic in competition for the network.

Cellular packet data networks can transfer packets between remote users. Host computers can also be connected and given network addresses. RAM, Mobile Data's Mobitex network, for example, supports host computer connections using X.25 or HDLC communications on leased or dial-in lines. For applications in which small amounts of data are transferred, a radio connection to your host computer can be just as convenient.

Because the packet data networks are designed specifically for data, they offer advantages that cannot be provided by a nondata network. The network can accept a packet from a sender and hold it until the destination is ready to receive it. This feature is called *store-and-forward*, and it can work efficiently because the packet size is limited. The Mobitex network provides the sender the choice of requiring an immediate delivery or storing the packet for delivery. In that system, the sender gets an acknowledgment from the network that includes delivery status. If the packet could not be delivered, and the sender did not allow store-and-forward operation, then the packet is returned to the user with the failure indication. Using store-and-forward can make the mobile user unaware of gaps in the coverage area, because packets that are stored are delivered as soon as the mobile unit reenters the coverage area.

Pricing on these networks is usage-based. Because cellular packet data is not being widely used yet, the costs per packet are still fairly high, on the order of one dollar per kilobyte of data transferred. Successful applications will take advantage of the following characteristics of the packet network:

- Small amount of data to be transmitted
- Guarantee of delivery before acknowledgment is returned
- Store-and-forward is available

Applications

The cellular packet data providers are concentrating their marketing efforts in two areas: large accounts with large numbers of mobile users (for example, package delivery tracking) and value-added resellers who will develop custom software and find buyers for it. RAM makes available a list of vendors for such systems, including a company that will deliver stock quotes to your Mobitex terminal. Both Mobitex and Ardis are supported by Radiomail, Inc., which sells an electronic mail delivery service, including connection to the Internet.

Because the packet network guarantees delivery of each packet to the mobile unit, it can be used as a reliable, two-way replacement for paging. UPS Canada uses such a system to dispatch their drivers, using the Motorola InfoTAC radio modem. Because the driver can send a response to the dispatcher, the dispatcher can know that the driver has the message and that the driver can act on it. This replaces a system using one-way paging, which required the drivers to find a pay phone and call in to confirm receipt of the page.

RAM Mobile Data has developed a prototype tracking system for vehicles as an example of what might be used for long-haul truck tracking. Their mobile unit consists of a PC laptop, an Ericsson Mobidem radio modem, a PC *global positioning system* (GPS) receiver, and custom software. The demonstration host is a Sun workstation with a graphic mapping database, also linked to the Mobitex network by a Mobidem radio. This demonstration allows the dispatcher to track the location of vehicles and exchange text messages with the drivers. A database keeps the location history of each vehicle so that its path can be examined.

Hardware

The interface hardware between the computer and the packet radio network is the packet radio modem. Unlike those modems used in conventional cellular, the unit includes the radio transceiver. Generally, the interface to the computer is a serial port, although cards intended for PCMCIA slots are just starting to appear. Manufacturers are having difficulty fitting the radio and associated electronics into the small PCMCIA form factor and are using "extended" cards, which stick out of the PCMCIA slot while in use. This can make it difficult to leave the card in your laptop all the time, as is often done with conventional PCMCIA modems. To provide enough power to run the radio transmitter, batteries are included in some PCMCIA designs, making the card extension bulge out even more. Because of these size problems, IBM announced modems for either Mobitex or Ardis that are designed to replace the floppy disk drive on some models of IBM notebooks.

The radio modems generally use a proprietary command set. Some models offer a variation on the AT command set used in conventional modems, but these variants are still loaded with nonstandard commands. Table 13-2 lists some available radio modems.

Unit	Manufacturer	Network	Format	Approximate price
InfoTAC	Motorola	Ardis or Mobitex	2×3×6 inches with pager-like display	$900
Mobidem M2100 "PIA"	Ericsson	Mobitex	PCMCIA type 3	$700
Mobidem	Ericsson	Mobitex	2×3×10-inch brick	$500

Table 13-2. *Typical Cellular Packet Data Radio Modems*

With Ardis and Mobitex, you will need to develop or purchase application software that supports the proprietary network interface required. The network providers have developed lists of application providers that use their network. These ready-made applications include electronic mail, stock market data delivery, credit card validation/billing, and vehicle location. The electronic mail applications support e-mail to and from the Internet and can show messages on the display of the Motorola InfoTAC unit even if no computer is attached to it at the time.

If you wish to develop custom software, the network providers can supply information about library subroutines that implement the network interface, so the user need not be concerned with the details. If you are developing software for sale to others, each of the network providers is anxious to recruit you for their developer's program.

A major advantage of packet data systems over conventional cellular is that the radio transmitter need operate only when there is actually a data packet or control message to be sent, rather than continuously as in analog cellular. This can increase the battery life of a portable unit substantially, compared to cellular telephones. You can expect about one day of operation from the battery in current packet data modems, unless you are transferring large amounts of data. If you were to transfer data continuously, you could expect about one hour of operation, similar to the performance you get with a cellular phone.

CDPD

CDPD (Cellular Digital Packet Data) is designed to share the frequencies used by cellular voice services, and it will be operated by the cellular operating companies. While this provides a funding base for wide deployment, it means that there is competition for radio channels, making data transmission difficult in congested areas, such as freeway intersections. Nationwide use may also be complex, because the local providers will need to provide roaming arrangements similar to their arrangements for their voice customers. At this time, roaming issues for CDPD have not been resolved.

The developers of CDPD have attempted to provide an interface that is more compatible with existing network software. The interface they have chosen is the IP protocol as used on the Internet, specifically SLIP (serial line Internet protocol), which is already being widely used by dial-in Internet users. This should make application development less difficult and allow existing applications to be used.

Unfortunately, pricing is still vague. The proposed pricing so far is based on packets transferred, similar to the competing packet data networks. Many existing IP applications you might wish to take on the road, including Internet mail and file transfer, appear to be more cost-effective when connecting by AMPS cellular phone and a modem than by CDPD. This will hinder acceptance.

Public Wireless Comparison

Table 13-3 lists advantages and disadvantages of various public wireless services, especially with regard to data communications. When making your own choices, be careful to request current pricing, because these prices are changing rapidly. Some providers who are just entering the marketplace are willing to negotiate price to secure demonstration applications.

Trunked Mobile Radio

An older voice radio communications technology that is still being used is *trunked mobile radio*. A radio system is called *trunked* when it has a mechanism to divide users into groups and cause messages only to be received by members of the correct group. For example, if a trunked radio system is used for police car dispatch, the dispatcher might direct one message to police cars assigned to just one district and another message to all cars everywhere. Transmissions like "Calling all cars!" or "Attention district two" do not have to be made. Trunked radio systems are controlled by a base station, which has a strong transmitter and high receiving antenna. Similar to a hubbed local network, the base station repeats the transmission of each mobile unit.

A trunked system can be privately operated, or the owner of the system may attempt to recover some of the costs by sharing the system with others. Sometimes police and utility companies will make some arrangement. There are also companies that will operate a trunked radio system as a business and make money by leasing access. The trunking feature, separating the calls so that only the intended groups receive them, keeps different users' messages separate.

Trunked mobile radio is primarily used for police and taxi dispatch, where its inherent broadcast capabilities are useful. Modern trunked radio systems use powerful radio transmitters, which require an operating license. Typical systems are capable of covering a metropolitan area and support up to about 250 different user groups having about ten

Service	Advantages	Disadvantages
AMPS Cellular	Cheap hardware Fairly low cost per minute Compatible with existing telephone modems	Low battery life when transmitting Required complex roaming arrangements when out of a given provider's area Subject to intercept and fraud
Digital Cellular	Claims better talk quality than AMPS cellular More difficult to intercept	Less available (but fallback to AMPS) Compatible modems less available Roaming still an issue for nationwide use
Packet Data (Mobitex and Ardis)	Long battery life in normal use Private networks reduce competition for network bandwidth Single provider covers USA	Small messages best Custom equipment and protocols
CDPD	Wider coverage expected Competition between providers *should* mean lower pricing eventually	Not really ready yet Roaming issues are not resolved for nationwide use Prices not lower yet

Table 13-3. *Public Network Data Communications Choices*

channels used. The number of channels limits the number of messages that can be transmitted, so messages must be kept short to keep channels free for others.

Trunked radio systems are capable of being connected to a telephone line, either directly to the public switched network or to a PBX. This feature cannot be heavily used, since it requires a channel for the entire time the call is in progress. The connection to the wired network is located at the base station. The mobile units that use this feature must be equipped with an optional tone-dialing keypad so that they can dial the call.

Most trunked mobile radios are designed for installation under the dash of a vehicle, and they draw power from the vehicle. A few hand-held units are available, in the form of large "walkie-talkies," because of the large batteries required to support a powerful transmitter. These mobile units cost from $400 to $1,000, depending on the number of channels supported and the transmitter power. Base station units cost from $5,000 to $10,000 per channel supported.

While trunked mobile radio systems are primarily used for voice communications, there are add-on devices to adapt these radios for data communications. Like public cellular packet data modems, these trunking data modems move data using packets of a few hundred bytes. The major application these systems are used for is delivery of license information to police cars. Like packet data modems, the protocol used is nonstandard, and custom software will be required to be written or purchased. Data rates of about 1,200 or 2,400 baud are typical of these units.

You are unlikely to want to use trunked mobile radio for data communications unless its voice mode of operation fits your business needs. The low volume of production for these units makes it difficult for them to compete with more popular, public systems.

Private Wireless Equipment

Most of the wireless communications described so far in this chapter have been between a network provider and you, the customer. The provider charges you for use of its infrastructure. It is much easier for you to operate wireless equipment in such a way that both ends of a connection are under your control, without any need to pay by the call or by the packet.

In the past, microwave links sold for voice or data communications operated at frequencies that required a license. While this reserved a frequency for you, it usually meant your license was tied to a specific location. The equipment vendor helped you apply for the license, but it was still additional paperwork.

Rather than use licensed frequencies, newer microwave communications devices use frequencies reserved for industrial, scientific, and medical equipment, known as the *ISM band*. There are ISM bands reserved at 902 MHz to 928 MHz, 2.400 GHz to 2.483 GHz, and 5.750 GHz to 5.825 GHz. Users of the ISM band under an FCC classification called Part 15 must meet power limitations and be tolerant of interference from other users of the same frequencies. One of the ways to keep power low and avoid interference is to use *spread spectrum* techniques, which involve moving rapidly between many different frequencies in the band. By switching pseudo randomly, a spread spectrum system will only interfere with another system when the two happen to land on the same frequency at the same time. It also allows them to avoid the powerful interference from equipment, such as microwave ovens and X-ray machines, which usually falls at one frequency for longer periods. By using some error correction and redundancy, an error-free path can be found through the crowded radio environment.

Optical communications, using light waves, is another wireless mechanism for communications. Lasers need only meet power limitations for eye safety. Other light beams, including infrared light, are not regulated at all. While subject to atmospheric interference, especially outdoors, optical communications can work very well. Ranges of up to one kilometer are cited by sellers of interbuilding infrared laser communications systems operating at 16 Mbps.

Microwave radio is only occasionally subject to weather interference. Ice forming on antennas can reduce signal strength and increase the number of errors in a wireless link. Neither microwave nor optical communications systems will operate with large physical objects between the two endpoints. This means that your signal path cannot be at ground level where it will be blocked by cars or pedestrians.

Point-to-Point Links

The easiest kind of wireless communications to imagine is a simple point-to-point link, replacing a wired line. Such links replace a voice telephone line, a serial data line, ethernet, or a T-1 line. Figure 13-2 shows a wireless ethernet bridge being used to connect local networks in two buildings. With directional antennas, the same kind of bridge can operate at distances of 15 or more miles, if you have the line-of-sight path that either a microwave or optical system requires.

The cost of wireless links of this type varies from under $3,000 to about $10,000, depending upon performance. Network bridges, with ethernet or token ring interfaces, are typically in the $5,000 to $8,000 range. You need to double this price, of course, as the link requires a unit for each end. Antennas and cables need to be selected appropriately for the building environment. If you are going a mile or more, you will usually want the equipment vendor to install the hardware, since link performance can depend on aiming if directional antennas are used. Shorter distance systems can be aimed by eye or by observing a signal strength indicator while moving the antenna.

Obviously, wireless is not cheaper than wire, for most applications. You will want to use a point-to-point wireless link when it is not possible to use a wired link. Examples of this include interbuilding or cross-town links, as well as applications where the equipment setup is temporary.

Table 13-4 shows some typical products and their performance specifications.

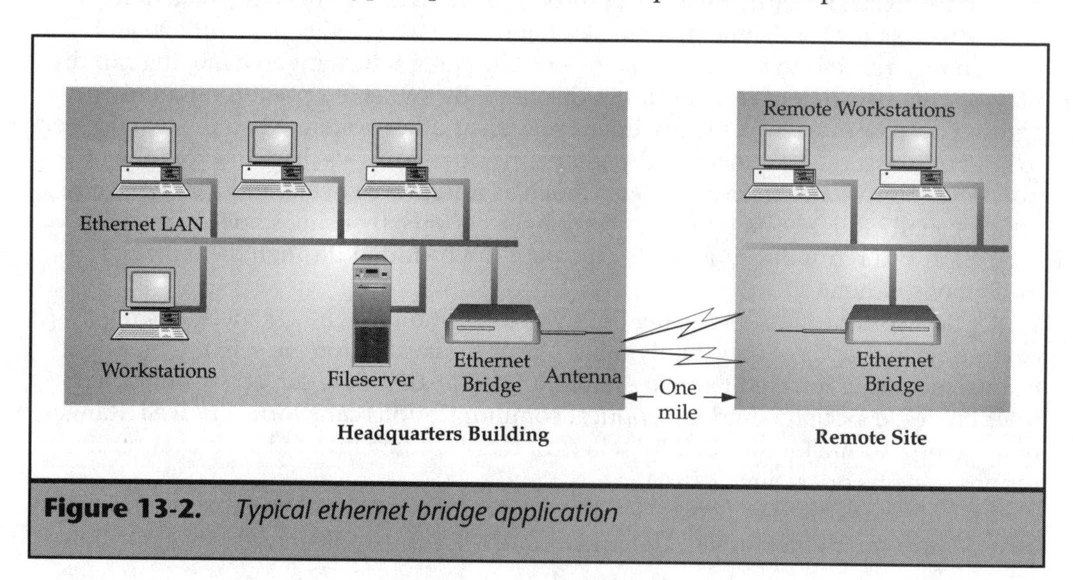

Figure 13-2. *Typical ethernet bridge application*

Manufacturer	Product	Interface	Technology Used	Data Rate	Distance
Laser Communications Inc.	LACE (Laser Atmospheric Communi-cations Equipment)	Ethernet or token ring bridge	Infrared laser	Up to 16 Mbps	1 kilometer with line-of-sight
Cylink Corp.	Airlink	Voice, RS-232, RS-422, V.35	Spread spectrum (ISM band)	64 Kbps asyn-chronous, up to 256 Kbps syn-chronous	Several miles, up to 15 or more with appropriate antenna
Persoft, Inc.	Intersect Remote Bridge	Ethernet or token ring bridge	Spread spectrum (ISM band)	2 Mbps	800 feet with omnidirectional antenna, 3 miles with directional antenna
Microwave Radio Corporation	Various	Digital line	22GHz Microwave (license required)	Up to 8.4 Mbps	Up to 10 miles
Motorola (Wireless Data Group)	Altair VistaPoint, VistaPoint LR	Ethernet link	18 GHZ Microwave (Motorola holds license and grants use to user)	5.3 Mbps	VistaPoint: 500 feet VistaPoint LR: 4,000 feet

Table 13-4. *Typical Wireless Link Product Specifications*

Multipoint connections can also be done without using wires. While some network bridge units can operate in a multipoint mode, debugging a multipoint system between several buildings or farther can be very difficult. It is much easier to find and fix problems if the long links can be diagnosed one at a time. For this reason, most longer-distance wireless links are used as point-to-point links.

Some factory data terminals are starting to be available with integrated wireless links. Currently, there are a few specialized units that incorporate an RS-232-equivalent link using either proprietary licensed or spread spectrum (ISM band) radio. These units will become widely available in one to two years.

Wireless LANs

There are shorter-distance systems that can implement a wireless network, and these can be of use to you. While wires are still cheaper, wireless in-building networks can be useful when the installation is temporary or changes very rapidly, or when wires cannot be used because of building limitations (for example, in a historic building). Wireless promoters have advertised the benefits of cable replacement, but this has not been demonstrated in real-world use.

Wireless LANs available today are generally suitable for a room or a few rooms. Most of the available systems use the 900 MHz ISM band, which will pass through a few of the concrete block walls found in most office buildings, but does not pass well through steel-reinforced floors. In large office areas with metal partitions, the radio waves will reflect off the partitions, limiting the distance they will travel.

Data rates on current wireless networks are lower than those available on wired networks. Typical wireless systems achieve 1 Mbps to 2 Mbps.

Wireless LANs either can use a *hub,* a special node that controls the operation of the network, or can be *peer-to-peer*, with all nodes communicating equally. While the hub provides the disadvantage of a single point of failure for the network, the hubbed design is often used because it allows the designer to reduce costs in the other nodes. In the wireless network, the hub must be centrally located, so that every other node has a good path to it. Locating the file server or backbone connection at the hub will make the network more efficient, since traffic between two other nodes will need to pass through the hub. In some wireless networks, the hub has an antenna that can be mounted high up so a clear path between the other nodes is easier to achieve. In Figure 13-3, the central hub provides a bridge between nodes at two ends of the room.

If the network uses a peer-to-peer design, then every node must be able to communicate to every other node. This can limit the size of the network compared to a hubbed system. Figure 13-4 shows a hubless network that will have problems because the two distant nodes cannot communicate directly. Some hubless networks avoid failure in this case by allowing intermediate nodes to retransmit the data.

A compromise design allows the network to dynamically select a hub node. This requires less design work to install, but the best performance will still locate the file server or other heavily used connection near the center, to increase the likelihood of the file server being selected as the hub.

Typical wireless LANs replace wired LANs in a way that is transparent to computer software. For PC networks, drivers are supplied for interface cards to support network operating systems such as NetWare, NDIS, or Windows for Workgroups. Table 13-5 shows some typical products and their performance.

Wireless networking hardware is substantially more expensive than wired equipment. Typical adapter cards cost from $400 to $2,000. Hub-based networks usually have cheaper adapter cards, but the hub will cost at least $1,000. You will find that wireless LANs cannot be justified by eliminating the cost of installing cables, as some early advocates claimed would be the case.

If you have an application for a wireless LAN, be sure to be conservative in configuring the network. Distances supported are often quoted under the best possible conditions. While you can gain substantial performance by raising antennas above cubicle walls and checking sight lines with improvised tools such as a laser pointer, you will get the best results if you derate the specifications.

Future Trends

You can expect to see new wireless products due to two driving factors: advances in miniaturization and availability of new frequencies for use.

The trend toward smaller portable electronic devices can be seen in the race to produce smaller cellular telephones. The same advances in technology are helping to reduce the sizes of packet data modems from brick-sized units to PCMCIA card size. Improvements in battery chemistry, which allow the same power output from a smaller volume, will help to reduce the size of hand-held and portable units.

Improvements in microwave technology are making it practical to use higher frequencies, which make more channels available for new applications. The newly allocated PCS frequencies are an example of this growth. Future allocations are likely to be determined by demand, as the governmental rule-making bodies (at least the FCC in the U.S.) consider the amount of current use when allocating new frequencies.

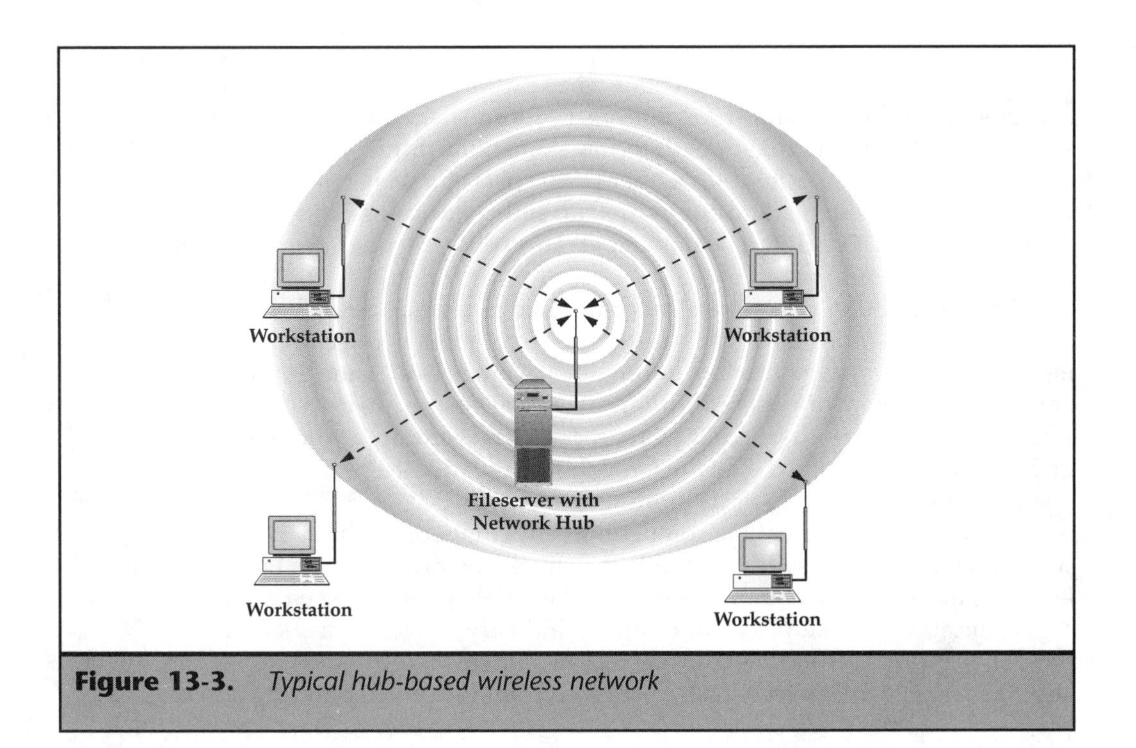

Figure 13-3. *Typical hub-based wireless network*

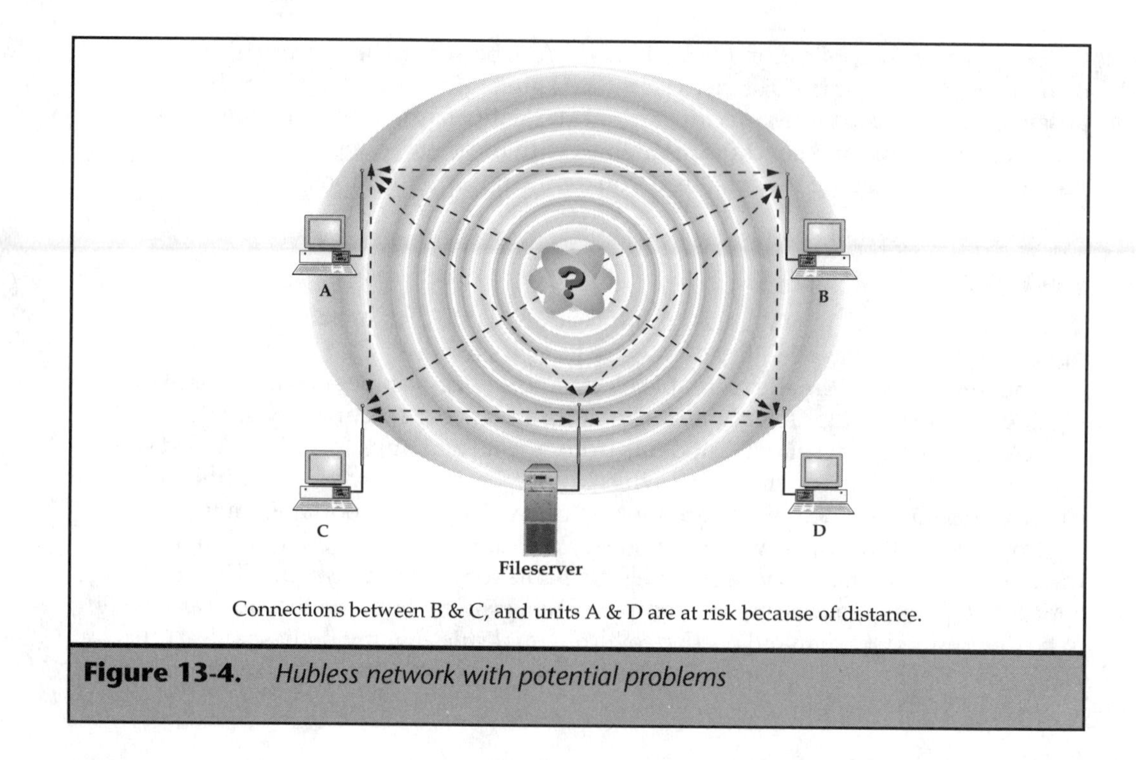

Figure 13-4. *Hubless network with potential problems*

Equipment manufacturers will offer wireless communications of various sorts, particularly for factory data and point-of-sale terminals. While existing units use a separate point-to-point link for each terminal, manufacturers are now developing units that will be compatible with either public packet data networks or private wireless LANs. These manufacturers are currently experimenting with PCMCIA slots in their terminals. The differences in the various software interfaces will be the factor that holds this development back.

The costs of using public wireless communications facilities will go down only when use increases. To provide a wireless service, whether voice or data, takes the installation of enough equipment to provide coverage over some geographic area. It is not possible for network operators to recover their costs of startup unless there is sufficient usage. Even then, additional user demands require additional cells so that a greater load can be carried. Only when volumes grow quite large can the network operator tolerate lower charges per minute or packet.

When there is incompatible competition, the competitors that do not achieve sufficient market share will not be able to compete on price and can be expected to leave the market. This kind of shakeout is likely to happen in the packet data communications marketplace, for example. Industry watchers currently expect that the CDPD protocol will eventually become dominant, because those operators will be able to subsidize their packet data operations from the profits of their voice cellular

Manufacturer	AT&T Global Information Systems (originally developed by NCR)	AIRONET Wireless Communications	Black Box Corporation	INFRALAN Technologies, Inc.	Proxim, Inc.	Proxim, Inc.
Product	WaveLAN	ARLAN	BestLAN2 Hub	INFRALAN	RangeLAN1	RangeLAN2
Hardware Interface	ISA, MCA, and PCMCIA cards	ISA, MCA, and PCMCIA cards; direct connection to ethernet	ISA, MCA, and PCMCIA cards; direct connection to ethernet	Adapters plug into standard IEEE 802.5 token ring cards, plus hub unit	ISA, MCA, and PCMCIA cards; parallel port adapter	ISA and PCMCIA cards; direct connection to ethernet
Network Compatibility	All major networks	Ethernet or token ring-equivalent	NetWare, LANtastic, Windows for Workgroups	Anything supported by the token ring card; all major networks	LAN manager, NetWare, NDIS, Windows for Workgroups; not TCP/IP	All major networks; includes routing software for nodes moving between networks
Technology Used	Spread spectrum (900 MHz ISM band); peer-to-peer	Spread spectrum (900 MHz ISM band); peer-to-peer	Spread spectrum (900 MHz ISMband); hub-based	Infrared light	Spread spectrum (900 MHz ISM band); peer-to-peer	Spread spectrum (2.4 GHz ISM band); peer-to-peer
Data Rate	2 Mbps	Up to 1.35 Mbps	2 Mbps	Up to 16 Mbps (if supported by token ring card)	242 Kbps	1.6 Mbps
Distance	Up to 800 feet indoors	3000 feet indoors	Up to 800 feet indoors	80 feet	Up to 500 feet indoors	300 to 500 feet indoors, up to 800 feet outdoors

Table 13-5. *Typical Wireless LAN Product Specifications*

businesses. If, however, the CDPD network operators find themselves unable to move funding from the voice cellular businesses—perhaps because there is too much competition from similar services in the new PCS frequencies—then CDPD is likely to fail, because it is last into the marketplace.

The wireless market has seen grandiose promises made for new products. This trend is unlikely to change. The best thing about wireless telephony is that it eventually delivers on its promises.

Conclusion

Wireless communications is not a universal solution. There are specific problems that wireless can solve, and it does solve them very well. You will find that the challenge of using wireless technology is usually in making the appropriate cost decisions rather than in making technical capability evaluations. The wireless marketplace is very active, and pricing changes rapidly. To fully understand how wireless communications take place, you would probably need to read thousands of pages of technical information and then completely digest it. Instead, you might be better off understanding the different wireless technologies and keeping on top of the operations and cost to determine where wireless telephony fits into your life.

Chapter Fourteen

Voice Processing Systems

Voice processing is a generic term that covers several industries in the telephony marketplace today. Voice processing, sometimes referred to as *call processing*, could be defined as any application that combines human or computerized speech and computer technology in order to automate a procedure or provide a service. You probably want to know what voice processing can do for you in the real world. In this chapter, you will learn about some of the more popular forms of voice processing and hopefully how you might use voice processing in your telephony environment.

You will find that you are quite familiar with a multitude of voice processing applications that you deal with on a regular basis. Here is a list of the more popular forms of voice processing:

- Voice-mail
- Audiotex
- Interactive voice-response (IVR) or voice response unit (VRU)
- Integrated voice and fax (fax-on-demand)
- Operator services
- Telemarketing
- Text-to-speech
- Speech recognition
- Multi-application platform (MAP)

One final component of voice processing that should be addressed is voice prompt editing. All of the forms of voice processing listed also require voice prompt editing to develop the application. Make sure you read the section on voice prompt editing later in this chapter so you understand the function of this vital tool.

In the current market, it is very difficult to distinguish between the different platforms of voice processing. In fact, the various voice processing applications have significant overlaps with one another and with other areas of telephony and computing, as depicted in Figure 14-1. In this chapter, you will learn basically how each of the major voice processing applications works and how they are combined to provide a multiapplication platform (MAP).

MAPs are voice processing applications that combine different voice processing technologies into one voice processing platform. A MAP always begins with a voice processing board and some computing device. From there, the voice processing system can be enhanced to include switching capability, fax capability, text-to-speech, and speech recognition. The application itself can be designed to incorporate any or all of the voice processing functions listed above and detailed below.

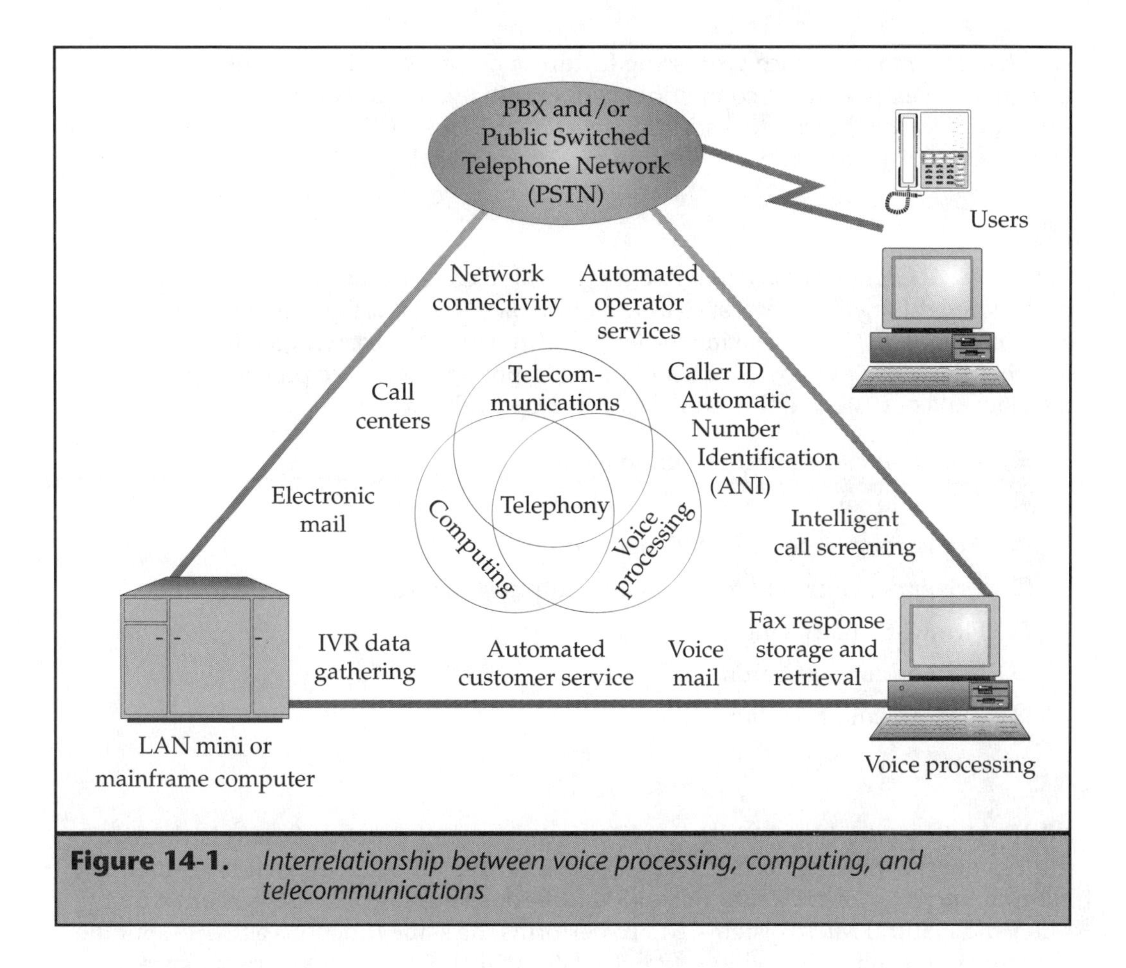

Figure 14-1. *Interrelationship between voice processing, computing, and telecommunications*

The Personal Computer and Voice Processing

You should understand first that there are many ways to configure a voice processing system. Unless your application is very large, a PC-based voice processing system can probably handle all of your needs. A PC-based voice processing system can be added to a LAN simply by installing your network interface card and properly configuring the operating system.

The nice thing about modern voice processing on a PC platform is that the architecture is open. Hundreds of organizations are out there designing and developing new ways to enhance voice processing features and services. The products that result from this development are compatible with existing voice processing hardware and software, because the vendors stick to the standards set forth. Two major initiatives are in place for voice processing development: SCSA and MVIP.

SCSA

SCSA (Signal Computing System Architecture) is a cross-industry initiative supported by hundreds of organizations, such as Dialogic, IBM, NEC, Northern Telecom, Siemens, and Tandem, with Dialogic taking the lead. The initiative has been successful in bringing together the complex aspects of telephony, computing, and voice processing. Some of the elements of system architecture SCSA is responsible for are:

- Adaptation to emerging technologies
- Alternate computing model standards
- Applications Interface (API) standards
- Driver development for input and output standards
- Firmware interfacing standards
- Networking standards
- Platform and Bus standards

MVIP

MVIP (Multivendor Interface Protocol) is the other voice processing standard used throughout the industry today. Some supporters of MVIP are: GammaLink, Mitel, Natural Microsystems, Picture Tel, and Voice Processing Corp. MVIP is headed up by Mitel and Natural Microsystems. MVIP performs the same functions as SCSA, but the structure of the alliance is a bit different. MVIP will allow new developers to use its architecture, but it requires that the specifications only be shared with other MVIP affiliates. SCSA does not require any nondisclosure. In addition, MVIP is centered around the MITEL ST-bus, which makes MVIP a little less open.

When choosing a voice processing platform to go with, you must choose between these two standards. SCSA has a little more backing than MVIP. Microsoft has been contributing to the SCSA standard to roll out the Phase III client/server version of TAPI (see Chapter 18). Both offer advantages based upon the telephony and computing environments in which the voice processing system will be used. Make your decision about which standard to go with after you have clearly defined the environment the voice processing system will work in and what your voice processing application will do.

Voice-Mail

Voice-mail, also known as voice messaging, allows you to leave a message for someone. Not just anyone, but a specific someone, and you can leave the message in your own voice. Answering machines were the very first forms of voice-mail systems, with prices starting around $40. Now, modern voice-mail systems range in price from around $100 (for a voice mail card and software to put in a PC) to over $200,000.

Modern voice-mail systems offer a great number of features that were previously not available on answering machines. Enhancing communication is the primary focus of telephony, and voice-mail has added a few new ways to help us communicate. Here are some of the features that you will find on a modern voice-mail system:

- Private voice-mailboxes
- Digitally recorded announcements and messages
- Message back up ten seconds and repeat function
- Message waiting notification (the voice-mail system lets you know you have a voice-mail message waiting)
- Message forwarding/redirection
- Message broadcasting
- Skip message
- Individual message deletion or save
- Scheduled message reminders (messages that you leave for yourself and are returned to you on a specific date)
- Online directory system

Voice-mail does have its abuses. It can be a bit intimidating to some. Other people hide behind voice-mail and never take calls or return them. You may have also experienced "voice-mail jail," where you are sent into the voice-mail system, never to be able to reach the person you want or talk to a human being. Most of these scenarios arise from poor implementation or intentional negligence. Try not to judge the entire industry by a few bad apples in the bunch.

If you are not a big fan of voice-mail, let's see if your mind can be changed. Take a look at some of the advantages voice-mail systems have to offer.

- 75 percent of all business calls are not completed on the first attempt. With voice-mail, you can leave a more detailed message in your own voice and intonations. The receiving party can return the voice-mail with an informed response, rather than trying to make the communication again and again or leaving a generic callback message with an operator or secretary, which may never get delivered.

- Two-thirds of all phone calls received are less important than the ones you are on. Having voice-mail can reduce the number of interruptions you get.
- When you leave voice-mail messages, the calls you make are invariably shorter. Live conversations include a great deal of chitchat. An average voice-mail call takes 43 seconds, while the average long-distance call is over 3.4 minutes. When used properly, voice-mail can be up to 80 percent faster.
- Time zone problems between bi-coastal or international businesses become less of an issue. Messages can be left for individuals, and the corresponding response can be made at a later time, possibly with another voice-mail message.
- Fewer callbacks. Voice-mail, many times, gets the information across the first time, thus reducing the number of telephone calls made by as much as 50 percent.
- 24-hour availability.
- Increased customer service (messages are alway delivered).
- Reduced labor costs (fewer receptionists are needed).
- You can use a voice-mail system to broadcast announcements throughout the organization or to specific groups of people.

These are just some of the benefits that voice-mail has to offer. Best of all, voice-mail can save a company between $50 and $150 per employee per month on long-distance charges.

What are the components that make up voice-mail? To begin with, you must have a voice-mail hardware platform. In today's market, there are three major hardware platform types that you may find yourself working with.

PC-Based Voice-Mail Systems

PC-based voice-mail solutions are primarily for the low-end market. They require that specialized hardware and software be used along with a personal computer. The ability to perform the voice-mail function has been included on some modem cards these days and can be used to answer a single line while distributing voice messages to multiple people. More advanced PC-based systems use specialized voice processing cards that have become a cross-industry standard, such as the Dialogic D/41 and D/42, which can answer up to four lines simultaneously, go on and off hook, initiate and terminate calls, and maintain mailboxes for hundreds of people. Cards with greater capacity are available as well. Systems may be placed either stand-alone or integrated with a PBX. Over the years, the PC-based voice-mail systems have grown in complexity, and now they perform virtually all of the same advanced functions as the larger proprietary platforms.

LAN-Based Voice-Mail Systems

LAN-based voice-mail systems offer a distinct advantage over their PC counterparts. They typically offer greater integration with the other components of the LAN, such as e-mail. Messages can be directed to e-mail and then retrieved over the LAN. Access to databases on the LAN is also made easier. The cost for implementing a LAN-based voice-mail solution varies a great deal depending on the LAN platform you are using. Some LAN-based solutions can start at over $40,000 for the basic hardware, while others can use the same voice processing cards as the PC-based voice-mail systems. As you learn more about MAPs, you will begin to recognize the advantages of a LAN-based voice-mail system that can grow into a complete MAP.

Proprietary Voice-Mail Systems

Proprietary voice-mail systems utilize custom-built computer-based hardware specifically designed for the voice-mail application. The voice-mail industry really does not have a single standard. All of the major players in the business, such as Centigram and Octel, have designed their own hardware and software to perform the voice-mail function. Costs for these systems are typically determined by the number of mailboxes they are designed to support and the amount of recording time available. The proprietary voice-mail system can be placed behind a PBX, or it can also be used as a low-level switching platform for answering, switching, and taking voice-mail messages.

At one time, the proprietary voice-mail platform was the only solution for a voice-mail application, regardless of the size of the organization. With the invention of the PC-based voice-mail platforms, the proprietary systems are probably only suited to large organizations.

Audiotex

Audiotex is an entire industry devoted to supplying information and services either for a small fee or as a public service. Audiotex is a glorified voice processing system that is menu-driven. Audiotex systems typically play prompts and pose questions. You make your selection by pressing digits on your touch-tone phone corresponding to the prompts given.

The costs associated with audiotex systems vary among service providers. If you are looking at audiotex as a potential addition to your business, look at spending around $5,000 dollars for a startup four-line system. Costs for larger systems can have extremely wide ranges. Check with some of the vendors listed in Appendix A for more detailed pricing. You will find you really just need a unique service that can be recorded into voice prompts to have a successful audiotex business. A 900 number can be used for the billing, credit card verification can be added, or you may simply want to give it away as an added service your organization offers either to your employees or to your customers.

As a consumer, you want to be aware of the charges that you might incur for using audiotex systems. Many systems inform you of their charges in their advertisements (if there are any) while others include a voice message that informs you of the rates being billed for the service. There are a few vendors that are not as forthcoming with the rates that they charge for their services. Beware! You should always know the cost of the audiotex services that you subscribe to before you use them. Nightmare telephone bills are born from user negligence.

Audiotex can be put to limitless uses with very little modification to the base audiotex design. Any information or service that can be recorded into messages can be turned into an audiotex application. Here are some of the more common applications of audiotex that you might be familiar with:

- Community bulletin board
- Get your horoscope
- Movie listings and show times
- Concert event line
- Recorded sex line
- Sports line

Figure 14-2 shows an example of a typical movie directory and show time audiotex application. The caller is greeted by the audiotex system, and a menu is presented. Additionally, preview clips of new movies can be advertised before the menu is announced (a source of revenue). The user may select the desired movie by entering in the first three letters of the movie on a touch-tone phone. The audiotex system contains a table of movie names and their translated three-digit codes. If two or more movies have the same three-digit code, the audiotex system presents another menu for the caller to choose from. The system then requests that your zip code be entered to focus in on the theaters in your area that are showing your movie. At this time, the audiotex system determines which theaters are in your area and showing your movie selection. Finally, a list of theaters and show times is provided. You get the picture (ha ha).

The example given is of a fairly complex audiotex application. Many audiotex applications involve only a list of menus and sets of voice messages. No table lookup is necessary. One thing that should be mentioned about audiotex applications is that the voice prompts usually require constant modifications. The information provided by audiotex is usually time-sensitive, although this is not always the case. If you are looking into audiotex applications as a business, investigate the amount of time that will be required to maintain the system before making a commitment.

Interactive Voice-Response

Interactive voice-response (IVR), also referred to as a *voice response unit* (VRU), is without a doubt the most flexible part of voice processing. Not only do IVR systems provide

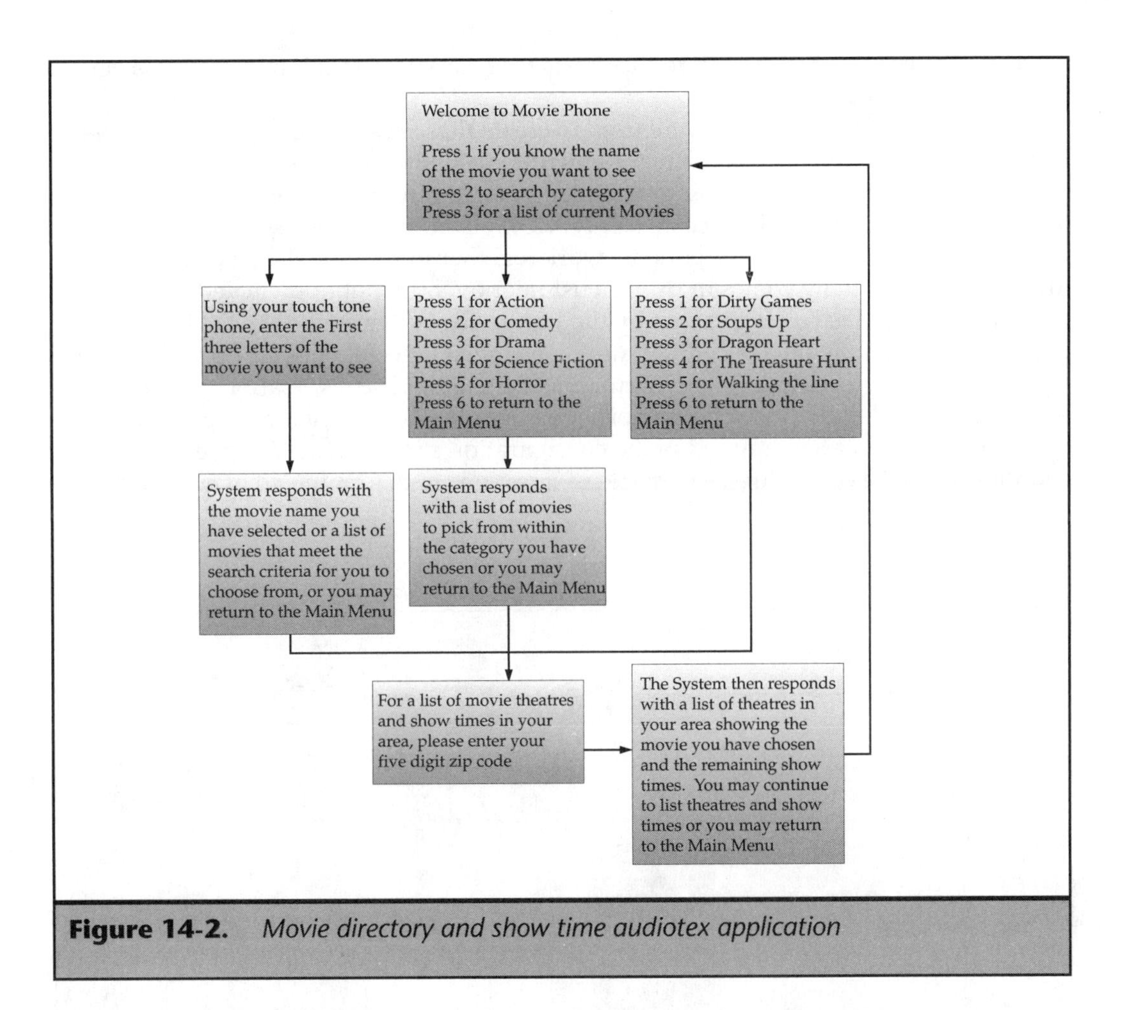

Figure 14-2. *Movie directory and show time audiotex application*

menu processing and voice messaging, but they also incorporate transaction processing and, potentially, transaction creation. IVR can perform almost any function that provides information from other computer systems and the information used in IVR systems can also be self-contained.

You are certainly familiar with some of the more popular forms of IVR systems:

- Banking by phone
- Cable pay-per-view
- Flight status
- Order entry and tracking
- Inventory control systems
- Traffic violation lookup and payment

All of these applications and hundreds more can be developed using IVR technology. The benefits of a system like this can readily be seen. IVR systems increase productivity, provide utility, and reduce costs for companies that implement this very powerful technology.

The equipment and the software used to create these systems is quite remarkable. Figure 14-3 shows a typical IVR system configuration. In the example shown, the IVR system sits behind a PBX switching system. Similar configurations to this could be used to provide a multitude of customer service-related functions.

Let's review how these components work together to provide the customer service function. A call initially comes in from a customer over the public switched telephone network (PSTN). The call is directed by the PBX to the IVR system. The D/42 voice processing cards shown in the configuration can emulate several popular PBX telephone stations and can process up to four simultaneous telephone conversations. Many variations of card and line configurations are possible when implementing an IVR application, so don't feel trapped in the configuration shown. The IVR greets the caller and then steps the caller through a series of questions to determine what customer service function is required.

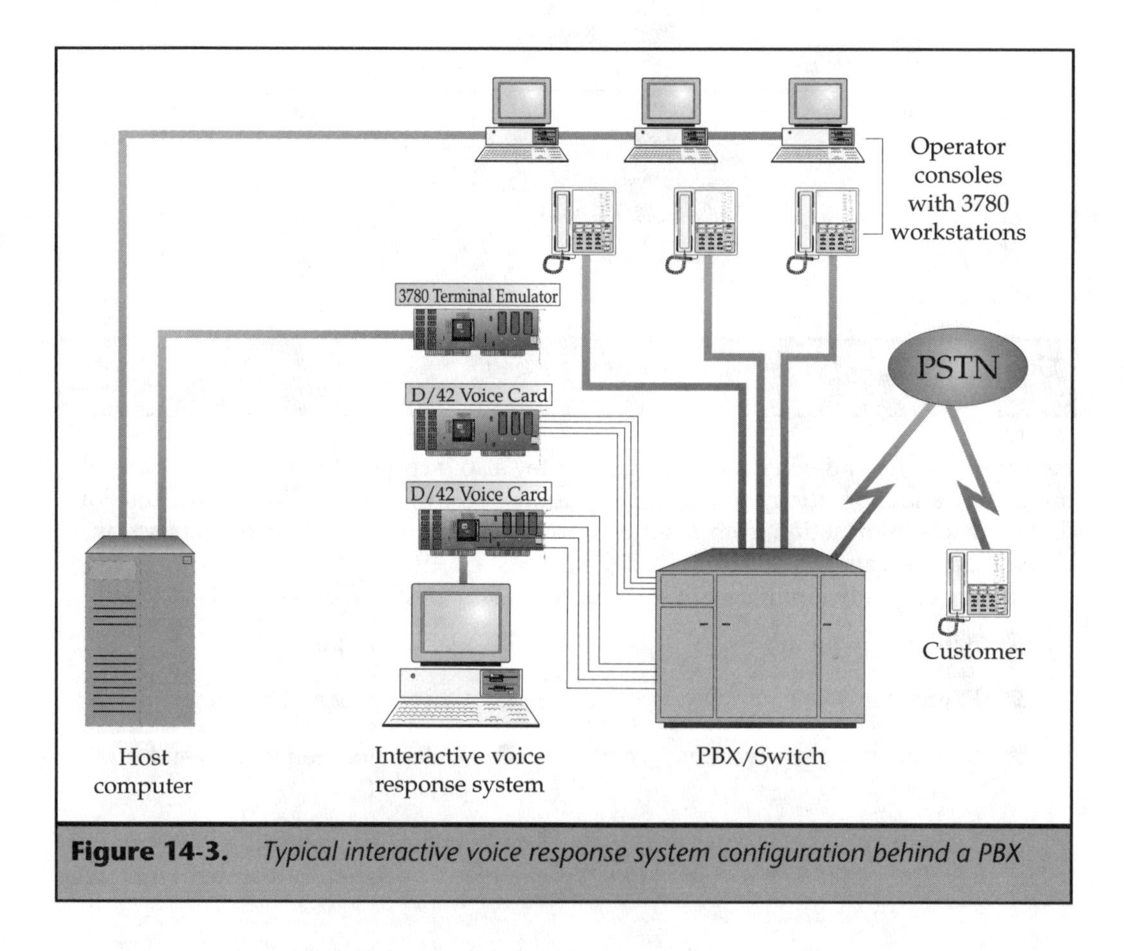

Figure 14-3. *Typical interactive voice response system configuration behind a PBX*

The eight IVR customer service lines shown in Figure 14-3 all share one 3780 terminal emulation card to interface with the organization's host computer system. The interface between the IVR platform and host computer system can be made in a number of different ways. Substitutions are available for a wide variety of terminal emulation cards. Other methods of interface can include standard RS-232, or the IVR system may be incorporated directly into a LAN environment.

The IVR uses the terminal emulation card to access information in the same manner that a customer service representative would using their operator console. Only a single line on the IVR system can take control of the terminal emulator at any time. A request is made to the host computer by the IVR for data pertaining to the information entered by the customer on a touch-tone phone. Once the screen's worth of information is passed from the host computer to the IVR system, the IVR stores all of the pertinent information on the screen in memory. The IVR then releases use of the terminal emulation port so that another voice processing port can take control. This entire operation usually takes place in a couple of seconds or less. Any information gathered from the lookup can then be reported to the customer. New requests can be made wherever the IVR application is configured to do so.

The ability of voice processing to emulate a PBX telephone station, rather than just being a dumb analog telephone, is a distinct advancement in IVR technology. Through the use of this feature, calls can be transferred to live customer service reps, message waiting lamps can be turned on, even automated voice-mail messages can be left. If the problem or inquiry cannot be resolved by the IVR system, the user can be transferred to a customer service rep or other department/extension for further assistance.

There are almost as many software packages available to generate IVR applications as there are IVR applications. IVR software generation packages come in three flavors:

- Programming language
- Text menu-driven
- Graphical user interface (GUI)

All three versions work admirably, but if you are not a hard-core programmer, you may want to look into one of the text menu-driven or GUI versions of IVR applications generators. Expert Systems, based in Atlanta, Georgia, is one of the few applications generator suppliers that offers both text menu-driven and GUI versions of their software. Even the programming novice can be up and generating working IVR applications in a matter of days with the powerful tools that are supplied in the enhanced applications generators available in the market today.

The text menu-driven and GUI applications generators have a gigantic suite of voice processing functions that you may call from a series of menus to create your voice processing application. The primary difference between the text menu-driven and the GUI applications generators is how the system presents the programming options to the user. One provides a set of menus and the other provides a graphical representation of the programming options. Both present their suite of functions clearly along with online help and a description of how to use the function. When you select a function, you may be asked a series of questions that you must answer to complete the different operations. Other operations, like answering the phone, gathering touch-tone digits, or transferring a call, are just a menu selection away. The applications generator then creates the source code from the information you enter, saving you tremendous amounts of development time.

When choosing your applications generator, try to keep in mind your upfront and recurring costs. The prices for development and run-time licenses vary a great deal between vendors. If you plan to resell the application you develop over and over, you may want to purchase an applications generator or programming language with little or no run-time license fee. However, if you are looking to use your IVR applications strictly in-house, go with the product that suits your needs the best. A list of IVR applications generator software providers is provided in Appendix A.

IVR systems are very cost-effective. Costs start at around $5,000 for hardware. The software and programming can bring the total average cost of an IVR solution to somewhere between $10,000 and $25,000. Adding features like credit card verification can make the price jump sharply. Depending on the number of simultaneous IVR telephone connections and the complexity of the application(s), an IVR system can run well over $100,000, but this is very rare. However, when all is said and done, IVR systems will either save you or make you money. A small IVR system can easily free up four employees from answering customer inquiries so they may be put to better use. Most IVR systems pay for themselves in a matter of months.

IVR technology unleashes the full power of voice processing. MAP systems focus their applications around IVR technology. New commands, features, and interfaces are being developed almost every day to further enhance the product offerings. As you look further into IVR technology, let your imagination run wild. You can change the way you do business through the use of advanced IVR technology and applications.

Integrated Voice and Fax (Fax-on-Demand)

Integrated voice and fax, more commonly referred to as *fax-on-demand*, is used by thousands of companies to provide enhanced service to customers, vendors, and distributors. Fax-on-demand combines the power of IVR technology with the convenience and speed of fax technology. Specialized fax modem cards are added to the IVR system along with enhanced IVR applications generation software, which allows processes to be passed from the voice processing portion of the system to the CPU and then to the fax modem to facilitate fax delivery. For a more thorough description of fax-on-demand, refer to Chapter 16.

Operator Services

Operator services is an area of voice processing that entails several different functions based upon how it is applied. Operator services are operations that have traditionally been performed by human operators, such as the ones described below. Operator services available through voice processing applications are used by many of us in our daily routines.

Automated Call Distribution

Call reception and distribution, commonly referred to as *automated attendant* or *automated call distribution or distributor* (ACD) is the answering of your general business lines by an automated attendant (the voice processing system). The ACD allows the caller to enter an extension number and/or presents a list of departments to be transferred to. Online name directories are also available on many systems. The caller also has the option to be transferred to a live operator, typically by pressing zero. These systems are usually used in conjunction with voice-mail systems or integrated in multiapplication platforms, discussed later in this chapter.

800 Collect

800 collect is one of the newer applications of operator services. The way it works is this: You call an 800 number, and the voice processing system asks you to enter the telephone number of the person you wish to call. The system asks you to speak your name. It then asks you to speak the name of the party you are trying to reach. The 800 collect system then places you on hold and calls the number you have entered. The system asks the answering party if they are willing to accept the collect call for (person you wish to reach) from (your name). If the answering party accepts the call, you are connected to your party, and a collect call is billed. All this is accomplished through an advanced voice processing system designed specifically to place collect calls. Costs are reduced and discounts are passed along to the customer.

International Callback

Would you believe that it can cost from 30 to 70 percent less to call from America to another country than it costs for the same call to go in the opposite direction? In fact, it can cost less to make two calls from the U.S. to two different foreign countries than it costs to place one direct dial call from one foreign country to the other. Why? Quite simply, the United States has the best telecommunications infrastructure of any nation in the world. In addition, deregulation has pushed down the rates that we are charged for our long-distance calling. In countries where monopolies exist for carrying outbound long-distance traffic, toll charges are absolutely ridiculous, but you are a captive audience.

With international callback, you can break free from your captivity. Europe, Asia, and South America are big users of international callback systems. The companies in the U.S. can charge full retail rates for both calls placed through the callback system, while they receive 50 percent or more discounts from the long-distance carriers they subscribe to. Savings are big to the user, and profits are equally as large for the operators of international callback systems.

How does a callback system work? Very basically, a user sets up an account with an international callback supplier. The user is then assigned a telephone number and an authorization code. The user also specifies a specific telephone number that the callback system will call when it is signaled. When the user wants to make a long-distance call, he or she first makes a call from the callback number earlier specified. The international callback system does not answer the call, but directly after the rings stop, it calls the predefined callback telephone number. The user answers the call and is prompted to input the authorization code. After verification, the user is supplied a dial tone from the U.S., at which time multiple calls can be placed to anywhere in the world at the reduced rate.

International callback comes in several different configurations with different options, such as variable callback telephone numbers and added security. Use of the service can be a bit cumbersome, but the savings are very real. If you or others in your organization travel abroad, you may want to look into an international callback account. If you are interested in getting into the business, research the competition and design your system the way you would want to use it. Customers will find you if you supply a quality service at a good price. Additional insight into the operation of this form of operator service can be found in Chapter 2.

Although international callback is an operator service of sorts, it is also a true example of a MAP. The system utilizes interactive voice, table lookup, data generation, and call switching to perform its function. As you read through these examples of voice processing applications, try to recognize when different telephony technologies are being used in conjunction with one another to provide new solutions, products, and services.

Credit/Debit Card Validation

Credit card validation and now debit card validation was one of the first applications ever developed for voice processing systems. In today's market, the proliferation of credit and debit cards has grown drastically through the use of voice processing. After you dial your network access number ("0," "10333," "800-XXX-XXXX," etc.), many of these voice processing systems begin with a bong tone. Actually, the type of prompt does not matter. If you do not know what to do, the system will speak to you in a human voice to explain what you need to do. The user then enters the credit or debit card number. The system first verifies the card against a database and then places the call. Billing is also generated by this application.

Directory Assistance

Perhaps the most common form of voice processing operator services used in the world today is half human/half voice processing. You know this application as directory assistance (411 and XXX-555-1212). The human operator ask you for the city and listing you are looking for. The person then looks up the listing and then passes the found record over to the automated operator to read you the telephone number. The human operator is then freed up to answer other calls. As speech recognition becomes more advanced, look for the directory assistance process to be completely automated in the near future. Pilot programs are already in place.

New applications that fall under the heading of operator services are being developed. As the multiapplication platforms become more refined by the programmers developing them, virtually all operator service functions that can be automated, will be.

Always keep in mind the main reason why voice processing is so popular. This technology saves or makes MONEY! Productivity can be increased, transaction auditing can be generated, and labor costs can be reduced. Who wouldn't want to implement one of these systems?

Telemarketing

Well, what can you say about telemarketing? It is a fact of life that telemarketers exist, and they exist in almost every industry in the world today, especially in the United States. Believe it or not, telemarketing is a two-way street. Call centers use both outbound and inbound voice processing techniques to accomplish their jobs. If your firm utilizes telemarketing, you should recognize the value of voice processing in this arena.

Inbound telemarketing uses voice processing typically in the form of automated call distributors (ACDs). As inbound calls are received, the ACD evenly distributes those calls to the call center operators. An additional function that can be added to the inbound voice processing application is statistical generation. The system can track and report on the number of calls handled by each operator, total time on the phone, average time per call, average time on hold, etc. This information can also be supplied through a telemanagement system (see Chapter 6) or through some PBX systems. If these resources are not currently available, adding this functionality to your telemarketing voice processing system is a great idea!

Outbound telemarketing is not done in the old impersonal style of a few years ago that may still ring in your ears. Completely automated outbound telemarketing, except for limited applications, was recently made illegal by the Telephone Consumer Protection Act. That is why you don't get those annoying automated telephone calls any more.

Today the outbound telemarketer is assisted by a much more sophisticated voice processing design. Outbound voice processing systems now use a technology known as predictive dialing. *Predictive dialing* gives voice processing the ability to call telephone numbers found in a database on the voice processing system or other computer platform that the voice processing system is integrated with. The voice processing system listens in on the telephone line and can detect busy signals, no answer, fax machines, the presence of an answering machine, and yes, the presence of a human voice.

If the human voice is detected, the call is instantaneously transferred over to a telemarketing operator who responds to the call just as though he or she had originally dialed the number in the first place. All of the other calls placed by the voice processing system are logged with their status but are not transferred to an operator station. The system can intuitively speed up or slow down the calling process to match the speed at which the operators are handling the calls. Your telemarketers are now able to do what you are paying them for—talk to customers. As much as 30 percent of a telemarketer's time is spent dialing and waiting for someone to answer a telephone. Through the use of voice processing, you can increase your employees' productivity by this same percentage while at the same time getting superior logging of the call disposition.

Text-to-Speech

Text-to-speech has been around for a few years. Although the technology has become fairly good, the primary applications for text-to-speech are found on voice processing systems. Text-to-speech is not a separate horizontal market in voice processing, but it deserves mention for its ability to enhance automated voice processing applications. Putting facsimile transmissions aside, there are two basic ways to report information to users of voice processing systems: through prerecorded voice messages or through text-to-speech technology. Text-to-speech gives your voice processing application the ability to say actual words, such as "Holly Road," that are read in through the database retrieval engine for reporting by the voice processing system. Specific recordings are not needed for every possible word. Depending on your application, you may find that text-to-speech can enhance the functionality of your voice processing system.

Speech Recognition

And now for the technology you have all been waiting for—may I introduce you to *speech recognition*. How long have you been dreaming about being able to talk to a computer and have it obey your commands? Probably since you saw your first episode of *Star Trek*. Well, ladies and gentlemen, you're not going to have to wait until the 24th century to get it. Speech recognition technology is on the horizon, and the impact on the computer industry, not just voice processing, is going to be immense. Speech recognition, for those of you who are not familiar with the concept, is the ability of a computer to understand human speech and act on the commands and processes it is instructed to perform by the speaker.

You may already be familiar with some of the recent incorporations of speech recognition with voice processing. Some of the latest operator services applications allow you to speak the words "Operator" or "Collect" to branch through a voice processing application, rather than using your touch-tone keypad. The latest directory lookup systems used by some bell operating companies collect the "City" information part of a directory assistance call through speech recognition before the call is passed to a live operator, thus reducing the amount of time required by the live operator to find the listing you are requesting.

Speech recognition technology has improved in performance dramatically over the past few years. Although speech variations can still throw this technology a few curves, it is showing tremendous promise. Computer applications have already been developed that are extremely accurate in taking dictation directly into word processing. Systems are advanced enough to know the difference between the words "to," "too," and "two" through the use of grammatical software. Speech recognition is a topic that goes beyond the scope of this book, but you may want to look further into this fantastic technology. The face of voice processing will change forever once this technology is refined and embraced by the voice processing community.

Multiapplication Platforms

What do you get when you combine all of the functions of voice processing into a single system? You get a multiapplication platform (MAP). As you may have realized by now, all telephony is interrelated. All of voice processing is interrelated. MAPs were a natural progression from stand-alone application platforms. All the functions being so close to one another has made MAPs a thing of reality.

MAPs no longer perform just the functions traditionally provided by voice processing. Other industries and functions have been added to the suite of offerings. Arguably the most notable application developed on a MAP is a product called Wildfire. This product actually goes over the edge into a new label called an *electronic assistant* or an *intelligent assistant*. It is, however, an example of an advanced voice processing multiapplication platform that combines all of the functions of voice processing today and at the same time provides a new level of computer telephony integration.

l 14-1

Let's step through an electronic assistant's capabilities and refer back to the technologies that are incorporated in a MAP. A person makes a call to his or her electronic assistant. Through the use of speech recognition, the system can recognize that its master is calling once the password is spoken. At this point, the power of the system is unleashed. An electronic assistant can perform all of these functions:

- Voice-mail storage and retrieval
- Call screening and announcement of callers on any phone
- Call routing, switching, and placement
- Intuitive response to spoken commands
- Fax storage, retrieval, and forwarding
- E-mail retrieval

Voice storage and retrieval is handled in the same traditional manner, but the user issues voice commands rather than touch-tone digits. The system can even sort your messages in a variety of manners.

Call screening is accomplished by making a verbal list of people you wish to talk to or a list of people you don't want to talk to. The electronic assistant requests that the caller's name be spoken. Depending on which list the caller is on, the assistant will either put the call through and announce the caller or put the caller into voice-mail. Callers not on either list can automatically be defaulted to being announced or being placed into voice-mail.

Call routing and switching is a remarkable feature. When you are out on the road, the system can be programmed to automatically forward callers to your location. The call screening function can be put on the front end of this application. The switching function can also be used to place calls. If you receive a message to call someone urgently, the assistant can automatically place the call and connect you, even if you are away from the office and calling in for your messages. If you have 800 access into your office and you place calls in this manner, you may actually save money on your long-distance charges over using a credit card.

The intuitive response to spoken commands is a complex series of predetermined responses to commands that might be issued by the user. Advanced speech recognition and voice processing is used to accomplish remarkable results.

Fax storage and retrieval incorporates another aspect of voice processing. Text-to-speech can be used to actually read fax transmittals to the user over the phone in an automated fashion. Of course, many things on a fax transmission may not be readable, such as illegible text, figures, diagrams, and pictures. To make the system more robust, a pseudo-fax-on-demand can be called upon to send the actual fax transmittal to your current location.

To top it all off, electronic assistants can actually read your e-mail messages to you over the phone. Again, text-to-speech is used to accomplish this function. E-mail messages can also be directed by the assistant to the fax server for hard copy retrieval.

All of these and several other functions are now being performed by electronic assistants using MAP technology. The electronic assistant is at the leading edge of what is rapidly becoming an entire industry unto itself. Costs for these systems are not cheap, but the savings are equally great. One day soon, you might be talking to a machine. It's not as far off as you might think. In fact, vendors are out there right now waiting to make it happen for you.

Voice Prompt Editing

So now you might be thinking that voice processing has potential, and you've begun to put together a plan to implement voice processing. You should also include provisions for *voice prompt editing*. There are two types of voice messages that can be called upon by voice processing applications. The first you are already familiar with: temporary recorded messages. These are the messages that are recorded on the fly, typically over the telephone. They are commonly referred to as voice-mail messages, but other forms of temporary recorded messages are also used in many voice processing applications. The second type of voice message is the voice prompt: these are the messages that are spoken over and over by voice processing systems. The numbers, dates, times, and other frequently spoken messages are other examples of voice prompts.

Voice prompts are stored in several different formats depending on the voice processing platform. Wave, Pure, and VOX are common voice prompt recording formats. The voice prompts are the most important part of your voice processing presentation after the application's functionality itself. Figure 14-4 shows the VFEdit voice editing toolkit being used to modify a voice prompt. As you are creating your prompts, it is important to obtain a high-quality recording. You must also assure that the volume of your prompts is consistent and that any leading and trailing silence around prompts is removed for optimal presentation. In essence, voice prompt editing turns your computer into a recording studio. Music with voice overlays, advertisements, echo, reverb, and a host of other functions are accomplished through the use of a voice prompt editor.

Once you have recorded all of your voice prompts, they are usually combined into one large file with indexes pointing to the beginning and end of each prompt. Your voice processing application may then call on these prompts whenever needed. Loading this voice prompt file onto a RAM-drive results in clean and immediate voice prompt playback by the voice processing system. The better voice prompt editors cost a few hundred dollars, but you will find it is a must-have when you are creating a professional voice processing presentation.

Conclusion

Just take a close look at the many different ways that voice processing applications might be used to provide greater productivity throughout your organization. Any process that you can automate will ultimately save or make money for your organization. Development of new and unique applications is not as difficult as you might think.

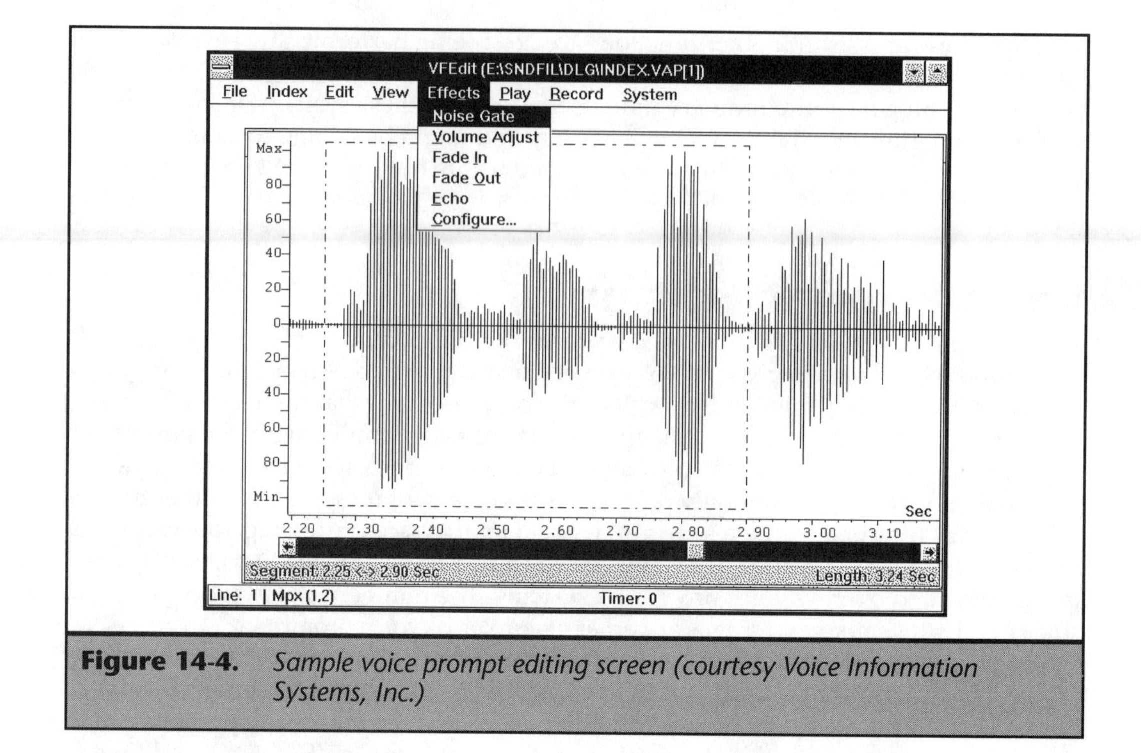

Figure 14-4. *Sample voice prompt editing screen (courtesy Voice Information Systems, Inc.)*

Whether you're starting from scratch or hiring a vendor to supply new voice processing services, you really can't go wrong. An investment today will bear fruit tomorrow.

Due to the limited scope of this book, deeper discussions about the technology that sits behind this industry were not possible. Try not to feel intimidated about what you do not know. Incredible advancements in hardware, applications generators, and voice prompt editing toolkit software make almost all of these voice processing applications rather simple to implement. Browse over Chapter 5 for information on how to create a strategic plan. You will learn how you can formulate your own plan to implement voice processing in your organization today.

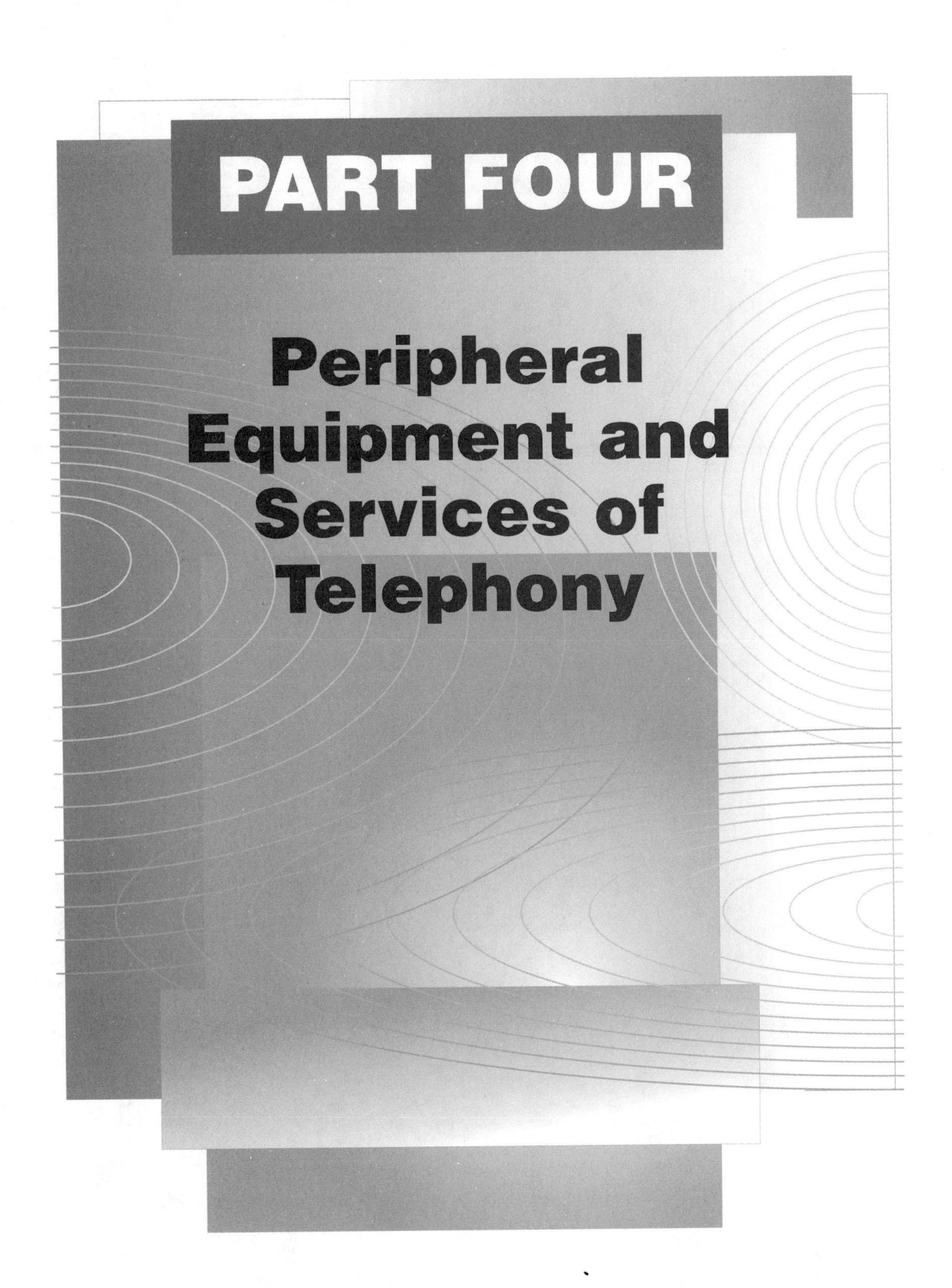

PART FOUR

Peripheral Equipment and Services of Telephony

Chapter Fifteen

Modems

Modem (*mo*dulation/*dem*odulation) is the common name used for communications equipment that modulates digital signal information received from a computer and turns it into an analog form, which can then be carried successfully on a telephone line. A modem also demodulates analog information that it obtains from the phone line back into digital signals before passing them to the receiving computer. Modems have been used for decades. The people who used them and used them well were few and far between. Because of varying line conditions, computer hardware, computer software, and modem specifications, reliable connections were as much an art as they were a science. The modem is considered to be a DCE (data communications equipment) device. The computer is commonly referred to as the DTE (data terminal equipment) device. These two components make up the starting base of a communications environment.

There are two main types of modems: synchronous and asynchronous. *Synchronous* modems are primarily used where there is a point-to-point connection between two locations. A point-to-point connection is like having two tin cans connected together by a piece of string. You can only reach the other can at the end of the string connected to your can. The primary advantage of synchronous modems is that they use separate pins to carry timing information, so fewer data bits are necessary to transmit the information from one location to another. Although synchronous modems are faster, the analog line quality must be superior in order to carry the additional clock signaling information.

Asynchronous modems are primarily used for connections made under dial-up circumstances, and they are the form of modem most widely used. The string on the end of an asynchronous can may be connected to any other can you know how to reach (the telephone number, in this case). Asynchronous modems use additional bits to identify the start and the completion of a byte/character. As a general rule, you can figure ten bits for every byte transmitted. To find out what type of modem you might already have, refer to the documentation that came with your modem. The actual number of bits transferred per byte will vary, based upon the modem settings, terminal emulation, and transfer protocol that are being used in the connection. These topics will be further discussed in this chapter.

Many of today's modems can operate in both synchronous and asynchronous modes, but the asynchronous applications are used far more often. The communications software that you configure and use will determine if the modem is set for synchronous or asynchronous mode. Some modems require that you change jumpers or dip switches to change from synch to asynch. Check your modem user's guide for more details. Since this book is about telephony and focuses on the most popular forms of communications, the remaining discussion in this book on modems will focus on asynchronous modem communications.

Asynchronous Modem Communications

For years, the standard speed for modems was 300 baud or bps (bits per second), followed by 1,200 bps, 2,400 bps, and then 9,600 bps. All through the '80s, the majority of modem users had 2,400 bps, and the elite few had the very expensive 9,600 bps modems. In order to obtain optimal throughput between two modems, users had to know precisely who and what they were talking to. Incorrect setting in baud, bit, parity, and terminal emulation would adversely affect the connection, if not make the communications attempt fail completely.

As you would expect, modem speeds continue to go up and up. In the early '90s, manufacturers came out with modems reaching speeds of 14.4 Kbps (14,400 bps) with maximum throughput speeds of up to 57.6 Kbps. In 1994, the modem world took a major step forward by releasing the V.34 protocol standard and a whopping throughput speed of 28.8 Kbps with a maximum throughput speed of up to 115 Kbps. The maximum throughput speed is never realized over dial-up connections, however. The incredible throughput speeds mentioned are made possible through the use of data compression and error correction protocols, which will be discussed later in this chapter. In order to achieve maximum throughput, not only do the modems have to be set up correctly and have similar features, but the POTS (plain old telephone service) lines must be crystal clear.

The more advanced modems will automatically adjust both up and down in speed based upon the line conditions. The speed at which the modems make the adjustment up and down is determined by the error correction protocols, compression protocols, and the overall quality of the modem itself. Don't count on achieving throughput any greater than the maximum standard speed of the slowest modem being used in the connection. If you are using a 28.8 Kbps modem, the maximum speed of the modem is almost never realized consistently.

In the past, to change the modifiable parameters of a modem, you had to open up the modem and physically set dip switches in order to achieve the desired configuration. In addition, some parameters could be changed through software settings. Some modems still use dip switches. If your modem does have dip switches, make sure that you understand all of their functions. Certain dip switches take precedence over software commands that you may choose to set later.

Modem communications standards have became more widely accepted and used. The majority of PC users now use what is commonly referred to as a Hayes-compatible modem, which uses the AT command set. Dozens of modem manufacturers use the AT command set as the core of their modem's programmable functions. Fortunately, modem manufacturers do add some enhancements to their product lines to set them apart from the others. Unfortunately, the AT command set is changed and added to slightly by each manufacturer to accommodate the special features that one modem might have over another. See Table 15-1 for a list of the most common and least modified AT commands and a brief description of each. Later in this chapter, we will discuss optimal settings that will help you get the best performance out of your modem.

NOTE: When purchasing a modem, make sure that you keep a copy of the exact command set being used by your modem, and try to understand as many of its functions and their effects as possible. You will most definitely be thankful that you have retained this information.

Communications Ports

Communications ports are sometimes forgotten when talking about modem communication. The port on your PC or LAN that connects to your modem is a vital

Command	Definition
AT	Attention command: must precede all other commands, except A/, A> and +++
AT/	Re-execute the last command entered
AT>	Re-execute the last command entered continuously
Dn	Dial the number that follows and go set the modem into origination mode. Use any one of the following options for dialing: P Pulse dialing T Tone dialing , (comma) Pause in dialing for two seconds ; (semicolon) Return to command state after dialing
En	Echo characters entered on the screen: E0 Echo off E1 Echo on
Hn	On/off hook control of phone line: H0 Hang up (go on hook) H1 Pick up the phone line (go off hook)

Table 15-1. *Standard AT Command Set*

Command	Definition
Ln	Speaker volume control: L0 Off L1 Low L2 Medium L3 High
P	Pulse dial
Sr=n	Set one of the S-register configurations; n must be a decimal number between 0 and 255
T	Tone dial
Vn	Verbal/numeric result codes: V0 Numeric result codes displayed V1 Verbal result codes displayed
Z	Software reset to setting stored in nonvolatile RAM
+++	Escape code sequence, preceded and followed by at least a one-second delay where no data is sent. The modem then returns to the command mode.
&Cn	Carrier detect (CD) operations: &C0 CD override &C1 Normal CD operations
&Dn	Data Terminal Ready (DTR) operations: &D0 DTR override &D1 On-line command mode with DTR toggle on/off &D2 Normal DTR operations
&F	Load factory default settings into RAM
&Hn	Transmit data flow control: &H0 Flow control disabled &H1 Hardware flow control (CTS/RTS) &H2 Software flow control (XON/XOFF) &H3 Hardware and software flow control
&V	Display current and NVRAM modem settings
&W	Write current settings to nonvolatile RAM

Table 15-1. *Standard AT Command Set* (continued)

part of the connection process and should not be overlooked. When putting together a communications link, you might view it as a relay race. The team members in the race are: the software package you are using, the CPU, the communications port, the communications port cable, the modem, and the telephone line. All components are important, but for now we are going to focus on the communications port. A number of different communications ports are available on the market today, and choosing the right configuration from the start will determine your success in the modem game.

8250 UART Chip

The majority of personal computers come standard with two communications ports that both use the 8250 UART chip. These communications ports are typically called COM1 and COM2. The 8250 UART chip was designed for a maximum throughput of 9,600 baud. Unless otherwise specified by the manufacturer, you probably have the 8250 UART chip in your IBM PC-compatible computer. If you have purchased your computer with an internal modem, the UART chip is built right into the modem card. (See the "Internal Modems" section later in this chapter for more information.)

If you are uncertain of what type of communications ports you have on your PC and you have Microsoft Windows installed, change to the Windows subdirectory and type **MSD**, then press ENTER. This runs Microsoft's diagnostic program, which will show you the existing system setting on your computer, including your communications ports, the type of port, the IRQ (interrupt request line), and the base I/O address for the ports.

16550 UART Chip

When configuring a communications port for an individual PC, the user should look for a controller card or a separate communications card that uses the newer 16550 UART chip. This chip boasts reliable throughput speeds of up to 115 Kbps. In addition, some communications cards can be configured with an additional 4K or so of RAM to be used as an additional buffer. For some unknown reason, this basic part of a reliable communications connection is frequently overlooked and, strangely enough, is not the easiest part to find in the secondary market. But, if you ask for the newer 16550 UART chip communications card when you purchase your PC, you will be thankful that you spent the extra time and the few extra dollars.

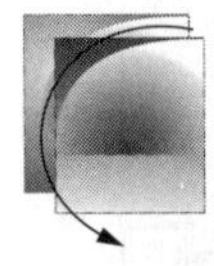

TIP: 8250 UART chip communications ports are still being sold today in the majority of PCs, but they are not desirable in the educated user's computer. Make sure when purchasing a new system that the configuration includes 16550 UART chip communications ports or better.

COM Ports and IRQ Settings

A standard PC can be configured with up to four communications ports; however, only COM1 and COM2 are present in a standard configuration. These ports come with their IRQs (interrupt request lines) preset at IRQ4 and IRQ3, respectively. These IRQ settings are also shared by COM3 and COM4 respectively when they are installed in a PC. Think of an IRQ setting as a radio station on your FM dial. Tune into the same IRQ that your communications card is set to, and you will be able to talk to one another. IRQ settings can be changed on many controller and communications cards. Try to avoid changing the standard settings of your COM1 and COM2 ports, as most software only looks for the communications port on a specific IRQ setting. The standard communications ports, IRQ settings, and base I/O addresses are listed in Table 15-2.

IRQ management becomes increasingly important as users add additional cards onto their motherboard slots. Almost all cards use an IRQ setting that addresses the card to the different software packages you might be using. If two devices are using the same IRQ setting, it is like having two people listening and talking to you at the same time—usually one of them is not welcome in the conversation. If for some reason your environment requires that you have more than two communications ports, avoid at all costs talking to two COM ports at the same time on the same IRQ setting. The result of this type of action can be highly undesirable.

Multi-I/O Cards

Multi-I/O (input/output) cards are single-slot cards that can provide 4, 8, 16, and up to 128 high-speed communication lines simultaneously. If you plan to have multiple users on your network share communications resources, then a multi-I/O is a necessary addition to your hardware configuration. Some multi-I/O cards support simultaneous use on all ports at 230 Kbps to 920 Kbps. Others have what is called an *aggregate line speed*—the total speed of all the ports running at a given time may not exceed the aggregate line speed, which is also usually between 230 Kbps and 920 Kbps. A number of multi-I/O cards are available on the secondary market. Choosing the right one for you, again, is an important step in establishing a solid communications platform.

Communications Port	IRQ Setting	Base I/O Address
COM1	IRQ 4	3F8
COM2	IRQ 3	2F8
COM3	IRQ 4	3E8
COM4	IRQ 3	2E8

Table 15-2. *Communications Port, IRQ Setting, and Base I/O Address Listing*

Choosing the Right Multi-I/O Card

Choosing the right multi-I/O card means that you will need to understand the LAN environment your company operates. This doesn't mean that you just go and find out if you are using Windows for Workgroups release 2.0, oh no. You need to know what operating system and revision you are running, what types of applications your people will be using over communications ports, what specific software they are using, and what types of hardware systems, applications, and software they will be connecting to. It also wouldn't hurt to find out how many users of communications applications you might have and the average amount of time they will spend online each week. Once you have gathered as much of this information as possible, it is time to shop for a multi-I/O card.

The first thing you want to make sure of is that the multi-I/O cards you are considering are compatible with your LAN and come with all necessary device drivers. There is nothing more frustrating than trying to make something work that you simply don't have all the pieces to. Check with your LAN provider for a list of any cards that would work particularly well in your environment. If the list of multi-I/O cards given is short or nonexistent, try not to get frustrated. Not many people know about multi-I/O cards, and even fewer know how to get them and what they can do.

Don't forget about the aggregate line speed. If you determine that you will have a lot of users making telephony connections all at once and at potentially high speeds, even though you may have enough ports on your multi-I/O card, you may run out of aggregate line speed. The solution to this problem is to buy two or more fast multi-I/O cards with fewer I/O ports on each card, or get a multi-I/O card that has the necessary simultaneous line speed for your environment. Make sure the specifications of the card you purchase meet the needs of the application. Some multi-I/O cards send weaker signals, which can result in data loss over long distances. Check Appendix A at the back of the book for a list of vendors that supply multi-I/O cards.

Replacing the Communications Ports on Your PC/LAN

Replacing the communications ports on your PC with a multi-I/O card is not an inexpensive upgrade, but it frees up valuable resources while at the same time providing greater performance and reliability. Prices range from around $400 to $3,000 and up, depending on the number of ports, the aggregate line speed, and the operating system you are running. That's a big jump from $10 to $50 for 8250 and 16550 UART chip communications ports. In order to install a multi-I/O card, you should disable the existing communications ports that came with your PC. You will probably need to change jumper settings on your PC's controller card, remove an auxiliary communications card, or both. (This is the time that you wish you had saved all of the documentation that came with your computer, or you hope that your hardware supplier also provides good customer service.) Once the existing communications ports have been disabled, you can install your new multi-I/O card by following the instructions that came from the manufacturer. You will find that having the right hardware for the job can make life a whole lot easier.

Communications Cabling

There are physically different types of communications ports. There are DB9 pin ports, most commonly configured as COM1, and DB25 pin ports, most commonly configured as COM2. However, the DB port you use to connect to your modem will vary from PC to PC. Make sure you know which port(s) are available and purchase the right communications cable configuration to connect your COM port to your modem. Generally speaking, a DB9 COM port is male, requiring a DB9 female connector to join the two, and a DB25 COM port is male, requiring a DB25 female connector to join the two. An external modem has a DB25 female connector requiring a DB 25 male connector to join the two. Don't forget to include a communications cable, usually called an AT/Hayes-compatible communications cable, when installing an external modem. Refer to Figure 15-1 for information on cable pin outs and their assignments.

External Modems

External modems are the quickest and the easiest type of modem to install on any PC. The user can look at, touch, and feel the communications gateway that is about to be put into operation. In order to install an external modem, the user needs five basic components:

- An available communications port on the computer that the modem will be installed on
- A communications cable to connect the communications port to the external modem
- An external modem
- A telephone line
- Communications software to issue commands to the modem

If the user has these five elements available and the communications software is good and easy to use, then the modem can be installed and ready for use in a matter of minutes. Of course, it always helps if you know which communications port your modem is plugged into. And, if you know how to use the communications software that is prescribed for the application, you will significantly speed up the installation process. Most modems sold today come complete with communications software at no additional charge. This software is usually a stripped-down version of a more robust communications software package. If you intend to be a serious modem user, invest in quality software. The communications software you choose should perform all of the functions that you intend to use it for, and if possible, the specific modem that you are using should be one of the preconfigured brands and models that the

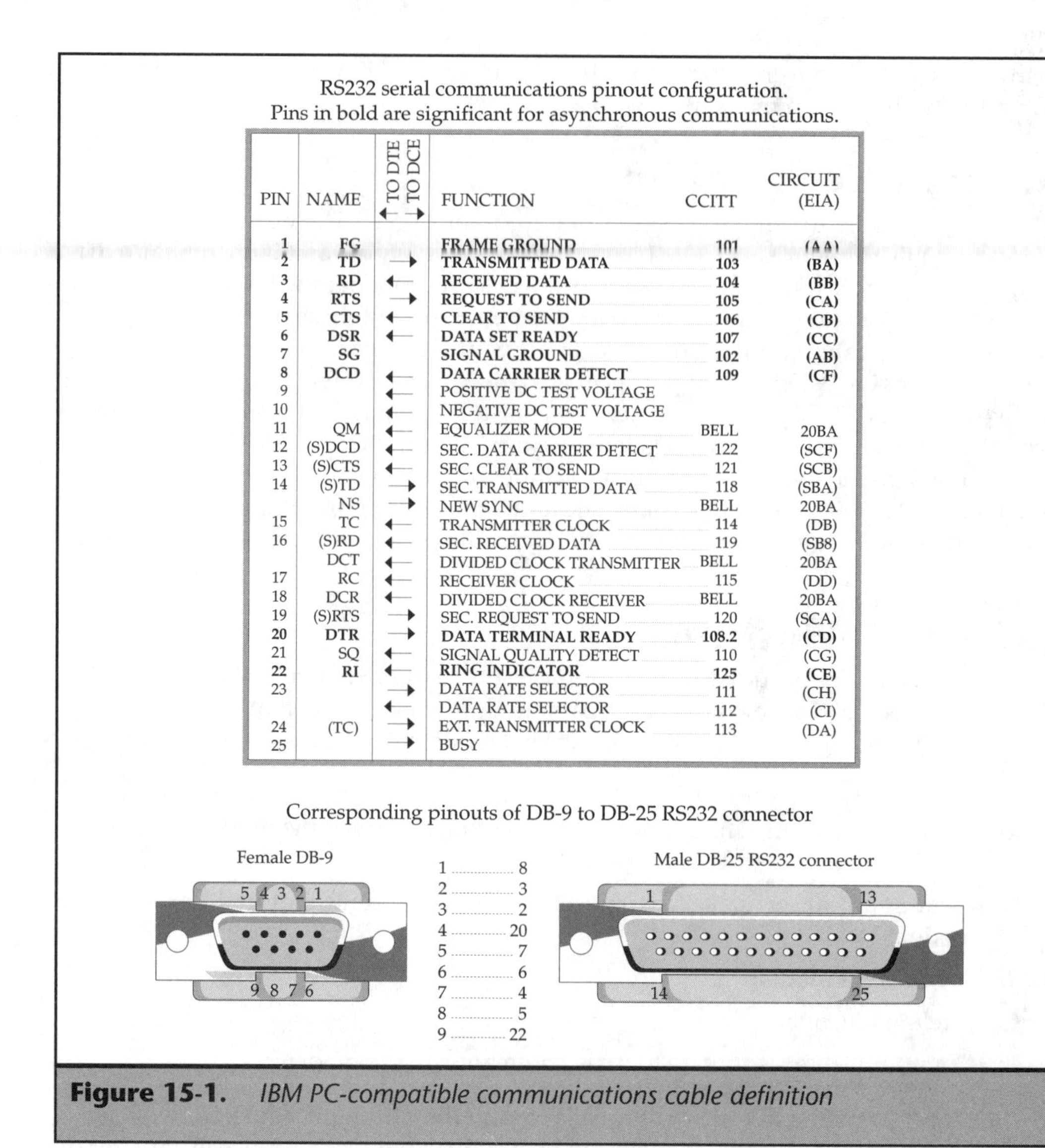
RS232 serial communications pinout configuration.
Pins in bold are significant for asynchronous communications.

PIN	NAME	← TO DTE / → TO DCE	FUNCTION	CCITT	CIRCUIT (EIA)
1	**FG**		**FRAME GROUND**	**101**	**(AA)**
2	**TD**	→	**TRANSMITTED DATA**	**103**	**(BA)**
3	**RD**	←	**RECEIVED DATA**	**104**	**(BB)**
4	**RTS**	→	**REQUEST TO SEND**	**105**	**(CA)**
5	**CTS**	←	**CLEAR TO SEND**	**106**	**(CB)**
6	**DSR**	←	**DATA SET READY**	**107**	**(CC)**
7	**SG**		**SIGNAL GROUND**	**102**	**(AB)**
8	**DCD**	←	**DATA CARRIER DETECT**	**109**	**(CF)**
9		←	POSITIVE DC TEST VOLTAGE		
10		←	NEGATIVE DC TEST VOLTAGE		
11	QM	←	EQUALIZER MODE	BELL	20BA
12	(S)DCD	←	SEC. DATA CARRIER DETECT	122	(SCF)
13	(S)CTS	←	SEC. CLEAR TO SEND	121	(SCB)
14	(S)TD	→	SEC. TRANSMITTED DATA	118	(SBA)
	NS	→	NEW SYNC	BELL	20BA
15	TC	←	TRANSMITTER CLOCK	114	(DB)
16	(S)RD	←	SEC. RECEIVED DATA	119	(SB8)
	DCT	←	DIVIDED CLOCK TRANSMITTER	BELL	20BA
17	RC	←	RECEIVER CLOCK	115	(DD)
18	DCR	←	DIVIDED CLOCK RECEIVER	BELL	20BA
19	(S)RTS	→	SEC. REQUEST TO SEND	120	(SCA)
20	**DTR**	→	**DATA TERMINAL READY**	**108.2**	**(CD)**
21	SQ	←	SIGNAL QUALITY DETECT	110	(CG)
22	**RI**	←	**RING INDICATOR**	**125**	**(CE)**
23		→	DATA RATE SELECTOR	111	(CH)
		←	DATA RATE SELECTOR	112	(CI)
24	(TC)	→	EXT. TRANSMITTER CLOCK	113	(DA)
25		→	BUSY		

Corresponding pinouts of DB-9 to DB-25 RS232 connector

DB-9	DB-25
1	8
2	3
3	2
4	20
5	7
6	6
7	4
8	5
9	22

Figure 15-1. *IBM PC-compatible communications cable definition*

software supports. This will drastically reduce the amount of knowledge you will need to have about your modem.

There are some distinct advantages to using an external modem. The first and most valuable reason to use an external modem is that you can see the status lights on the front of the modem. These status lights can tell the novice and the advanced user a great deal about the performance of their modem. Refer to Table 15-3 for a list of

Indicator	Definition
HS	High Speed: when lit, indicates that the terminal connection is set at the maximum speed of the modem operating in standard, noncompressed mode
AA	Auto Answer: when lit, indicates that the modem will answer on a specified number of rings set in the S0 register of the modem. When not lit, the modem will not be able to receive calls
CD	Carrier Detect: when lit, the modem believes that there is a modem tone on the other end of the telephone connection
OH	Off Hook: when lit, indicates that the modem has the receiver off hook and is attempting to communicate on the line. When not lit, the phone line is on hook
RD/RX	Receive Data/Transmission: when flashing, indicates that the modem is receiving data from the remote modem
SD/SX	Send Data/Transmission: when flashing, indicates that the modem is transmitting data to the remote modem
TR	Terminal Ready: when lit, indicates that the DTE device (via pin 20) is in the active state (usually controlled by the communications software)
MR	Modem Ready: when lit, indicates that the modem is in the ready state
RS	Request to Send: when lit, indicates that the DTE device (via pin 4) is telling the modem that it is ready to send data to the remote modem.
CS	Clear to Send: when lit, informs the DTE device (via pin 5) that the remote modem is ready to send data to the local modem

Table 15-3. *Standard External Modem Light Indicators*

standard status lights and their functions. Another advantage to using an external modem is that you can easily move it from one PC to another. Although most users don't have their modems migrating from PC to PC, it sure is a nice option when you are trying to diagnose a problem and you think the modem might be the cause. Swapping the questionable modem for one that you know works will be a sure way of telling if the modem is the cause of the problem.

NOTE: Your modem may not have all of the lights listed in Table 15-3 or they may be labeled a bit differently. Refer to your modem user's manual for more information on external modem lights.

There are two distinct down sides to using an external modem. External modems rely on the communications port they are connected to for the data they send and receive. If the communications port's speed is not equivalent or is superior to the modem's potential speed, then the application will suffer. (See the discussion on 8250 and 16550 UART chips earlier in this chapter.) In addition, external modems are a bit more expensive than their internal counterparts, although the extra money spent is usually worth the investment.

Follow these steps to ensure a safe and successful external modem installation:

1. Determine which communications port(s) is/are available for use. In a standard-configured PC, the COM1 port is used for the mouse, and COM2 should be available for connection to the external modem cable.

   If you are uncertain which communications ports are available on your PC and you have Microsoft Windows installed on your PC, change to the Windows subdirectory and type **MSD**, then press ENTER. This will run Microsoft's diagnostic program, which will show you the existing system setting on your computer, including your communications ports, the IRQ, and the base I/O address for the ports.
2. Once you have determined the COM port you will be using, power off your PC and unplug the power cable from the back of the PC.
3. Plug the communications cable into the available communications port determined in step 1 and secure the locking screws.
4. Plug the communications cable into the back of the external modem and secure the locking screws.
5. Plug in the telephone line that will be used with your modem. There are usually two jacks on the back of an external modem: one for the telephone line cord and one for a possible telephone jack to be connected. Always connect the telephone line coming from the wall directly into the line jack on the back of the modem.
6. Plug in the external modem and the PC power cord, then turn on both devices.
7. Install the communications software that you have chosen to use. Be sure to remember the COM, IRQ, and base I/O address in case the software asks you questions about your communications settings. Also, try to configure precisely the correct make and model of your modem in your communications software if at all possible.

Your modem should now function properly. Try to open the communications port and type **AT**, then press ENTER. The modem should respond with "OK." Or you may try to run one of the diagnostic tests that the communications software might have or try to call another modem that you are hopefully familiar with to ensure a successful installation.

Internal Modems

Internal modems are becoming more and more popular with PC users today. The internal modem has virtually the same functionality as an external modem, with some minor differences. If you are a more experienced user and you intend to use your modem exclusively on a single PC, then an internal modem might be the right choice for you.

Internal modems have three advantages over their external counterparts. First, if a particular modem brand and model comes in both an internal and an external model, the internal version of the modem will always be less expensive. Second, an internal modem has its own communications port built right into the card, so you don't have to worry about the type of communications port you are connecting to. And third, you won't have another piece of equipment plugged into the back of your PC and cluttering up your work area. Instead of a big communications cable, a modem, and a phone line, you will just have the phone line plugged into your internal modem's phone jack.

All of the same signaling and operational parameters that apply to an external modem also apply to an internal modem. The main difference in functionality between the two is that you get very limited feedback from an internal modem. There are no status lights to indicate what state your modem is in. Some communications software, such as ProComm Plus for Windows and Delrinas Communications Suite, will show you a picture of status lights onscreen, but there is no true substitute for the lights on an external modem. Only your level of understanding of the communications software and what you are telling the modem to do will compensate for the reduced amount of feedback provided by an internal modem.

Follow these steps to ensure a safe and successful internal modem installation:

1. Determine which communications port(s) is/are available for use. In a standard-configured PC, the COM1 port is used for the mouse, and COM2 should be available for use by some other serial communications device.

   If you are uncertain of which communications ports are available on your PC and you have Microsoft Windows installed on your PC, change to the Windows subdirectory and type **MSD**, then press ENTER. This will run Microsoft's diagnostic program, which will show you the existing system setting on your computer, including your communications ports, the IRQ, and the base I/O address for the ports. You must at this time decide either to disable one of the existing communications ports that already exist on the PC, or if available, plan to use one of the unused communications ports.

REMEMBER: Having two active COM ports on the same IRQ setting can cause very undesirable results. So try not to configure more than a one-to-one ratio between COM ports and the IRQ setting, unless you are a more advanced user and will consciously not use two COM ports on the same IRQ at the same time.

2. Once you have determined which communications port(s) is/are currently in use, you should unplug all cables from the back of the PC. If you are unfamil-

iar with the locations of all cables plugged into the back of your PC, it is a good idea to label where each cable goes.

3. Remove all necessary screws on the case to remove the cover and expose the cards in the computer.
4. If you have determined that you need to disable one of the existing COM ports (usually COM2) for modem installation, check your controller card user's manual for the correct jumper settings to disable the appropriate port. Write down the starting setting of all jumpers and dip switches before you begin. If you are not 100 percent certain of the correct changes you need to make, you can always return the settings to their original positions.
5. Check the manual that comes with the internal modem for how to set the communications port and IRQ settings. Make sure that your internal modem is set to the COM and IRQ settings that you determined in step 1.
6. Choose an available slot on your PC's motherboard and remove the screw that holds the cover plate in place.
7. Install the internal modem in the chosen slot and press firmly down until the card is properly seated. Reinsert the screw.
8. Replace the cover on the case, return all screws to secure the cover, and reconnect all cables to their proper locations on the back of your PC. If you are comfortable with leaving the cover off and testing the system, there is nothing wrong with temporarily not replacing the cover—the choice is yours.
9. Plug in the telephone line that will be used with your modem. There are usually two jacks on the back of an internal modem: one for the telephone line cord and one for a possible telephone jack to be connected. Always connect the telephone line coming from the wall directly into the line jack on the back of the modem.
10. Install the communications software that you have chosen to use. Be sure to remember the COM, IRQ, and base I/O address in case the software asks you questions about your modem configuration. Also, try to configure precisely the correct modem make and model in your communications software if at all possible.

Your modem should now function properly. Try to run one of the diagnostic tests that the communications software might have or try to call another modem that you are hopefully familiar with.

Fax Modems

One of the added features of modems in the 1990s is the incorporation of faxing ability into the modem card. What this means is that you can now send or receive facsimile transmissions directly from or onto your PC. A fax modem should be able to communicate with any fax machine that you can send or receive from using a regular fax machine. Purchasing a fax modem over a standard modem adds very little to the price of the modem, if anything. As you become more familiar with the communications environment you operate in, you will find that having the convenience of faxing documents directly from your desk is a nice little added bonus.

In order to make your fax modem work as a facsimile machine, you will need software for your PC that is capable of sending and receiving faxes using your brand of fax modem. Almost every modem sold today includes communications and faxing software at no additional charge. These software packages are usually stripped-down versions of more robust packages or they are very remedial software packages to begin with.

The software that you use to control your fax modem is extremely important. After all, this is where you interface with your fax modem. If at all possible, try to purchase a full version of PC faxing software that is compatible with your computer system and which hopefully has support for your specific fax modem. If your specific brand and model of fax modem is supported by the software that you purchase, the amount of knowledge you need to have is greatly reduced. For more information on fax modems and computer faxing applications, refer to Chapter 16.

Modem Pools

A *modem pool* is a group of modems hooked up in a series to answer telephone calls from multiple modem users at the same time. All of the modems in a modem pool are usually hooked up to the same computing device. The initial connection is usually to a multi-I/O card, which in turn is plugged into a PC, LAN, mini, or mainframe computer. When you call your favorite bulletin board service or access your LAN from home, you are calling into a modem pool.

Modem pools can also be accessed in the reverse direction. Multiple PCs on a LAN can have a few modems available for outdial connections over the switched telephone network. The reason for modem pools in an outdial configuration is to cut back on the number of modems needed to support the user base. Since the cost of modems has dropped drastically over the years, the majority of modem users have their modem connected directly to the back of their PC. Therefore, the outdial modem pool is becoming a thing of the past. Security is an issue with modem pools, depending on the information that is accessible through the modem pool. Proper security measures should be implemented when necessary. (Network security can be highly complex, is unique to every computer network, and needs to be addressed with the specific user's requirements in mind.)

The telephone line configuration for a modem pool deserves some mention. The phone lines can be set up in a number of different ways. The most common configuration of the phone lines is to have one main telephone number that rings on any one of a number of physical telephone lines. The calls are spread out evenly over all of the phone lines in the group. This type of line configuration is known as a *universal call distribution group* (UCD), *call pickup group*, or *longest idle*.

Another configuration used for modem pools is to have one main telephone number that rings on the first available phone line. The call will always be answered by modem number one if it is not busy. If modem number one is busy, modem number two will always answer the next call unless it is also busy, and so on. This type of line configuration is known as *asequential* or a *hunt group.*

The last form of phone line configuration used for modem pools is one in which you have multiple telephone numbers that ring on separate or multiple lines, and they are all connected to the same modem pool. This configuration is used to give the system administrator greater flexibility over the access to modems in the pool. Special users can be given a special phone number (possibly an 800 number), which accesses specific modems that are reserved. If those modems are all busy, the phone system can still hunt the call over to the standard modem pool lines. This method of line configuration can combine the use of both configuration methods described above.

Modem Gateways

Modem gateways perform the function that the name suggests: they allow the user to have a gateway between two or more systems. Quite often, a modem gateway performs some kind of data translation from one protocol to another, and once the gateway is accessed, the remaining links to other systems are usually fixed. A user on a LAN-based platform may need to access a mainframe, a PC, or another LAN-based system. A modem gateway gives the user the ability to access a wider variety of systems from a single exit or entry point. Modem gateways can work in both directions.

Let's say a user is working on a home PC and needs to access some information available on the mainframe system back at the corporate office in Dallas. The PC user could dial a local telephone number that is connected to a modem gateway. The user would then enter in the proper password or entry path. The modem gateway can connect the user via a *wide area network (WAN)* connection from the local access point to the mainframe computer in Dallas, where the information can be accessed. The same modem gateway given a different password or entry path might connect the user to the local mainframe, another network computer, or a LAN—the list goes on and on.

The secret to a modem gateway is knowing *what* can be accessed through the gateway and *how* to get to it. The nice thing about modem gateways is that they take care of the various connectivity and protocol issues. The entry point protocol and configuration is the only thing that the user needs to know in order to make a successful connection. The modem gateway is a valued asset to the people who use them to their fullest capabilities, but they are also doorways into your company from the outside world. Unlike modem pools, where the modems direct the user, usually, to one specific system, modem gateways link many systems to one access point. Security measures must be properly implemented at all times.

Gateways do not always involve modems. Quite often, your LAN or WAN is acting like a gateway as well, connecting multiple systems together with sometimes little or no barriers. Configuring your bridge, router or software with even a small amount of security and/or auditing is a wise decision.

Baud, Bits, and Parity

Baud, bits, and parity are three things you need to know about the modem equipment you will be communicating with. These three items vary consistently from one system to another. Baud, bit, and parity can easily be changed through your communications software, providing you know what the proper settings should be and you have the right communications port for the job. If the baud, bit, and parity are not set correctly, you may see all sorts of garbage characters on your screen, your keyboard may not respond, you may get partial pieces of information, or you may not be able to make a connection at all. When you are the caller, it is your responsibility to make sure that your communications software and modem settings are identical to the called modem's baud, bit, and parity.

Baud

Baud is a measure of transmission over standard telephone lines. Baud rate is often confused with bits per second (bps). Based upon the number of cycles per second available over voice lines, the theoretical maximum baud rate would be 6,000 baud. But modems today are much faster than this. Although baud rate has a physical limitation over POTS, the number of bits per second is still on the rise. Baud rate and bits per second used to be equal measures of transmission, but the actual bits-per-second rate has continued to rise well above the maximum baud rate of 6,000. Modem communications standards have been developed over the years that encode more bits in a single baud, allowing greater amounts of data to be sent over an analog line in a given instant. Some of the more recent standards used by modems are V.32, V.32bis, and the latest V.34 ITU-T standard, which allows for the current 28,800 bits per second maximum transmission rate. Today, modem speeds are quoted in bps, but sometimes you will hear the words "baud rate" uttered. Don't worry; even though there is a technical difference between the two, in common usage, the two terms are synonymous.

Bit

Data begins with a little thing called a bit. A *bit* is, quite simply, a 0 or a 1. Computers combine either seven or eight bits to form one byte or character of information. Seven-bit data is in ASCII (American Standard Code for Information Interchange) form, where up to 128 different characters of information can be exchanged. Eight-bit data is in EBCDIC (Extended Binary Coded Decimal Interchange Code) form, where up to 256 different characters of information can be exchanged. The majority of computer data can still be represented with the seven-bit ASCII standard, so whenever possible the seven-bit format of data transfer should be used, to reduce the number of bits necessary to transfer the information and therefore increase the overall throughput speed. The actual interpretation of the bits/bytes of information being received is determined by the terminal emulation that is being used for display information. Refer to the "Terminal Emulations" section later in this chapter for more information.

Parity

Parity is an extra bit of information attached to every byte of information that is sent between two modems. The parity bit is used to determine if any of the information has been dropped or changed from one end of the transmission to the other. Both modems involved in the transmission and reception of data must have the same parity definition in order to obtain a reliable connection. There are five different settings of parity that modems can use to verify the transfer of information. *Even parity* is when the modem adds up all the bits to be transferred in a byte, and if the total of those bits is an odd number, the parity bit is set to 1, or the parity bit is set to 0 if the total is an even number. Conversely, *odd parity* is when the modem adds up all the bits to be transferred in a byte, and if the total of those bits is an odd number, the parity bit is set to 0, or the parity bit is set to 1 if the total is an even number. *Space parity* always sets the parity bit to 0, while *mark parity* always sets the parity bit to 1. The last setting for parity is *no parity*, where the parity bit information is ignored.

Terminal Emulations

Terminal emulation is how a computer interprets the information that is received from a remote source. The remote source could be a network or a modem connection. Users need to be concerned with terminal emulation whenever they are communicating between two different systems. Users can get into trouble simply by running different applications on their own LAN that require specialized terminal emulations or, in some cases, specialized hardware. Just because you can connect to the application you want to run doesn't mean that you can communicate and work with it. The modem or the network simply provides the connection between the systems. Terminal emulation is how the data received is interpreted and presented to the user.

Literally dozens of terminal emulations are used on a wide variety of hardware and software platforms. The majority of communications packages off the shelf come with a number of terminal emulations built in as part of their features. However, there are several packages that are strictly for terminal emulation. ADM 3A, ANSI, AT&T UNIX, Interactive UNIX, SCO UNIX ANSI, DG Dasher 100, DG Dasher 200, Hazeltine 1500, IBM 3101, IBM 3780, Solaris Console, Unixware Console, Televideo 912, Televideo 920, Televideo 925, VT-100, VT-102, VT-220, VT-52, Wyse 50, Wyse 60, and X/Open base ANSI are just a few of the more common terminal emulations that are used in the industry today. By far, without a doubt, ANSI (American National Standard Institute), which uses the ANSI eight-bit character set of 256 characters, is the most common form of terminal emulation used today. Almost all bulletin board systems support the ANSI terminal emulation for access to their systems. If you are not sure what terminal emulation you should use for a connection, start with ANSI and see what happens. It may not be the ideal emulation for the application, but it works in a lot of situations.

Whenever possible, try to find out what the required or desired terminal emulation should be for optimal performance. Improper settings can cause a great number of problems, ranging from unwanted characters being displayed on the screen to complete and total failure of the application. Knowing in advance what terminal emulations you will require from your communications software is a tremendous asset. No one communications software package supports all terminal emulations, so knowing the emulations that you will need in advance of selecting a communications package (or, if necessary, a stand-alone terminal emulation package) is a distinct advantage. One other thing to keep in mind is that some emulation packages aren't 100 percent compliant with the original terminal specifications they claim they support. Try to make sure that you can return the product easily in case your application does not work well with the emulation package you are using.

Error Correction Protocols

Error correction protocols verify that the information transferred between two modems has been received without any transmission errors. If errors are detected in the transmission, the receiving modem will request that the information be resent. Error correction protocols are increasingly important as information is condensed and compressed and then shot over your voice lines at the incredibly high speeds at which modems operate in today's world. Your overall throughput goes down when packets of data are reshipped over the connection, but corrupt data is even worse than no data. Without getting into a deep discussion of the intricacies of each and every error correction protocol that has been and is being used in modem communications today, you can easily make an educated decision about error correction and your modem. Make sure that the modem you purchase supports as many error correction protocols as possible.

The more popular error correction protocols used in asynchronous communications today, in order of increasing sophistication are V.22bis, MNP 1, MNP 2, MNP 3, V.32 terbo, LAPM, MNP 4, and V.42. Virtually all modern modems will support these and maybe a few other error correction protocols. Both modems must have the protocol in order for that form of error correction to be performed. For the novice, or even the intermediate user, error correction is not a major concern. The modems handle this function automatically, so your level of knowledge does not have to be high. Simply understanding what error correction does for you is enough.

Compression: The Key to Speed

Compression of data is how modems can send greater quantities of information over the same telephone lines in the same amount of time. The actual CCITT (Consultative

Committee on International Telegraphy) and ITU-T (International Telecommunications Union-Telecommunication sector) modem standards that have been set down over the years are a form of compression. After all, the V.21, V.22, V.22bis, V.27, V.32, V.32bis, and the most recent V.34 standards all allow more information to be sent at once, even though the data is not technically compressed, so the net effect is more information being sent in a shorter period of time.

There are, of course, protocols that have been developed in addition to the modem standards that preprocess the data and compress the same amount of information into a smaller number of bits. You may be familiar with PC products such as Stacker, Double Space, and PKZip, which all compress data for storage on your hard disk. When a modem compresses data, it performs a similar compression function on the fly. As data is sent to the modem, the modem compresses the data down and sends the compressed information. The receiving modem, using the same compression protocol, expands the information back into its original form so that the data can then be either displayed or captured to the hard disk or the printer. Microprocessors are extremely fast, while data transmission over voice lines is relatively slow by comparison.

Even though additional functions are being performed by the modems when using compression protocols, the net result can sometimes more than double the throughput of your modem. When looking for a modem, remember to get one that has as many compression protocols as possible. The two most popular and standard compression protocols are MNP5 and V.42bis. MNP5 compression usually yields a 200 percent increase in overall throughput. V.42bis can yield anywhere from between 300 to 500 percent compression of information being sent. If your modem supports these two compression protocols, then your overall throughput will be greatly increased.

It should be mentioned that compression protocols work with varying degrees of success, based upon the type of data that they are trying to compress. If you are trying to compress and send a data file that was already compressed using PKZip for example, you may actually experience a drop in overall throughput. The reason for this is that the modem may be trying to compress data that cannot be compressed any further. The actual number of bits being transferred remains the same, but the compression processing time is still added to the transmission.

File Transfer Protocols

File transfer protocols give the user the ability to upload and download data files from one computer system to another. A file transfer protocol provides data compression and error correction while a file transfer is in progress. In order for you to use a file transfer protocol, both systems must have the same file transfer protocol available for use. ASCII, RAW ASCII, CIS-BT, XMODEM, XMODEM CRC, XMODEM-1K, XMODEM-1K-G, YMODEM, YMODEM (batch), YMODEM-G (batch), ZMODEM, SUPER ZMODEM, and KERMIT are some of the more popular file transfer protocols in use today. They are listed above, for the most part, in the order of increasing speed and sophistication. You will find that KERMIT is probably the most widely available file transfer protocol in use on the widest variety of computer platforms. ZMODEM, SUPER ZMODEM, AND KERMIT all have an added feature that allows a file transfer session to be restarted even after a communications link has been lost. This is an especially handy feature when you are transferring large amounts of data. In most cases, these protocols should be used whenever both ends of the connection have these more advanced file transfer protocols available.

The file transfer protocol discussed here should not be confused with FTP, which also stands for file transfer protocol. FTP is a file-sharing protocol that allows for a LAN-based SLIP or PPP connection file transfer of layers five through seven of the Open Systems Interconnection (OSI) model. FTP can also move files from computer to computer, but it performs this function at a different level than the file transfer protocols mentioned above.

DTE and DCE

Modem communications is made up of many different components. This chapter has focused on modem-to-modem communications. Computer-to-modem communications should not be overlooked. DTE stands for data terminal equipment, or simply put, your computer. DCE stands for data communications equipment, or simply put, your modem.

The speed at which you set your communications software is the speed at which your computer talks to your modem, and it is the maximum speed at which your modem can communicate information to another modem, regardless of how fast your modems can connect to one another. The DTE-to-DCE speed should be set to the maximum speed that you can hope to achieve between the two DCE (modem) devices. Any faster, and you are just blowing smoke. Any slower, and you might not be realizing your modem's maximum throughput. There are circumstances where you intentionally want to drop down the DTE-to-DCE rate. If your modem is constantly renegotiating the line speed and you are experiencing poor performance out of your modem, then you may want to drop the DTE speed back a bit. Yes, you won't be transmitting at the maximum rate, but you will optimize your connection by reducing the amount of time wasted with modems retraining their connection to each other.

Making a Reliable Connection

It is extremely important that you know who you are talking to if you want to make a reliable modem connection. At a bare minimum, you need to know the telephone number of the modem you are calling and these four pieces of information:

- Baud rate or bps
- Number of data bits

- Parity being used
- Terminal emulation

With these four pieces of information, you can be off and running rather quickly. Modems will drop down to the proper speed needed for the connection, based upon the line conditions and the modem on the other end of the connection. Communications software will allow you to define different configurations of your modem. You might have one configuration to call a friend who has a 2,400-baud modem, another for your local BBS, and a third for your modem pool that gets you into your company's mainframe computer system. Make your configuration match as many of the four points mentioned above as you possibly can.

Optimal settings for a modem connection are not always as simple as one might hope. Modems have their own language made up of error correction protocols, compression protocols, and modem standards. You must also keep in mind the terminal emulation and the specific software applications. With all these different conversations going on at the same time, it's not difficult to see how something can get lost in the translation. Ideally, you want to turn on only the features of your communications software and modem that the other side of the connection is capable of taking advantage of. By disabling and enabling the proper features of your modem and communications software, you can cut down on unnecessary modem negotiations and invalid data interpretations.

Gathering the hardware and software specifications of all the modem connections that you make is not always possible. In fact, the modems and communications software will negotiate the majority of the pertinent settings and capabilities themselves. But if your modem settings are configured specifically for the connection you are trying to make, you will be assured of a more reliable connection.

Chapter Sixteen

Facsimiles and Fax-on-Demand

You need it right away? OK, I'll fax you a copy. Sound familiar? Facsimiles have become a common part of our everyday lives in the last ten years, both in business and, increasingly, in the home. Fax-on-demand is a new application of telephony whereby you are able to request information in the form of a facsimile to be sent to you in an automated fashion.

A fax machine actually acts as three machines, because it has to scan, transmit, and reproduce information. As a scanner, it can take any image on paper and convert it to binary data. The binary data is then transmitted via a modem connection across switched telephone lines to another modem on another fax machine. That machine then converts the binary data back into an image that can be stored or printed on a piece of paper. Of course, there are variations in how the information is sent and received, but the three functions of scan, transmit, and reproduce are at the heart of fax technology.

The Evolution of Fax

Fax machines have been going through the same product life cycles as the calculator 20 years ago and the VCR a decade ago. Both products were available many years before they became an integral part of our everyday lives. It was the improved technology associated with these products that enabled them to become affordable and settle in as a common part of the home.

The fax machine is proving to be no different. In 1991, the market for facsimile and fax services was $115 million. In 1994, it jumped to $775 million and is expected to reach over $1.2 billion in 1995. By the late 1990s, fax machines will be as common in homes as VCRs were in the 1980s. Lower prices and a fledgling resale market are enabling these mainstays of the office to become affordable for home use. Experts agree that residential applications will open an enormous new market for fax-on-demand. Fax machines for consumers are already selling for well under $500. Two million stand-alone fax machines are expected to be sold in the U.S. in 1995. Experts predict that 62 percent of those machines will be for home offices, work-at-home, or other residential settings. As fax machine prices continue to fall, people will be better able to shop and to do research without leaving the comfort of their den or living room.

With the invention of the fax modem and advanced PC faxing software, the shift to paperless faxing is beginning to become a reality. Advances in fax technology have also brought new tools to the business user, tools that can provide information faster and save time, money, and labor. New ways to serve your customers and exciting business opportunities have been born through the proliferation of fax machines, particularly with the recent integration of voice-processing technology (see Chapter 14) with fax technology.

Types of Facsimiles

You may be bewildered by the wide variety of fax machines available today. They all do about the same thing, right? Like any other product, though, fax machines vary in quality, features, and functionality. For example, in addition to the traditional fax machine that stands alone, sending and printing faxes as necessary, a whole new breed of fax modems integrates fax functionality into the computer. When you decide to enter the world of faxing, or if you are in the process of adding new facsimiles, you must select the facsimile(s), and possibly software, that are right for you and the applications that you will use it for. Here is some information to help you make your decision.

Thermal Bond Fax Machines

At the low end of the stand-alone fax machine market is the thermal bond fax machine. Ranging in price from about $200 to $600, these machines use special (annoying) paper that comes on a roll. As faxes are received, the image is transferred to the page through heat. No ink or developer is used in the printing process. The main drawback to thermal bond fax machines is the paper. Not only does it have a strange feel, but—because it comes on a large roll—the sheets tend to curl (though some of the newer models have a decurling feature). In addition, the fax machine must cut the sheets of paper (some machines require you to cut the pages apart), which it does not do consistently, leaving some sheets longer than others.

In defense of thermal bond fax machines, they do have several advantages. They are typically low maintenance, with less down time than other stand-alone fax machines. They are less expensive to purchase and operate. For basic use, particularly in the home, you may find that a thermal bond fax machine is the right choice for you.

Plain Paper Fax Machines

Plain paper fax machines are extremely popular in the business world today. With prices ranging from $500 to $2,000, they are less popular in the home, though SOHOs (small office/home office) may use them for the professional appearance of the resulting fax transmission.

These machines produce an image in a number of different ways. As the name indicates, the plain paper fax uses standard paper, such as the kind you might use in your photocopier or laser printer. The image can be transferred to the page using one of the following methods.

INK JET This is the newest generation of plain paper fax machines. They are primarily for the low-end market. They use basically the same technology as ink jet printers. The image generated is good, but the print time is usually a little slow. The maintenance needed for these machines is just a little more than for an ink jet printer, but they are not designed for high volumes of traffic. These machines are capable of producing a gray-scale image instead of straight black and white, which, as you can see in the following example, is a distinct advantage over comparably priced products:

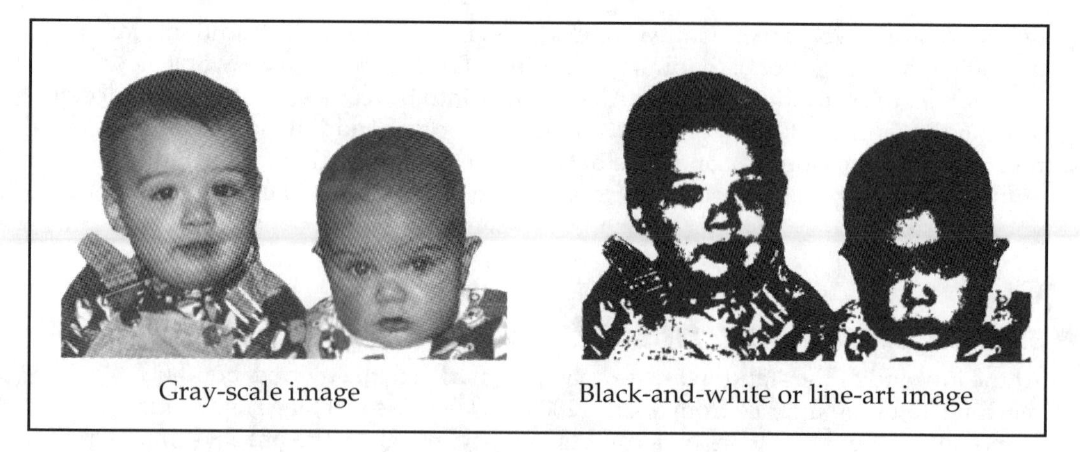
Gray-scale image   Black-and-white or line-art image

Usually, these models do not offer some of the higher-end functions of stand-alone fax machines, such as fax broadcasting (see the "Fax Broadcasting" section later in this chapter) and multiple programmable speed dialer buttons.

CARBON FILM TRANSFER A carbon film transfer fax machine uses a roll of inked film to transfer the image onto the paper. These fax machines produce a nice image and are usually more reliable than the ink jet models. However, they do not produce gray-scale images very well and require significant maintenance proportional to the amount of use. These machines do have their place in business. They are typically designed for high volumes of use and usually include the majority of functions found in stand-alone fax machines.

TONER These are the highest quality fax machines available. They work in the same fashion as a laser printer or a photocopier in order to transfer the image to the page. These machines combine the best of all the plain paper fax machines. They can produce a high-quality gray-scale image, which results in higher quality overall. These fax machines are made for high volumes of use and should stand up to the abuse. Maintenance is about the same as a laser printer or photocopier in terms of down time and cost. Virtually all of the functions found on stand-alone fax machines should be found on a toner-based fax machine.

Fax Modems

Fax modems are growing rapidly in popularity. The cost to add faxing ability to a PC modem card is minimal. In fact, most modems today include faxing capability as a standard function with no effect on the modem function at all. Fax modems can cost as little as $50, yet their performance is admirable. Most come with faxing software, however basic, at no charge. Faxing using a fax modem does, however, rely on at least three and perhaps four components: the computer, the fax modem, the fax software, and (arguably) the printer.

A fax modem relies on the power of the personal computer and software to break apart the three primary functions of a fax machine. A fax modem cannot scan a page of information unless the PC is equipped with a scanner, which is an expensive piece of equipment. But a fax modem can transmit any document that is already on a computer. This is accomplished through specialized printer drivers that redirect the computer print output to the fax modem for transmission. The fax transmission function of a fax machine is performed by the fax software and the fax modem itself. Lastly, received facsimile transmissions can be printed by the fax software and the printer that is connected to the PC, or the fax can be viewed onscreen, thus avoiding the printing process entirely.

Fax Software

When all is said and done, the true power of PC faxing lies in software. WinFax Pro by Delrina is by and large the industry leader in PC faxing software, and justifiably so. The fax software provides a number of high-level functions with starting prices at around $50.

Network versions of fax software have all the functionality of stand-alone fax software, but they also allow the fax modem to be used as a fax server for everyone on the network. Network fax software starts at around $100 and goes up based upon the number of users. Network faxing software allows transmissions to be queued, which means no more waiting in line for the fax machine. Incoming faxes can be distributed automatically by the system, and unidentifiable faxes may be directed to a specific person or automatically printed.

Fax software also provides a high degree of functionality with regard to phone number storage and retrieval. The personal computer does not have the memory limitations of stand-alone fax machines. Fax broadcasting is extremely easy to set up and to manage. One strong advantage of computer faxing is the ability of the software to archive any and all faxes that are sent and received by the software. A copy of the fax transmission can always be retrieved, and you don't have to worry about losing your paper transmission log, as you would with a regular fax machine. Other services, such as off-site fax mailboxes and broadcasting, are offered by many of the major fax software companies.

If you are moving in the direction of the more economical fax modem, invest wisely in the fax software you choose. The fax software is your interface into the world of faxing.

Fax Capabilities

The functions modern fax machines offer are extremely diverse. Some of the features you may take for granted or never use, while others may bring new functionality to your operations. Here are some of the features you may want to evaluate in a fax machine:

- Group 3 or Group 4 support
- Class 1 or Class 2 support
- Fax broadcasting
- Directory of frequently called names and numbers
- Automatic transmission retry
- Transaction logging
- Number of pages held in memory
- Fax scheduling
- Gray-scale transmission and reception
- Support for Error Correction Mode (ECM), a data correction enhancement to Group 3 facsimiles to request retransmission
- Baud rate supported (14,400, 9600, 4800, etc.); see Chapter 15
- Support for Caller Station ID (CSID) which allows the fax machine to detect the telephone number of the calling party

Group 3 and 4 Fax Support

Group 3 facsimiles are the industry standard for facsimile transmission. Group 1 and Group 2 are older standards that supported slower transmission rates. Group 3 facsimiles support the slower Group 1 and 2 standards, but they are only used when line quality is poor. Group 3 transmissions are done over regular POTS lines. Group 3 fax transmission supports two resolutions: standard (203×98 dots per inch, or DPI) and fine (203×196 DPI); Group 3 enhanced also supports superfine (203×391 DPI).

The new standard for digital faxing is Group 4 and is supported only over 56/64 Kbps lines, i.e., ISDN lines (see Chapter 11). Where Group 3 fax transmission on POTS lines on average takes about 20 seconds per page, Group 4 transmission takes less than six seconds per page. Group 4 currently offers 400×400 DPI resolution, which is high quality, however the primary advantage is in the speed. Group 4 faxing is very rare; few people or organizations have ISDN circuits, and even fewer have Group 4 fax machines. Wide acceptance of this technology is still three or four years away.

Fax Modems and Classes

Fax classes are something to keep in mind when choosing a fax modem. The fax class only applies to fax modems, not fax machines. Fax modems that use Class 1 (also called Class 1/EIA-578) are based on the Hayes AT command set (see Chapter 15) that can be used by software to control fax machine functions. Class 1 fax modems are the most common type of fax modem used in the world today.

Class 2 (also called Class 2/EIA-592) is a new standard used between facsimile software and fax modems. A Class 2 modem places more of the emphasis on the hardware for the connection while leaving the transmission of the data to the software module. Typically, Class 2 modems also support Class 1. The Class 2 standard is still under study by the EIA (Electronic Industries Association), and further revisions are expected.

Fax Broadcasting

Fax broadcasting, one of the more popular features of modern fax machines, lets you send a fax to several recipients. Some fax machines will allow you to broadcast to a limited set of numbers stored in memory in a specified order, but fax broadcasting is primarily the domain of computer-based fax software (it can also be contracted through a fax service bureau). Having the ability to send specialized fax information to a number of people with a touch of a button and to schedule the faxing process to occur during off-peak hours is a tremendous cost saver. New ways to use your system's fax broadcasting ability should be explored by your marketing, promotions, and public relations departments. Your costs will drop and your productivity will rise.

Fax-on-Demand

A new technology called fax-on-demand (also referred to as interactive fax, fax back, or fax retrieval) allows anyone with a touch-tone telephone and a facsimile machine to request documents and receive them immediately. With the marriage of interactive voice and fax, this versatile technology is proving its worth as a direct marketing tool. Using this tool places greatly enhanced business opportunities at your fingertips. The need to travel around the world or even across town to communicate effectively with customers and prospects has been greatly reduced. Instead, new and more sophisticated methods of getting messages to your audience, no matter how specialized or targeted, have been developed. At the same time, fax-on-demand can reduce your costs for mailing and printing, and the labor of hand-feeding your own fax machine.

The residential segment will certainly need access to a broad range of information. And just think of all the types of information families would want. Information could range from getting restaurant menus and movie reviews to the more complicated, such as referencing databases to help children do their homework.

Making Your Information Accessible

To be accessible to the public, the faxable information must be digitally loaded into the fax vault. A *fax vault* is typically a personal computer used to store digital fax information for later retrieval by the fax-on-demand system. Some companies make hardware specifically designed for this function.

The information is normally set up into different documents by document number. A caller dials the fax-on-demand service number and is voice-prompted to enter a document number. If the document number is unknown, the caller is usually given the option to receive a directory of documents. Many companies publish their directory in a full-size ad when advertising this service. After selecting from this directory, the caller enters a document number and is voice-prompted to enter the fax number to which they want the information sent.

Many fax-on-demand services are on an 800 number, whereby either the information is free to the caller or the caller is charged a fee, which requires the caller to enter a credit card number and expiration date. When the caller is charged for the information on an 800 number, the call is referred to as a "pay-per-call" service, and the caller is charged a flat rate for the call.

For example, when you call the 800 number, a voice-prompt will ask you to enter your credit card information by using the keypad on your touch-tone phone. You will then be asked to enter the document number for the information you are requesting, followed by your fax machine phone number. You will be charged a flat fee for the information, which is billed to your credit card. This option is becoming common practice for businesses such as magazines and newspapers, which provide reprints of past articles and issues to callers. Setting up a fax-on-demand service on a 900 number, also a "pay-per-call" service, is a trend that is also becoming popular with companies that choose to charge for the information, such as magazines and newspapers, because it costs them money to supply this information.

This service can be put into your business by promoting it in two different ways. You can add your fax-on-demand service number to any and all advertising you are currently running. Like many companies, you can simply place the number at the bottom of your ad, or you can place an entire ad focused towards your new service, as *Fortune* magazine has done with their current ad, "5 Years of *Fortune* in 5 Minutes Flat." Some businesses choose not to advertise their service in this fashion; instead, when the company receives a request for information that is frequently requested and faxed out, they then refer the caller to their fax-on-demand service number.

How Fax-on-Demand Is Used

Many companies are turning to fax-on-demand to help control their mailing, postage, staff, and printing costs. Instead of having employees tied up on the phone accepting requests and faxing information over and over again, companies are referring those calls and requests to their fax-on-demand service number. This way, staff members are free to use their time more wisely, and callers receive their faxed information in minutes. Examples of companies already taking advantage of this new technology include the following:

Acer: (800) 554-2494
Hewlett-Packard: (800) 946-0714
Microsoft: (206) 635-2222
Detroit Free Press: (313) 886-6685
IBM: (800) IBM-4FAX
NEC: (800) 366-0476

Try some of these fax-on-demand service phone numbers to get an idea of how fax-on-demand works. You will be able to immediately obtain a good amount of information about the products and services that these companies offer through their fax-on-demand systems.

Many variations of the fax-on-demand technology can be readily implemented. Here are some ideas that have been set up using a fax-on-demand service:

- A magazine can use fax-on-demand to provide reprints of articles and reader service.
- A catalog merchant can use fax-on-demand to provide access to detailed product information, including instructions, manuals, and manufacturers' warranties.
- Real estate listings can be made available via fax-on-demand. A simple set of voice-prompts allows callers to select properties by price, location, and major features, such as number of rooms.
- Restaurants can use a simplified fax-on-demand service to dispense a single document—the menu or menu of the day.
- A fax-on-demand system can be used as a literature dispensary by the sales force of a company. Instead of faxing materials directly, the sales force calls the fax-on-demand system, which automatically sends the information requested. Customers and prospects may also use it directly.
- Banks are beginning to offer statement information using fax-on-demand. Customers can call in and request a current statement at any time. For security, statements are transmitted only to a predetermined number, addressed to the account holder.
- Colleges can use fax-on-demand to supply students with updated information on classes and schedules, recreational activities, sports events information... and the list goes on.
- Fax-on-demand can be used as a direct revenue generator. For instance, *Consumer Reports* and the Auto Club for a small fee will allow you to obtain actual factory invoices (the price the dealer paid) for any new car you might be thinking of purchasing. Valuable information at reasonable rates.

Using a Service Bureau

A fax-on-demand service can be set up in one of two ways. You can either buy all the equipment yourself and have an on-staff programmer load all the information into the fax vault, or you can work through a service bureau that specializes in fax-on-demand. A fax-on-demand service bureau will provide all equipment and technical expertise to get your project off the ground in a professional and timely fashion.

Three-quarters of all businesses offering a fax-on-demand service in the United States choose to work through a service bureau. The entry cost for fax-on-demand can be high, depending on the number of documents offered through the fax-on-demand service and the complexity of the scenario. Using a service bureau is less expensive initially, and the service bureau will do all the work for you, which is very important to businesses new to this technology.

Choosing a Service Bureau

Several service bureaus offer a full-service customer support staff, programming staff, and technical consultants. They will work with you to create a customized service and setup on an 800 number, toll number, or even a 900 number. Here are some service bureaus that offer fax-on-demand (see Appendix A for a list of addresses and telephone numbers):

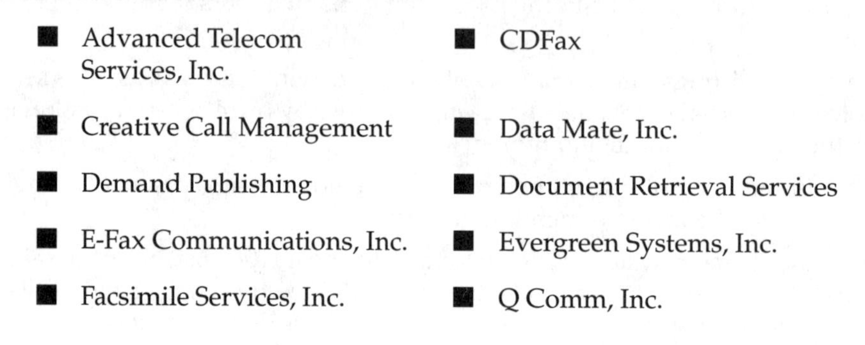

- Advanced Telecom Services, Inc.
- CDFax
- Creative Call Management
- Data Mate, Inc.
- Demand Publishing
- Document Retrieval Services
- E-Fax Communications, Inc.
- Evergreen Systems, Inc.
- Facsimile Services, Inc.
- Q Comm, Inc.

Something worth considering when choosing a service bureau is the assistance you will or will not receive during and after your setup. Research shows that many service bureaus do not offer unlimited customer service. In many cases, customer service from your fax-on-demand provider is the difference between success and failure. When getting involved in such high-technology services, you need to make sure that the service bureau you choose is going to be there to assist you with any changes or additions to your service.

Options relating to your fax-on-demand service are also important when choosing a service bureau. As mentioned earlier, most fax-on-demand services are currently being set up on 800 numbers. As this industry is growing, many large companies have moved to the use of a 900 number in order to charge for the information. Fax 900 charges start from about one dollar and go up, depending on the information being provided. You need to know in advance if the service bureau you choose has the capability and experience to offer these different options.

COST Pricing is extremely important. In addition to a setup fee, service bureaus may charge by the minute or page for additional documents or pages, for monthly storage, and for changing a document. Not all service bureaus charge for each of these services, but most do.

SETUP FEE This fee is normally a one-time fee that is charged for setting up your information in a fax vault, delegating document numbers, and setting up the voice prompts. This fee can range anywhere from $300 to $2,000 depending on the service bureau and the scope of work. When speaking to service bureau providers about this fee, it is important to have them list everything that is included. Some include a designated 800 number, while some do not.

ADDITIONAL DOCUMENT FEE This fee refers to any additional documents you may add after the initial setup fee. Some service bureaus limit the number of documents that are set up as a part of the initial setup fee. For instance, if the initial setup fee includes the setup of three documents, any document after the first three will be charged for. This fee can range from $25 to $75, depending on the service bureau.

ADDITIONAL PAGE FEE

This fee works similarly to the additional document fee. There is usually a specified number of pages per document that is allowed before a fee is charged. For instance, if your service bureau allows four pages per document and your document consists of five pages, you would be charged for that additional page of information. This fee can run up to $50, although some service bureaus do not charge at all for this service.

PER-MINUTE/PER-PAGE FEE This fee is charged for the information that is actually faxed out. The average fax-on-demand service faxes out two to three pages per minute. A service bureau will charge somewhere between 40¢ and 60¢ per minute. If they charge on a per-page fee basis, the average price would be between 20¢ and 30¢ per page. Make sure you understand what service you are contracting for. Your bill will definitely be affected by this decision.

MONTHLY STORAGE FEE This fee is sometimes charged by service bureaus to keep your information on their fax vault. It works like paying rent. It is a monthly fee that must be paid, or your information is taken off the system. Some service bureaus do not charge for this storage, but many do. The average monthly storage fee is $300. If the service bureau you choose charges for the monthly storage of your documents, be sure to find out when this fee is due and how long the grace period is before termination of your service occurs. This is a prime example of the cost factors that can be involved in providing fax-on-demand. Watch out!

CHANGE TO DOCUMENT FEE Nearly every business supplying information to its customers has the need to update or change certain parts of a document at some time. Some service bureaus take advantage of this need and charge anywhere from $25 to $75 to make these changes. Others do not charge for this necessary service, but all service bureaus need to work with you to know exactly what changes need to be made. Normally, it would take a programmer only minutes to make a change, so you should be leery of such charges if it is only a minor change to your document.

Using a service bureau is definitely a smart way to go unless your business is technically advanced, can afford the expense of setting up its own fax-on-demand service, and intends to use fax-on-demand for a long time to come. Just make sure the service bureau you work with has your best interests in mind and is reliable. Checking references before making a fax-on-demand business decision is a good idea whether you are going to set up your own system or get involved with a service bureau provider.

Frequently Asked Questions

Following are the seven most frequently asked questions that experts in this industry come across.

Q: I do business with customers all over the country, sometimes overseas. With all the different time zones, what if my office is closed when someone wants information?

A: A fax-on-demand service never shuts down. It's available 24 hours a day, seven days a week. Your customers and prospects can make requests any time of day or night, even if you are not available. This is one of the biggest advantages of fax-on-demand.

Q: Information that I distribute changes almost daily. How can I be sure that all the information being faxed is current?

A: It's easy. You can change information as often as you'd like, at any time. Simply fax, mail, or modem the new information to your service bureau or in-house programmer, and the information is updated in a timely fashion. You can also get copies of your information faxed to you.

Q: What about fax quality? Most faxed data comes out smeary and hard to read, especially on the thermal paper that most machines use.

A: You've just hit on one of the most important reasons to use fax-on-demand rather than hand-feeding your own fax machine. The digital equipment used for fax-on-demand directly faxes your documents out from the computerized fax vault after the information has been input digitally into high-resolution mode. The document sent is the highest quality possible.

Q: What about costs? How much is all this high-tech equipment and service going to add to my overhead?

A: If you choose to use a service bureau, you will see a savings on overhead immediately. The savings in printing, personnel handling, and postage alone is considerable. Plus, just think, no more "couldn't get throughs," no mail carrier foul-ups, no unopened envelopes, etc.

Q: Will I need to add a phone line or change my phone number?

A: Not if you go through a service bureau. You have no need to buy any additional equipment. The service bureau will do all the work and supply everything you need to start and maintain a successful and profitable information delivery system.

Q: I advertise in several different publications. How do I keep track of my responses from each of these for comparison purposes?

A: Simple. You can assign a different extension to your fax-on-demand service number for each publication and combine the results into one report for the period in question. Normally, a service bureau will do everything except calculate your cost per response, since they will not know your cost for advertising.

Q: I'd like to create a database of the people who call to request information. How can I do this?

A: Your fax-on-demand system can be configured with *called lead retrieval,* which for an additional charge gathers the caller's name, address, title, department, and phone number. This is extremely useful when placing sales information on a fax-on-demand service.

Of course, there are hundreds of possible questions about fax-on-demand, especially from individuals who are unfamiliar with how it works, but as you can see, the answers are usually very clear-cut and easy to understand.

Conclusion

There are an unlimited number of ways you and your business can use fax technology. It is suspected by many that faxing, along with e-mail and other computer services, may one day replace the United States Postal Service.

The increasing penetration of fax machines into the workplace and now into the home, coupled with the omnipresence of touch-tone phones, is expanding consumers' options for how and when they receive information. Consumers are proving themselves willing to pay premium prices for the convenience, ease, and immediacy of faxing. If you can find your faxable niche audience, you are on your way to expanding the value of your information resources in ways not possible even a decade ago. Just as computers have become a part of our everyday lives, so has the fax machine. It is a tool that most of us would not dream of doing without.

PART FIVE
Emerging Technologies

Chapter Seventeen

Videoconferencing

Videoconferencing is a technology that allows users to communicate interactively in real time through the transmission of images and sound between two or more sites. Videoconferencing gives an individual the power to be in a remote location without actually leaving the comfort and convenience of the home or office. The value of this communication medium lies in its ability to enhance productivity and save time. Today, placing a videoconference call requires special equipment and a reliable connection, and once these are obtained, communicating is as easy or easier than placing a regular phone call. This chapter will cover some of the basic aspects of the technology and its applications.

Emerging Technology

Videoconferencing is redefining the way the world meets. Business, government, and education are using videoconferencing to bridge time and space and enhance productivity and communication. Much has changed since the early 1980s, when this industry was getting started. The costly boardroom facilities have given way to less expensive roll-about systems, and recent advances in technology are shifting the direction of videoconferencing towards the personal computer. Desktop videoconferencing is the latest trend in price reduction and system flexibility.

Several factors are responsible for the rapid growth of this industry:

- Improvements in compression technology
- The decline in communications services prices
- The decline in computer and visual communications equipment prices
- Interoperability of visual communications equipment
- Desktop computers: an increase in power, yet a decrease in price

Group Systems and Personal Systems

Group systems are primarily designed with group meetings in mind. The equipment may be fixed, as in a boardroom, or may use a mobile housing for more flexibility. The systems use one or more large monitors, thus enabling easy viewing and allowing numerous meeting participants on-screen simultaneously. On the other hand, personal systems are designed for use on a personal computer and are best suited for one individual on-screen at a time. Frequently, audio and video quality in group systems is better than in personal systems.

Despite their differences, group and personal videoconferencing systems have a similar functionality. Live video and audio signals from a camera and microphone enter your machine through video capture and sound boards. Several additional computer components are also used in the transmission of voice, video, and data (see Figure 17-1). Information is digitized and compressed (*compression* is the process of reducing the information content of a signal so that it occupies less space on a

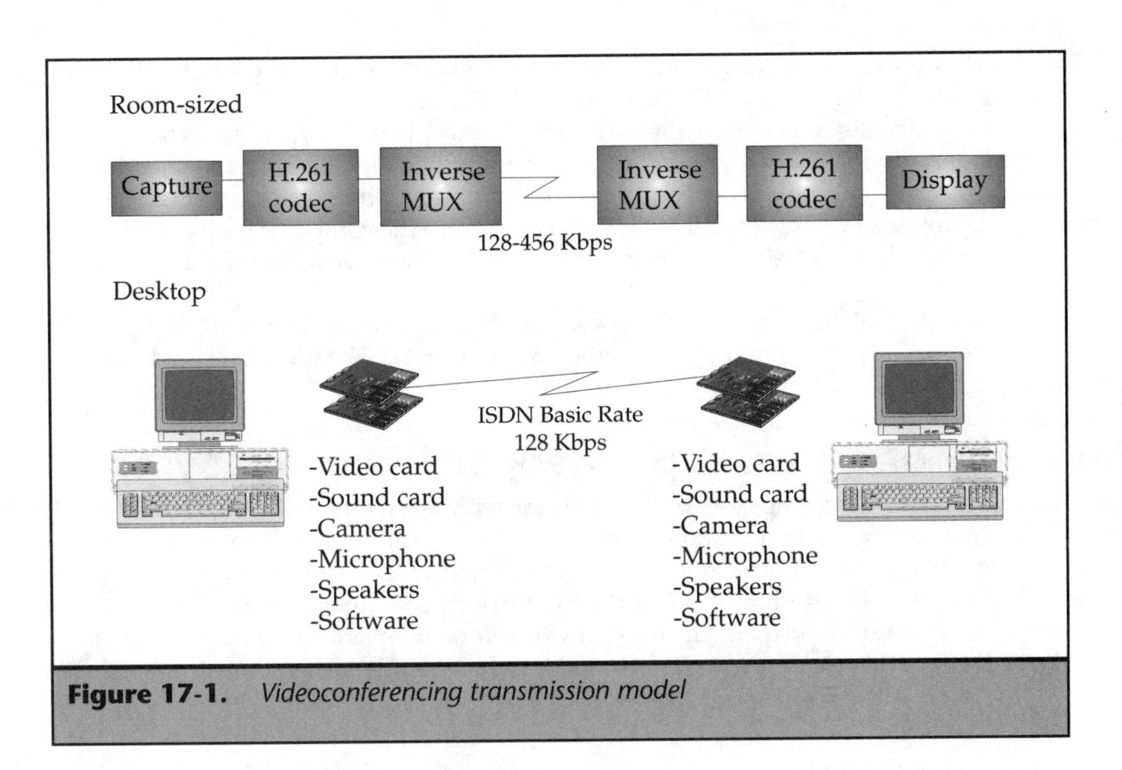

Figure 17-1. *Videoconferencing transmission model*

transmission channel or storage device) by a codec, then transmitted to a remote site via a communications interface such as an ISDN adapter card or an ethernet network interface card. A *codec* (a contraction for coder/decoder or compress/decompress) is a sophisticated digital signal-processing unit that takes analog input and converts it to digital on the sending end. When the data reaches its destination, the process is reversed, and the digitized signals are converted to analog format and output to a monitor and speaker. This technology is discussed in greater detail later in this chapter.

In addition to using comparable technologies, the applications of group systems and personal systems are similar as well. Whenever two or more people cannot be at the same place at the same time, then videoconferencing is the next best thing.

Desktop systems can dramatically improve personal communications. Also, most systems today have data-sharing capabilities—a critical element in collaborative and groupware applications. Furthermore, desktop videoconferencing can run on LANs (which the computers are already a part of anyway) or over circuit-switched networks like most group systems. *Circuit-switching* is the process of establishing a connection for the purpose of communication in which the full use of the circuit is guaranteed to the parties or devices exchanging information. Examples of circuit-switched networks are ISDN (discussed in Chapter 11) and POTS (standard telephone lines). The picture quality of POTS-line desktop videoconferencing is not too great, but it is adequate for the job. ISDN provides a high-quality video image, but it is more costly. The section on infrastructure selection covers these distinctions in greater detail. An example of a desktop videoconferencing system is provided in Figure 17-2.

The primary difference between the two options is price: Group systems range from $15,000 to $30,000 for entry-level systems (including all hardware and software) and can cost much more depending on features and capabilities. Desktop videoconferencing systems are typically sold separately from the computer itself and range in price from $2,000 to $7,000. As always, additional features increase the final cost of the system.

It is not difficult to see why desktop videoconferencing is attracting so much attention. For the price of two average group systems ($20,000 each), you could get ten desktop systems!

Anatomy of a Desktop Videoconferencing System

Desktop videoconferencing systems usually come with several separate components: a video camera, a speakerphone or microphone, video and compression boards, network interface cards, and application software. Desktop videoconferencing solutions are available for PC, Macintosh, Sun, and virtually any other computer platform. Some products are platform-independent, thus providing for cross-platform communications. Figure 17-3 supplies examples of platform-independent peripheral equipment.

Photo courtesy of AT&T

Figure 17-2. *Sample desktop videoconferencing system*

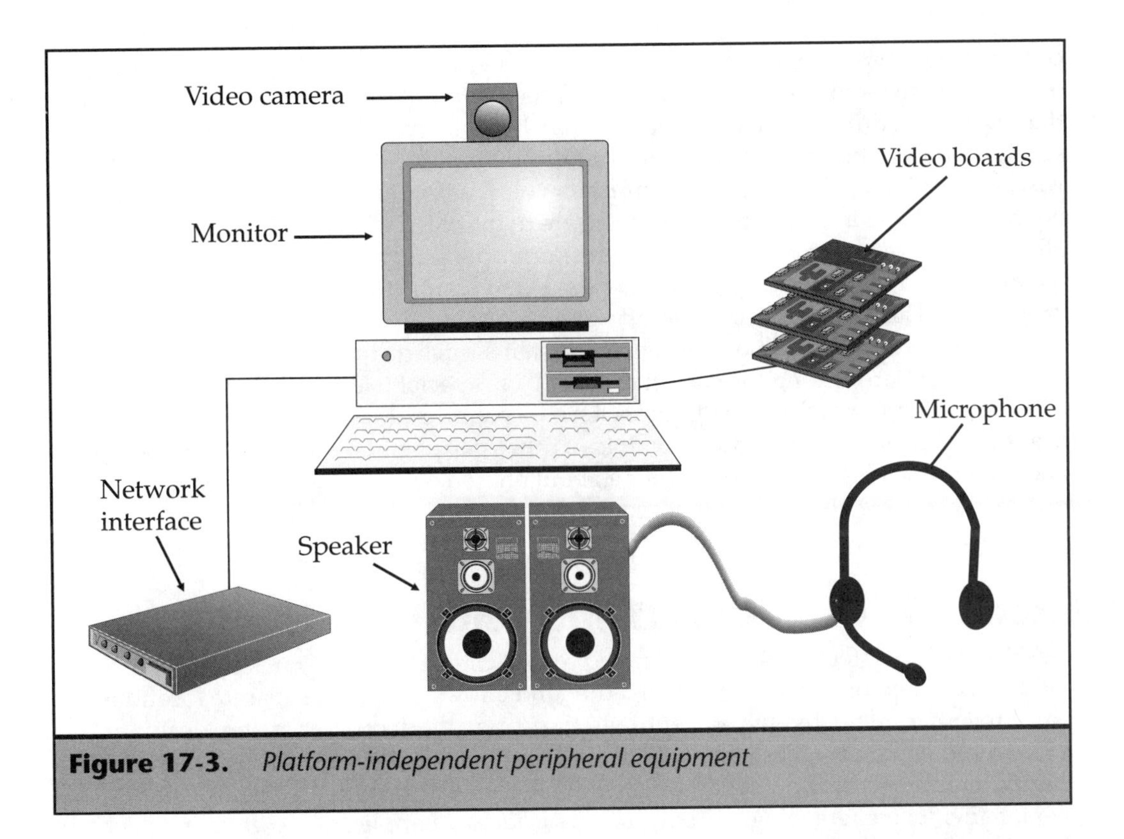

Figure 17-3. *Platform-independent peripheral equipment*

Setting up a PC for videoconferencing can be a complex proposition, particularly for the computer novice. Most products require the installation of at least two add-in cards, the video camera, and the microphone—not to mention several megabytes of software that change the configuration files. Most vendors have been responsive to this, though, and are providing clear installation documentation or sending engineering personnel on-site when multiple units must be installed.

The trouble many people have with ISDN desktop videoconferencing is proper configuration of the ISDN interface, which often requires coordination between the product vendor, the local phone company, and the long-distance carrier to get the system working. Installation of ISDN also requires specialized CPE (customer-provided equipment), such as terminal adapters, power source, and wiring. Once your desktop videoconferencing system is installed, though, it's as easy or easier to use than a telephone. Many companies integrate automatic dialing into their interfaces with electronic phone books or icons that launch the calling procedures. The real-time interactivity of voice, video, and data make you feel like the other person is virtually in the room.

Something to watch out for when you are selecting an ISDN solution is the ISDN capabilities of the system. Many of the ISDN cards included with ISDN videoconferencing kits can only perform the videoconferencing function. Though this may be adequate for some, a more desirable and flexible solution would be an ISDN card that performs additional data connections such as fax/modem capabilities at various speeds. One kit you might want to look at is the Vivo320. This product uses the IBM waverunner for its ISDN card and supports voice, video, and data in one integrated board. The Vivo kit uses specialized software to interface with the ISDN card. This allows for more information to be sent in shorter periods; however, it is highly recommended that you use a Pentium-class PC to handle the additional CPU processing. This is a very fair trade-off nevertheless.

The POTS desktop videoconferencing systems, such as the PC3000 made by Creative Labs, are among the easiest to install. No special telephone lines are required, and the software practically configures itself. If your desktop PC has the right configuration and is not weighed down by lots of other peripheral devices, you can be up and communicating with a colleague in about an hour. The sound, whiteboard, and document sharing all function well, so users should be able to acclimate themselves to the new technology rather quickly.

Desktop Videoconferencing and Multimedia

Desktop videoconferencing is a multimedia application that integrates computer, video, and telephony technologies together and allows people in separate locations to work together interactively. Conceptually the same as videoconferencing, desktop videoconferencing enables participants to see each other in a window on their computer screen, carry on voice conversations, and share onscreen data. Most personal videoconferencing systems include *whiteboard* application software, which lets two or more users simultaneously annotate graphics, text, or image files on screen. Some also support Object Linking and Embedding (OLE), which ties pieces of information throughout different applications, background file transfer, and application sharing, which lets participants jointly run third-party Windows packages.

Desktop videoconferencing is multimedia in its richest form. Other multimedia applications include presentations, groupware, and computer-based training. Multimedia can integrate voice, video, data, music, still images, graphics, and animation to provide a truly unique, educational, and entertaining experience.

Real-Time Versus Store-and-Forward

At this point, it should be noted that interactive visual communication can be separated into two main categories:

- *Real-time* is based on interaction occurring at the same time at different locations. This generally requires bandwidth-intensive infrastructures that provide the platform for applications such as live television or single- and multipoint videoconferencing. *Bandwidth* refers to the information-carrying capacity of a channel, i.e., its throughput. In analog systems, it is the difference between the highest frequency that a channel can carry and the lowest, measured in hertz. In digital systems, the unit of measure of bandwidth is bits per second (bps).

- *Store-and-forward* is used for different time at different location interaction. A stored multimedia solution relies on unique infrastructure attributes, such as extended storage capabilities and commonality of multimedia hardware and software components installed on each computer.

Focus on Applications

Videoconferencing has emerged as a cost-effective tool for enhancing productivity and interpersonal communication, which are mission-critical objectives in most organizations today. Originally developed to reduce travel and preserve the subtleties of face-to-face meetings, videoconferencing has greatly expanded its functionality. Videoconferencing can save money in reduced travel expenses, but conserving the time and productivity of key personnel and accelerating the speed of business processes are equally if not more important to most organizations considering the use of this technology.

In addition, videoconferencing provides a means for conducting virtual meetings that would not have been possible if travel were involved. One practical way of looking at it is this: if it takes more than two hours to get to your destination, it's probably worthwhile to use videoconferencing. Time and travel expenses become even more critical at the international level. Figure 17-4 illustrates how videoconferencing can become an extremely valuable global communications resource.

Business Applications

Videoconferencing is having a major impact on the way we communicate in the workplace. Most of us are not accustomed to "acting" on camera, but comfort and familiarity come quickly. The addition of an interactive visual element to the communications process has spawned numerous business applications:

Face-to-Face Meetings: Far and away the most popular application of videoconferencing, personal and group meetings bring people from remote locations together in a virtual conference room. The benefits of face-to-face communications have been proven over and over again by interpersonal communications experts. Seeing the person or persons you are meeting with in real time adds a dimension of comprehension that cannot be replicated simply by voice or static images.

Telemedicine: Hospitals and health care organizations are using videoconferencing technologies to cope with shrinking budgets and to increase access to subject-matter experts. Telemedicine uses a variety of high-quality imaging techniques. Teleradiology, telepathology, and teledermatology are examples of how doctors and nurses can work in multiple locations simultaneously, often in emergency procedures while accessing high-tech medical information simultaneously. Rural hospitals and clinics find videoconferencing particularly useful, because of their shortage of local expertise,

Photo courtesy of AT&T

Figure 17-4. *Multiple-location international videoconferencing*

which requires the consultation and diagnosis of medical professionals who are usually located in larger metropolitan areas.

***Telecommuting*:** Working from home or any remote location has numerous advantages: flexible work schedules, reduction or avoidance of commuting time and travel stress, improved quality of life for employees, and many more. However, this paradigm shift in the working environment raises a number of challenges, such as measuring productivity, employee isolation, remote supervision, and a basic shift in traditional business practices. Videoconferencing capabilities will help address some of these challenges. Communicating with coworkers, managers, and customers via video will help bridge the gap in physical presence.

***Project Management*:** Videoconferencing is an excellent tool for managing projects over distance. Team members can engage in *telecollaboration*—enabling geographically dispersed work groups to collaborate together electronically with greater speed and flexibility. This could take place with two or any number of individuals working on a project. Meetings occur interactively with greater frequency and can even include outside help such as subject-matter experts. The likelihood of reaching unanimous decisions and building team synergy is greater when the decision-makers are present and can see, hear, and share data with each other.

Design Teams (engineering and product development): Architecture and engineering firms are finding many uses for videoconferencing technologies. Particularly on large projects where numerous contractors and subcontractors are involved, designs can be discussed and changed interactively. Whiteboard and application sharing are very useful features to these professionals.

***Distance Learning (training and education)*:** The term "distance learning" is typically used to describe video-enabled instruction in a college, university, or K-12 environment. Group videoconferencing, as depicted in Figure 17-5, is used extensively in the delivery of educational services and programs. In the mid-1990s, tens of thousands of students receive their education via interactive video. Business schools across America are able to bring high-level executives as guest speakers into the classrooms. The executives never have to leave the office, and the students get the unique opportunity to speak with extremely busy organizational leaders. Video training can have an organizational or corporate focus as well.

***Customer Service*:** Customer service is the reigning motto of corporate America. What better way to serve customers than to give them instant visual access to the friendly customer service representatives. Help desks can now see what you see and diagnose problems much more quickly and efficiently.

Photo courtesy of VTEL

Figure 17-5. *Corporate training environment using desktop videoconferencing*

Other applications include:

- Intracompany communications
- Sales and marketing
- Presentations
- Legal depositions
- Surveillance

Which Videoconferencing Solution Is Right for You?

Many factors must be carefully evaluated when selecting a videoconferencing system. The difficulty in the decision-making process is augmented by the large number of systems currently available. This personal videoconferencing market segment will continue to grow as computer manufacturers and software developers integrate numerous third-party components into their own solutions. You might think that you're buying from a single vendor, when in fact the desktop videoconferencing system is an integrated product from several manufacturers. Also, some manufacturers advertise turnkey solutions even though their product may still be in the beta testing process or demo stage.

The team responsible for implementing desktop videoconferencing should do its homework before approaching vendors. Following are some factors to consider:

- Application(s) (current and future)
- Computer platform and operating system
- Physical network available (ISDN, T-1, SW-56, etc.)
- Number of users; multipoint capabilities
- Operation over LAN or WAN (circuit-switched network)
- Standards-compliance (international)
- Track record of vendor/integrator
- Service and technical support
- Features (e.g., audio peripherals, whiteboard, and application sharing)
- Encryption and security

Once the basics are covered, you can begin to assess factors such as video picture quality, frame rate, audio clarity, application interface, and data-sharing tools. A *frame* is an individual television, film, or video image. Television sends images at 30 frames per second (fps), while film (movies) uses 24 fps. Videoconferencing systems typically send between 8 and 30, depending on the transmission bandwidth offered. Frequently, a software application that comes with the system will control the frame rate to optimize system performance as well as control traffic flow on networks. Frame rate is important, because it determines how smooth or choppy the motion video will be. Smooth motion requires less compression, and consequently more bandwidth. Camera upgrades are one of the easiest and most common ways to enhance personal videoconferencing systems. If you expect to hold meetings that require paper documents or three-dimensional objects, a *document camera* is a good solution (see Figure 17-6). These devices have lenses, controls, and mounting mechanisms designed for viewing documents and objects resting on a flat, lit surface.

Each product has its own answer or solution to these and other factors. One of the best things you can do is talk to current users. Some vendors will release names of companies as references. The only problem with this is the vendor and customer may be "in bed together." Trade shows, such as TeleCon, ITCA, and DeskCon, are excellent opportunities to get the latest information and to network with the end user community. Some publications, such as the *Desktop Video Conferencing Report* (Applied Business teleCommunications, Livermore, California), contain comprehensive product evaluations and matrices to assist with comparative analysis.

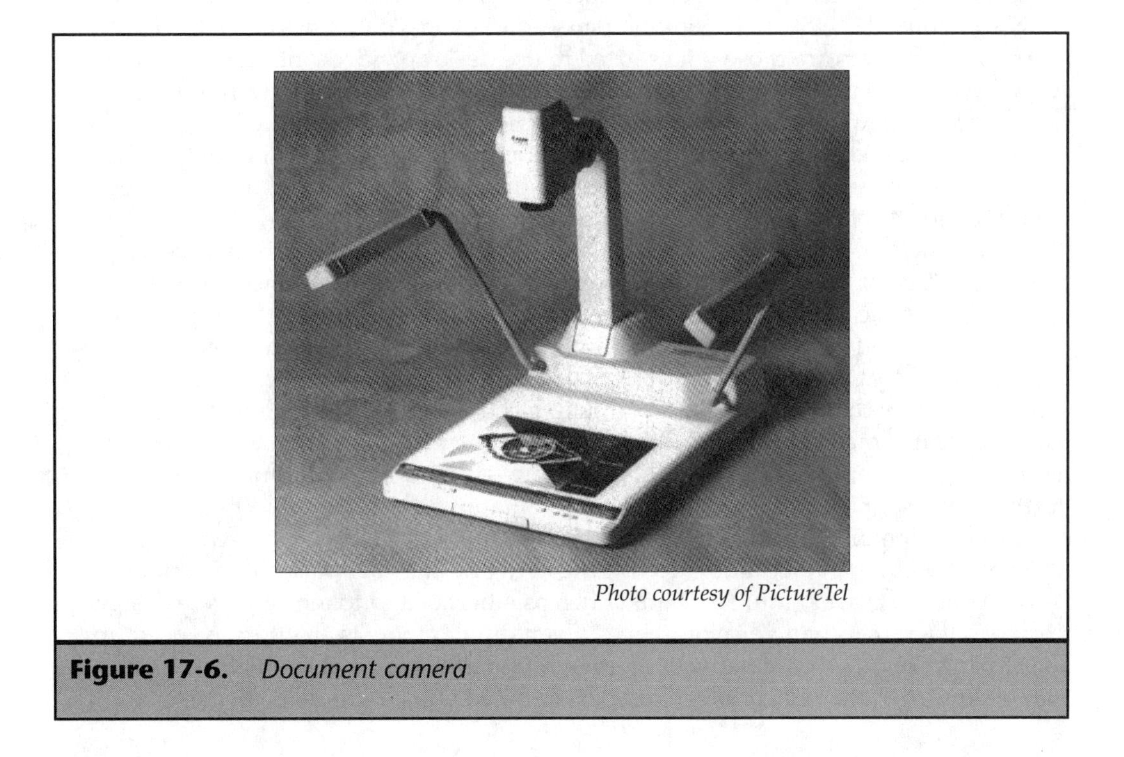

Photo courtesy of PictureTel

Figure 17-6. *Document camera*

Selecting an Infrastructure

Probably the single most difficult decision when selecting a desktop videoconferencing solution is whether to use the LAN infrastructure, a circuit-switched network such as ISDN, or just your normal POTS lines. The only way to bridge these transport mechanisms today is with specialized hubs and switches. This cross-transport communication should improve as the phone companies, groupware/network software developers, and desktop videoconferencing manufacturers form strategic partnerships. The quality, of course, will vary as well.

Local Area Network

It's inevitable: sooner or later, networks will have to be upgraded to accommodate voice and video. Several technologies, such as asynchronous transfer mode (ATM), promise to help with this endeavor, but they are either too expensive to implement or require extensive network retooling.

The use of different transport media is related to how the users got into desktop videoconferencing. That is, some of the users were introduced to desktop videoconferencing as an alternative to room-sized systems that were used for corporate meetings over distance. These users are likely to have configurations that are based on WAN solutions like switched 56 kilobit per second (Kbps) data lines or ISDN. The other common users are those who started to use desktop videoconferencing as a part of groupware and multimedia. These users are likely to have configurations based on LAN solutions like ethernet, token ring, or FDDI. The newest users are being introduced to desktop videoconferencing without prejudice and want both LAN and WAN access. The vendors are responding, but the user is not able to switch between the LAN and WAN alternatives as easily as they might like.

Despite network managers cringing at the thought of using their networks to distribute video or voice, executives are realizing the benefits these media can bring to organizational communications via the network. Data is losing the spotlight to more dynamic and richer forms of communication, such as *full-duplex* voice (an audio channel that allows conversation to take place interactively and simultaneously between the various parties, without electronically cutting off one or more participants if someone else is speaking), still images, and interactive video, because these media are approximations of the real thing—face-to-face communication. Furthermore, sound and images enhance presentations, training, project management, and product demonstrations.

So what has kept voice and video off the network thus far? Basically, computer networks are bursty in nature and links such as ethernet and token ring were designed to carry data, rather than voice or images. Even though there is an effort to speed up and expand the capacity of networks with new technologies such as 100Base-T and 100VG-AnyLAN, the networks remain packet-based and continue to have trouble providing the continuous uninterrupted bandwidth required by voice and video. Therefore, network managers need solutions that integrate the concepts of telephony, or switched networks, with packet data networks.

One such solution is an emerging communications standard called *isochronous ethernet*. It will allow broadband networks to support real-time video, as well as data and voice, while enabling low-cost access to ATM backbones. Isochronous ethernet involves a broadband architecture that includes a 10-Base-T, 10 Mbps data channel plus 6.144 Mbps of bandwidth supporting 96 ISDN B channels. Channels can be used in multiple combinations of Nx64 Kbps, supporting any video, voice, or other application. Using this technology, an enterprise network can support multiple video applications along with its data and voice applications.

Isochronous ethernet can connect to the WAN by adapting standard networking hubs or through a new generation of distributed PBXs, which are being designed to deliver computer telephony integration to the desktop. Typical ethernet traffic for file services, print services, and client/server applications can be kept completely separate from the channels carrying voice, application sharing, videoconferencing, graphical interfaces to the Internet, and any other ISDN applications. The importance of ISDN as the basis for isochronous ethernet lies in the fact that ISDN is the predominant network infrastructure that supports application sharing (T.120), videoconferencing (H.320), and voice (G.711), which are the standards set by the International Telecommunications Union (ITU-T) to ensure interoperability among videoconferencing systems. Standards have been critical to the success of this industry and a more detailed discussion of standards is included in this chapter.

Another alternative for personal voice and video applications on the LAN is *switched ethernet*. This type of network provides each desktop with 10 Mbps of uncontended bandwidth by microsegmenting the LAN to give each user his or her own segment. Segments are connected to a switching hub in a star configuration using 10Base-T twisted pair. Desktop videoconferencing applications run well over switched ethernet (despite not supporting isochronous traffic) because of the 10 Mbps of uncontended bandwidth available to each user.

The long-term solution is asynchronous transfer mode, a scalable network protocol that supports speeds ranging from 25 Mbps to 2 Gbps. ATM is voice and video "friendly" for two main reasons:

- ATM cells sizes are small (53 bytes) and can travel through a network very quickly and with very low latency. Therefore, video and voice don't suffer the delays larger packets on a packet-based network often experience.
- ATM cell transfers resemble a circuit-switched network such as ISDN or Switched-56. A computer sending voice, video, or data makes sure there is an open, "dedicated" path to the destination, which is similar to placing a phone call. This guarantees the continuous stream of information so critical for accurate reception of voice and video.

Wide Area (Circuit-Switched) Networks

Teleconferencing networking can use switched circuits at rates ranging from 56 Kbps up to T-1 (1.544 Mbps). In the U.S., the key data rates used in videoconferencing include 56, 112, 256, 384, 768, and 1536 Kbps. The basic concept is to provide connectivity to the network "cloud" and through switching systems in the network's central offices. Meanwhile, the videoconferencing signals are routed to the appropriate destination prior to the start of the conference, and once the connection is established, the circuit is kept open as in a regular phone call. Once the conference is concluded or disconnected, the circuits are once again freed up for use by others. In the early 1980s, most teleconferencing was done using satellites. Today, terrestrial networking via high-capacity fiber optics is the primary transport mechanism.

Personal videoconferencing systems are usually used for *point-to-point* conferences, meaning one location establishes a call with one other location. However, studies show that multipoint capabilities are high on the list of desired features. The ability to assemble numerous participants from around the globe into one virtual meeting place is one of the most unique and powerful features of videoconferencing. When linking more than two sites, a multipoint control unit (MCU) is required to bring together the various locations into a single conference. VideoServer and Teleos are two providers of multipoint control units. A diagram of how Teleos' MCUs work is provided in Figure 17-7. This device bridges together multiple inputs so that more than three parties can participate in a videoconference.

When a multipoint conference is set up, the circuits from all participating conferencing systems are routed to an MCU in star fashion. The MCU then demultiplexes the individual bit streams and bridges, or mixes, the audio and video for each location participating in the conference. Most MCUs provide multiple approaches for the participants to select whose video or graphics are viewed. The most commonly used is voice-activated switching mode—the person who is speaking is viewed by all. Current MCU technology is responsive enough to provide voice-activated switching that accurately follows the pattern of discussion between participants in small- to medium-size teleconferences. Multipoint connections can also be controlled by a single user or "chairman" who selects who has the floor. MCUs can be costly, but this type of service can be easily obtained through any long-distance carrier and provides greater functionality.

Most videoconferencing networks, including group systems and desktop systems, utilize narrow-band ISDN because of its cost-effectiveness and ability to support acceptable sound and image quality. *Narrow-band ISDN* employs two "B" and one "D" channels (2B+D). The B refers to bearer channels of 56 or 64 Kbps. These channels carry the digitized and compressed voice, video, and data information. The D channel is a separate 16 Kbps channel used for signaling information on the network. This version of ISDN is also called basic rate interface (BRI), and it is quickly becoming available across the country.

If video quality is a critical issue, as in telemedicine for instance, greater speeds can be used for conferencing. Primary rate interface (PRI) (discussed in Chapter 11)

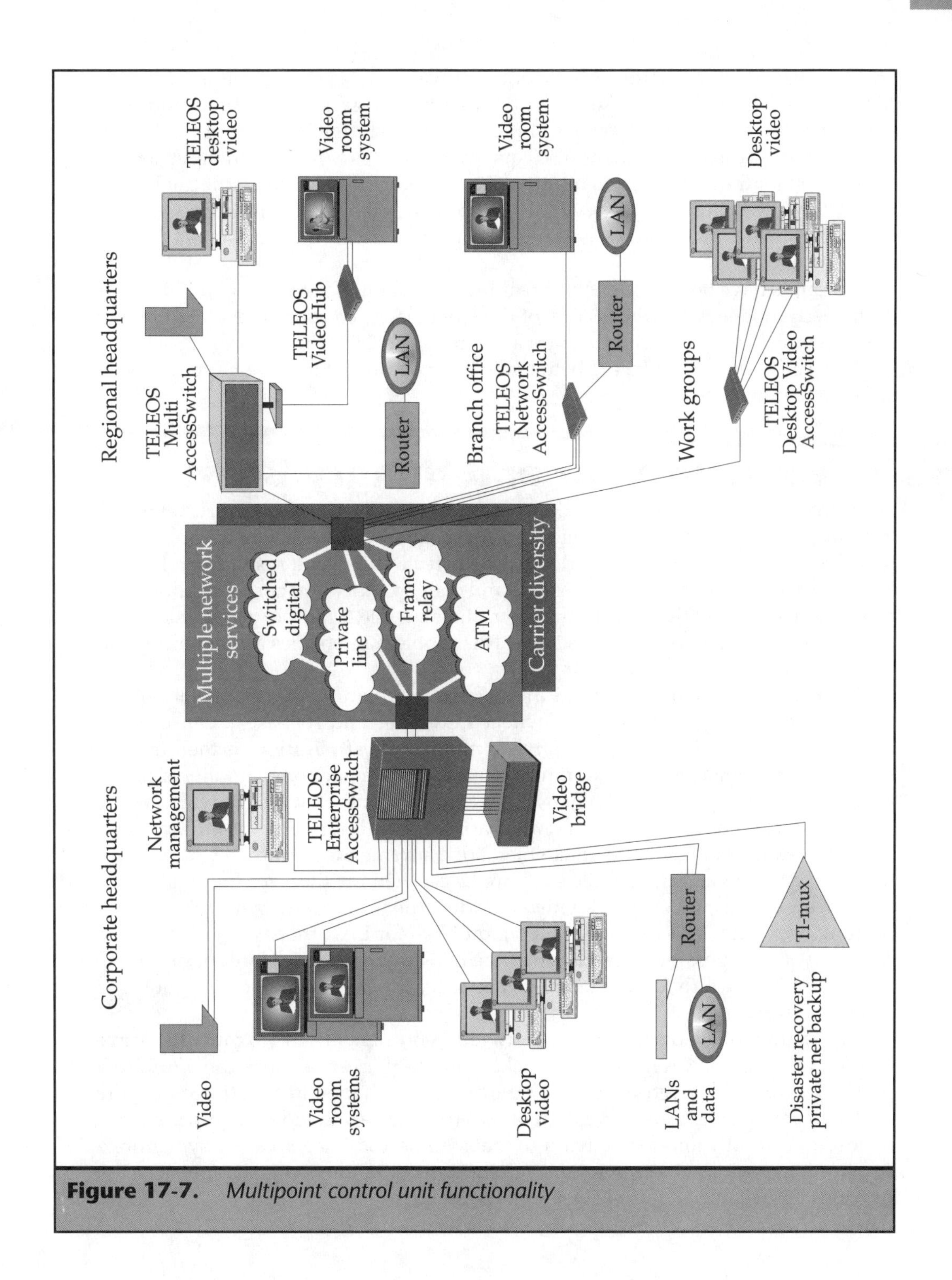

Figure 17-7. *Multipoint control unit functionality*

and T-1 (discussed in Chapter 10) are both equivalent to 24 digital channels, or 1.544 Mbps. These channels can be segmented into different sizes for flexibility and cost containment. An *inverse multiplexer*, or I-mux, is the most common piece of hardware used for bandwidth management. The equipment receives a high-speed input and breaks it up for the network into multiple 56 or 64 Kbps signals so that it can be carried over switched digital service. It also provides the synchronization necessary to recombine the multiple channels into a single integrated transmission at the receiving end.

Attempts have been made to put real-time video on the Internet and POTS using high-speed modems. But the quality of the video has been poor, and a lot of motion detail gets lost. Keep in mind that you are comparing 28.8 Kbps versus 128 Kbps (ISDN). These analog media are better suited for audio, data-sharing applications, or still-image transfers.

Video: A Peculiar and Hungry Beast

Performance of the network is the key reason fast LAN technology is crucial for multimedia applications such as videoconferencing. Multimedia chews up megabandwidth. Sluggish or overloaded LAN technology will result in broken, hard-to-understand voice transmissions and choppy video imagery. Nothing frustrates users more than twiddling their fingers while a complex image slowly paints itself onto a workstation's screen. This problem intensifies as the number of multimedia users on a LAN grows.

Many network managers are addressing the bandwidth issue by exploring new technologies, such as FastEthernet, Switched Ethernet, and 100Base-T. One drawback most of these technologies have relative to multimedia applications is their inherent transmission scheme, which can transfer frames of data packets in nonlinear order before reassembling them at the network reception point. That design works well for file transfers and other store-and-forward applications but poorly for multimedia.

Video traffic, or even large image files, adds a tremendous burden to a LAN's traffic capacity, because image file sizes are larger than the file sizes for which standard LANs were designed. When live video is added, the traffic in the network becomes a continuous stream. This can present a big problem for LAN managers, who must contend with the possible breakdown of the LAN under the pressures of providing continuous bandwidth for a video file while still carrying varying loads of bursty data-only traffic.

In addition to video's intensive bandwidth requirements, video must also travel without interruptions. Regular data packets on an ethernet network can afford to be split into chunks or be retransmitted if they don't reach their destination on the first try. However, the digital transmission of real-time voice and video requires a continuous or carefully timed bit stream so that sounds and images can be synchronized. Applications such as videoconferencing rely on isochronous transmission since it provides predictable and minor delays in packet delivery.

What occurs during conferencing over fully used packet ethernets is that the network loading causes video frames to be dropped and even lost completely. This could take place in the hub, switch, or router, and this delay or disruption in the delivery of the frames will cause severe degradation in voice/video quality. The frame rate can become so variable that words can lose synchronization with the speaker's lips, making communication irritating and possibly unacceptable. Clearly, adding voice and video capabilities to the network is a tricky proposition, unless your network is using the right technologies.

Video Compression

At the heart of videoconferencing products is compression technology that lets them send huge streams of information such as voice, images, and video over digital phone lines. Compressing video means transmitting signals at a greatly reduced bit rate by reducing the representation of the information, but not the information itself. Real-time video's huge bandwidth requirements make video compression a vital part of this process. Compression is necessary because video has excessive requirements in bandwidth, storage capacity, time, cost, and redundancy.

The fundamental goals of compression algorithms are the following:

- Reduce redundancy in each frame. Capitalize on the fact that the current frame of the picture is very similar to the last frame.
- Use "motion prediction" to avoid resending all the information for sections of the picture that have only moved slightly. Complex algorithms are used to analyze information within a frame and predict movement based on similarities within that frame.

Compression allows us to reduce the bandwidth of analog broadcast quality video from 90 Mbps to 56 Kbps. Compression ratios range from 10:1 at the low end to 200:1 at the high end. Digitally-based compression algorithms are used to transmit changes in information by eliminating identical bits. One popular way of shrinking video is by sending only changes that occur between frames, also known as *interframe compression.* Small movements, such as a person gesturing with arm movements, will degrade video quality and reduce frame rates. Therefore, there is a very direct correlation between the amount of movement in a frame and the quality of the video. The less information that has to be updated and retransmitted, the more attention that can be devoted to clarity of images by the compression algorithms. Static backgrounds and slow deliberate movements are highly recommended when videoconferencing.

Another type of compression is *intraframe compression*, a compression technique whereby redundancy within a video frame is eliminated. Video information is shrunk by selectively discarding redundant or hard-to-perceive information, such as monochrome backgrounds.

To further illustrate these processes, think of cartoons and how they are made. Cartoons are primarily hand-drawn images on cels, and motion is simulated by rapidly seeing each individual cel in sequence. Most of the time, the background scenery remains unchanged and only the portions of the cartoon that are moving must be redrawn to simulate that motion. Similarly, when an image is transmitted in a videoconference, only the moving parts of the picture must be updated. So if you are sitting behind your desk in your office, everything that is motionless, such as the walls, tables, papers, etc., only need to be transmitted once. On the other hand, motions such as your facial expressions or hand movements must be constantly updated on the other end.

Codecs

Codecs are the heart of videoconferencing systems, and they may use any one or a combination of methods to reduce video bandwidth. The primary elements of video that are reduced are resolution, frame rate, color information, and redundant information. These are all critical elements of crisp, clear video, and reducing any one or more of these elements will have a negative impact on picture quality. Therefore, there is an inverse relationship between compression and quality. A codec's task is to balance these two characteristics to provide an acceptable image quality.

The key component that makes videoconferencing possible is the codec. Without the codec to translate high-bandwidth analog video signals into compressed digital data, video communications would be practical only over satellites or fiber-optic cable.

Just a few years ago, codecs were the size of computer towers and cost tens of thousands of dollars. Technology has advanced very rapidly in this industry through enhanced microprocessor power and compression techniques. Today, most desktop videoconferencing systems squeeze an entire codec into a single full-length add-in board. A second card is usually required as a communications interface, such as an ISDN adapter or ethernet interface. Some systems are even taking advantage of the newer and more powerful CPU chips to handle most of the compression, leaving the remaining functions, such as application interfaces and decompression, to software. This latter scenario is a unique solution, because it does not require physical space on expansion slots in your computer. Still, the quality of software compression is directly related to processor speeds, and commercially available chips have not reached an adequate performance level for this application.

Compression brings up the issue of video quality in desktop videoconferencing. Even the most expensive systems won't approach the standard NTSC television rate and resolution of 30 frames per second (fps) at 640-by-480 resolution. For reasonably smooth perception of video, rates of 12 fps and 15 fps are a minimum. Window size and monitor size are also important considerations, since they affect resolution.

Interoperability in Videoconferencing

Interoperability is synonymous with compatibility. Communications devices from different manufacturers can "talk" to each other, provided there is a common format or standard. These standards may be formalized, as in the case of NTSC for broadcasting, or they may be informal, or de facto, as in the RS-232 data transmission standard. The importance of standardization can be illustrated by two technologies: facsimiles and HDTV (high-definition television). The first developed a common protocol and quickly became enormously successful. In contrast, HDTV continues to struggle due to the lack of an international standard. Naturally, standards alone do not guarantee the success of a technology, but in the marketplace, particularly in large organizations, compatibility is more of a necessity than a luxury.

Telephony technologies live and die based on their ability to gain market acceptance. This acceptance depends on a number of factors; however, the most important factor is standardization of the technology. Users want "seamless" communication regardless of who manufactured a piece of equipment. In the videoconferencing industry, the recent adoption of interoperability standards has spurred a tremendous growth in the use of this technology. Industry analysts predict the visual communications market will reach $6 to 7 billion by 1997—a growth of 90 percent annually.

Visual communications are made possible by the combination of complex technologies such as audio and video compression, bandwidth management, and system control. Each manufacturer/developer of videoconferencing equipment has created its own unique way of bundling these communications protocols. These proprietary solutions help each company distinguish itself from the competition.

As videoconferencing became increasingly popular in the 1980s, several manufacturers were selling a variety of systems, and users began demanding complete and seamless end-to-end interoperability. This meant making the equipment as simple and user-friendly to use as a fax or telephone (sometimes referred to as "plug-and-play"). In order to make a videoconference call as simple as making a phone call, equipment must have shared technologies. Therefore, the communication components had to be standardized individually before they could be brought under a comprehensive package.

International standards were finally ratified a few years ago by the International Telecommunications Union (ITU-T). The ITU-T, formerly known as the Consultative Committee on International Telegraphy and Telephony (CCITT), is an international treaty organization and is perhaps the most important international standards body in the world. The group developed a worldwide standard of interoperability for videoconferencing, labeled H.320.

H.320 Family of Standards

H.320 is a set of videoconferencing standards developed by ITU-T. Formally called Narrow-Band Visual Telephone Systems and Terminal Equipment, the H.320 standard is currently made up of nearly a dozen standards that establish design parameters for manufacturers. Table 17-1 provides a brief overview of current videoconferencing standards and their functions.

Standard	Purpose of Recommendation	Year Approved
ITU-T G.721	64 Kbps PCM digital audio from 3.3 KHz analog.	1980
ITU-T G.722	48, 56 or 64 Kbps PCM digital audio from 7 KHz analog audio.	1986
ITU-T G.728	16 Kbps digital audio for low bit rate video. Samples 3.3 KHz of analog audio.	1993
ITU-T H.200	Audio-only and still graphics with audio. T-120 will expand the standard.	1990
ITU-T H.221	Frame structure, protocol, and audio/video multiplexing technique. Specifies BAS codes used to combine B channels.	1990
ITU-T H.230	Multiplexing frames of audio, video, user data and signal information onto a digital channel. Includes BAS escape codes.	1990
ITU-T H.231	AV.231 multipoint control unit description.	1993
ITU-T H.233	Encryption technique used to protect the confidentiality of H.320-compliant exchanges.	1993
ITU-T H.242	Protocols for setup and disconnect; inband information exchange, recovery from faults and channel management.	1990
ITU-T H.243	Control procedures between H.231 MCU and H.320 codecs using ISDN B channels.	1993
ITU-T H.261	Provides a uniform process for a receiving codec to interpret a compressed video signal. Establishes QCIF and CIF display formats.	1990
ITU-T H.KEY	Describes how encryption keys are protected, managed and authenticated.	1993
ITU-T H.261 ANNEX D	High quality graphics system which doubled the FCIF horizontal and vertical resolutions.	1993

Table 17-1. *ITU-T H.320 Family of Standards (Source: Trowt-Baynard)*

H.261

Of the standards comprising the H.320 family, H.261 is the most prominent. It defines what an H.320 codec should do by recommending a set of algorithms for video compression, and it is the algorithm used when two or more different codecs have a need to interoperate. H.261 can also be a video system's only compression method and/or a supplementary algorithm. The standard is based on discrete cosine transform, Dpulse code modulation, and *motion compensation* techniques—in a motion sequence, each pixel can have a distinct displacement from one frame to the next. Motion compensation is a compression technique that exploits similarities in neighboring pixels, divides the image up into rectangular blocks and uses a displacement vector to describe movement of all pixels within the block.

H.261 is frequently called "px64" because it defines what steps should be taken to compress and decompress video transmitted in multiples of 64 Kbps, where p=1,...30 (e.g., 64 Kbps to 2.048 Mbps). The H.261 standard format is depicted in Figure 17-8.

H.261 was adopted to ensure compatibility for the differing television display formats found around the world. For instance, North America's and Japan's television equipment abide by a standard called NTSC (National Television Standards Committee). In contrast, Europe's predominant standard for color television is called PAL (phase alternating line system).

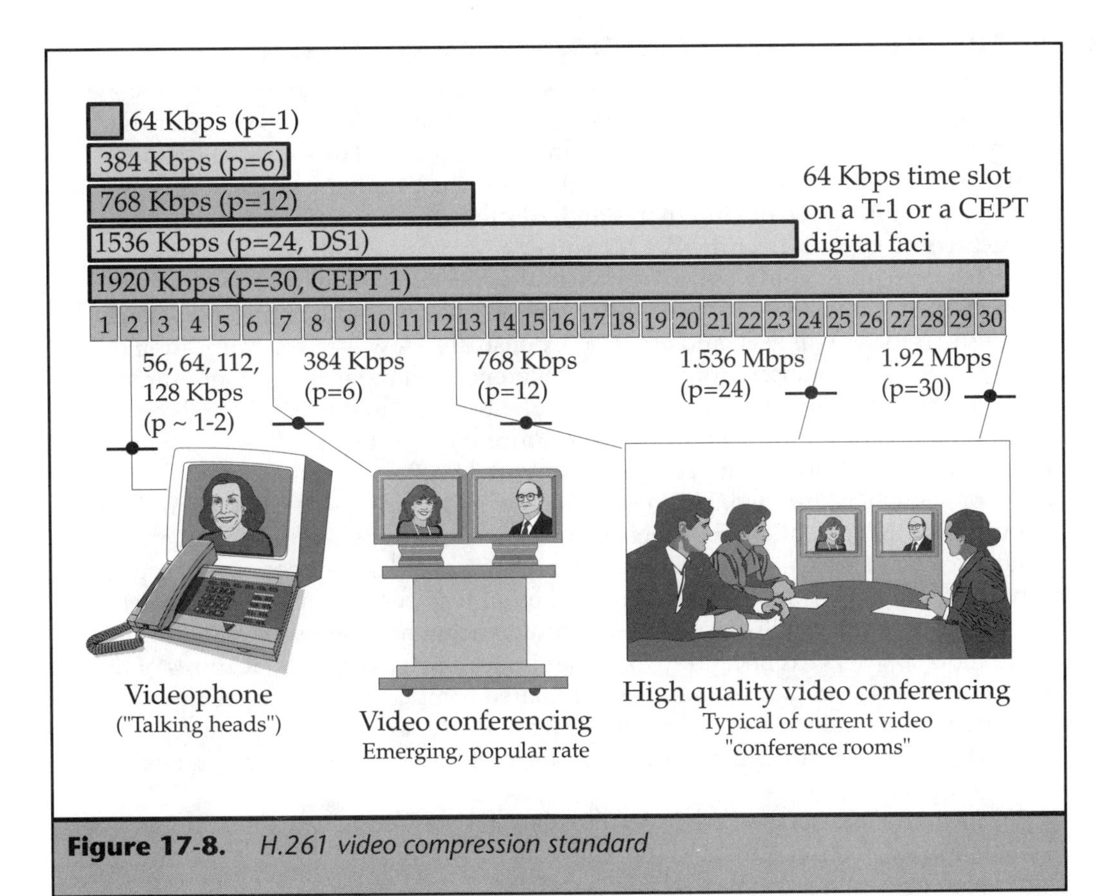

Figure 17-8. *H.261 video compression standard*

Under H.261, the ITU-T agreed to a new television display format, which any nation building codecs could use. Titled the common intermediate format (CIF), it is a compromise between the formats mentioned above.

CIF and QCIF

The common intermediate format (CIF) and the quarter common intermediate format (QCIF) are compromises that use the frame rate of North American TV (30 fps) and a resolution technique more compatible with European television. CIF displays video frames containing 352 pixels in each of 288 noninterlaced luminance lines. QCIF displays 144 luminance lines, each containing 176 pixels.

Although H.261 does not make CIF compliance mandatory (QCIF *is* mandatory), it is frequently found in group systems since it is best suited for large screens (over 20 inches). QCIF, on the other hand, fills one-quarter the spatial resolution of CIF and is usually found on desktop video systems, making for a much smaller picture on your video screen.

Proprietary Versus Standard Protocols

Prior to the adoption of the H.320 family of standards, videoconferencing systems communicated with proprietary protocols. Most LAN-based desktop conferencing systems still use proprietary protocols, mainly because standards have not been formalized until recently and, more often than not, the proprietary algorithms provide a better audio and video quality than standard protocols. Therefore, even if LAN products could connect over the WAN to other LAN products (from different manufacturers), most could not share data and applications. This means proprietary videoconference transmissions are mostly limited to intravendor communications.

These protocols still exist and are used extensively. However, the ratification of the ITU-T H.320 family of international videoconferencing standards and their implementation by equipment manufacturers worldwide have largely eliminated the debate over standard versus proprietary protocols. Proprietary codecs are totally different and are thus incompatible with each other or with H.320.

As previously mentioned, H.320 is a family of standards needed to design a piece of equipment that will allow for video compression and transmission. The major elements covered are: audio and video compression and decompression, encryption, graphics, multipoint videoconferencing, and other features such as far-end camera control. The interoperability issue in videoconferencing has been well planned and is allowing users to concentrate on other issues, such as cost and features of the systems. Likewise, standardization has stimulated competition among manufacturers and brought equipment prices tumbling down. Clearly, we can begin to see the advantages that standardization brings to a technology. Still, the H.320 standard has its imperfections.

The technical elements of the H.320 standard are very complex, and manufacturers have developed and implemented it in different stages. In addition, the complexity is augmented by continuous improvements and expansions of the standards. Therefore, users frequently find themselves having to upgrade their systems to the latest standards. The only catch with standards is that they are worthless unless the entire industry adheres to them in a joint effort.

Video and Computers

The H.320 series of standards was created with group systems in mind. Therefore, as new products, services, and applications arise, the family will expand to accommodate new standards. The next major effort in the standardization of visual communications is for desktop videoconferencing. The original videoconferencing standards were created for large-scale corporate uses, such as group meetings; personal videoconferencing was not taken into consideration. The standards outlined in the H.320 series are not compatible with application sharing or document conferencing. Consequently, several proprietary desktop videoconferencing standards are evolving to include cross-platform videoconferencing.

T.120

T.120 is a standard for audiographics exchange. While H.320 does provide a means of graphic transfer, T.120 will support higher resolutions, pointing, and annotation. Users can share and manipulate information much as they would use a whiteboard if they were in the same room, but they will be working over a distance and using a PC platform.

Intel Corp. has developed a proprietary algorithm that combines similar functions to T.120, but their solution has caused great concern in the videoconferencing industry since it challenges the issue of system compatibility. The rest of the industry is counting on H.320's and T.120's increasing acceptance to reduce other proprietary solutions' chances of survival.

We could soon see video compression handled by software rather than hardware. Though the technology is still being perfected, software control would eliminate codec interoperability problems. The software would recognize protocols from the sender and match them up at the receiving end with a lookup table. In the meantime, this solution is hampered by the heavy reliance on microprocessor power as the computer performs the compression "in house."

For the time being, interoperability is still the driving force behind the success of videoconferencing. The standards are fairly recent, but the industry players have gone to great lengths to demonstrate their commitment to users. At trade shows such as ITCA and TeleCon, the largest equipment manufacturers have set up their systems and simulated intervendor transmissions. It is collaborative efforts like these that give the marketplace confidence in the technology's longevity.

Lowest Common Denominator

Standards are a form of compromise. The H.320 standards are the initial step in the interoperability of dissimilar codecs. Picture and sound quality are generally lower than those delivered by proprietary algorithms running at the same speed. However, H.261 has the flexibility to evolve. Virtually every videoconferencing system sold today (except for desktop products) supports the H.261 algorithm, either as an option or as part of the system's standard package. Some have dropped proprietary algorithms altogether.

The industry is responding to the customer request for interoperability. Now, it is up to the customers to keep the momentum going by using the standard, feeding back information on where it needs to be improved, and insisting that manufacturers provide inexpensive ways to continue to add enhancements to existing systems.

Another common problem encountered when integrating personal videoconferencing systems is the incompatibility among long-distance services. Although you should be able to mix even Switched 56 and ISDN lines on the same conference, you're better off matching services as closely as possible. The quality of the phone line you get in the low-end POTS videoconferencing applications can have a tremendous effect on the application. Differences as subtle as inconsistent line-signaling schemes cause some systems to refuse to connect. Furthermore, a few videoconferencing products have features that are tightly coupled with PBX or telephone company central-office switching equipment.

Compatibility issues are even more critical for wide-area multipoint conferencing, in which the complexities of routing multiple lines through a multipoint control unit (MCU) can make spontaneous meetings nearly impossible. Furthermore, multipoint sessions are conducted at the lowest common speed among participants, much like modems. It is always best to select a long-distance provider with experience in these issues, such as AT&T's Global Business Video Services or Sprint's Meeting Channel.

The Videoconferencing Marketplace

Group systems have been available for over ten years. Substantial increases in quality and portability have been followed by decreases in pricing. Today, prices for a turnkey solution start between $20,000 to $30,000. Since they are rather costly and are usually devoted for use in one primary location, companies usually install one unit per site.

On the other hand, most employees in medium to large corporations have a personal computer at their desk. Therefore, the proliferation of desktop video-conferencing will be much greater. Many government, commercial, and educational organizations have set up pilot programs to test the technology. These test beds usually use from two to 20 units costing $2,000 to $6,000 per seat and may include a number of different vendors. Acceptance is growing quickly, though. Some large corporations have embraced desktop videoconferencing as an essential productivity enhancer and timesaver for collaborative applications and are deploying thousands of desktop videoconferencing units to employees and even to customers.

The general public, including government, education, and business, has a limited perception of the desktop videoconferencing industry because the technology is so new and changing so rapidly. The average person gets information from feature articles in telecommunications trade publications or sometimes even a television program with a future technology focus. Still, the industry is so dynamic and advancing so rapidly, it's hard to keep up. This is a common occurrence with emerging technologies.

Conclusion

One could argue that the videoconferencing industry is becoming a part of the computer industry. The growth in popularity of personal videoconferencing systems is logical when you factor in cost and convenience. Once considered a separate technology, videoconferencing has attracted the interest and investment of computer industry giants such as Intel Corp. and Microsoft Corp. The purpose of the computer industry is to convert the desktop PC into the ultimate communications tool by integrating voice, video, and data.

The desktop videoconferencing market is relatively easy to get into. The abundance of codec, sound board, camera, and data-sharing software developers and manufacturers is attracting a new breed of systems integrators. In the near future, desktop videoconferencing will be an off-the-shelf computer peripheral similar to scanners or modems. Prices are plummeting. There are excellent systems for ISDN applications in the marketplace today from companies such as PictureTel and Intel priced at $2,000 to $2,500. And in the low-end market, companies such as Creative Labs and Toshiba offer proprietary starter kits for the POTS line users. Much like other computer technologies, prices quickly drop with improvements in technology and a competitive environment.

In the group videoconferencing sector, PictureTel (Danvers, Massachusetts) has been the leader for several years. Approximately two-thirds of room-sized systems in use today are PictureTel units. The other major players in this sector are Compression Labs (San Jose, California), VTEL (Austin, Texas), and British Telecom (Herndon, Virginia). In the desktop videoconferencing sector, it is estimated that 20,000 to 30,000 units were shipped in 1994. That figure is estimated to more than triple in 1995. Sales were led by AT&T, Intel, and PictureTel. Look for even more competition to enter the arena as the standards start to take hold and the struggle for market share heats up. See you on your desktop.

Photo courtesy of PictureTel

Chapter Eighteen

Computer Telephony Integration

Now, more than ever, businesses are focused on customer service as a competitive advantage. Companies are striving to provide accelerated and uncomplicated methods for their customers and employees to obtain access to critical information resources. The computer and the telephone, the two most commonly used business productivity tools, are the primary vehicles used for acquiring and disseminating information. Advancements in the features of each technology have been so significant that little attention has been paid to the idea of combining their inherent strengths—until now.

Since the late 1980s an architectural framework has been unfolding for linking computer and telecommunications platforms. This architectural framework is known as computer telephony integration, or CTI. Early CTI solutions combined mainframe platform and proprietary telecommunications application programming interfaces (APIs). With the shift toward distributed computing and open telecommunications APIs, CTI is evolving from a limited solution to a widespread technology for integrating data and voice platforms.

This chapter will examine the historically proprietary nature of the telephone and the movement toward open systems, the integration of the computer and telephone, and the telephone as a vehicle to process information.

Messaging Standards

In the late 1980s, driven by customer demand for third-party applications, the European Computer Manufacturers Association (ECMA) initiated the movement toward a standard for communications between computer and telecommunications platforms. Since then, two CTI standards have evolved: the Computer Supported Telephony Applications (CSTA) and the Switch-Computer Applications Interface (SCAI).

CSTA

In response to the ECMA initiative, a committee was formed representing major PBX manufacturers, including AT&T, Northern Telecom, Alcatel, Siemens, IBM, HP, and others. The committee developed the Standard ECMA-179 *Services for Computer-Supported Telecommunications Applications (CSTA)*, and the Standard ECMA-180 *Protocol for Computer-Supported Telecommunications Applications* for OSI Layer 7 communication between computer and telecommunications networks.

CSTA defines architectural frameworks, requirements, and protocols for integrating computer and telecommunications platforms. This technical standard emphasizes flexibility, bidirectional communications, and a distributed model for both computing and switching. Aimed initially at private networks, CSTA focuses on switch-based objects such as telephones, trunks, and queues. CSTA services were designed to be independent of the switching platform. CSTA has no knowledge of the specific details of how the switching platform accomplishes requested CSTA services.

The first edition, now known as the 1992 specification, defines the following:

- Call control services, including events such as answer call, clear connection, hold call, and transfer call
- Device services, including phone set features such as do not disturb, forwarding, and message waiting
- Status reporting services
- Routing services
- System status and escape services

Publication of the second edition of CSTA, now under development, is expected in early 1995. The second edition will expand the definitions of various services, including:

- Call control services: single-step conference, single-step transfer, and park call
- Device services: control of microphones and volume
- Voice unit services: play message and record message

ECMA is now a worldwide standards body. In the future, it will expand its coverage to include the public network. It will also review specific application program interfaces such as the telephony server application programming interface (TSAPI), jointly developed by AT&T and Novell, for official standard status.

SCAI

The Switch-Computer Applications Interface (SCAI) standard was developed by the American National Standards Institute (ANSI). SCAI specifies an architecture and OSI application layer protocol for peer-to-peer data communication between computer and switch applications, thereby enabling the functional integration of computer and telephone switching platforms.

SCAI was influenced by other standards, including Integrated Service Digital Network (ISDN) and Intelligent Networks (IN).

The first version of the SCAI standard is targeted toward call center applications in public and private networks.

Unlike CSTA, SCAI emphasizes communications integration more than application integration. Since CSTA provides more event-reporting capabilities, it has attracted more application developers.

Versit

The geometric growth in computer and telecommunications applications and hardware platforms, and the undisciplined implementation of the technologies, has made it hard to achieve interoperability in and between many companies. The rapid growth also makes it difficult for any one company to dictate an interoperability standard.

Versit was formed by four leading computer/communication vendors—Apple Computer, AT&T, IBM, and Siemens Rolm—to endorse standards that allow computers, telephones, personal digital assistants (PDAs), and other network applications to work together. The consortium will certify that products bearing its seal of approval will interact.

Versit is also targeting areas for which to define specifications, and it has selected some existing standards on which to base the specifications. Novell's TSAPI, the switch-normalization scheme in IBM's CallPath, first-party call functions, and a set of object classes for the next generation of telephony applications will be supported by Versit.

Disagreement over the physical interface between the phone and PC threatens to split the CTI industry into two camps. Versit has selected Apple's GeoPort as the physical link from the PC to the phone. Intel has proposed an alternative interface, called high-speed serial interface, and has the support of Compaq Computer Corp., Microsoft, and Northern Telecom.

Intel and rival Versit are racing to complete a standard set for multimedia communications, which, if adopted by the standards bodies, will have far-reaching consequences. Both sides are initially using the same macroarchitecture, with a 5 Mbps serial bus at the top layer for isochronous data and layers of API and format specifications above this to provide data to various PC applications.

Architectural Implementations

Today there are two architectural implementations of CTI. The first implementation is the first-party connection, shown in Figure 18-1. In this environment, the desktop computer is directly connected to an analog or digital telephone. The computer emulates the telephone by sending the same type of proprietary signals to the switch that the telephone would send. The first-party connection is implemented with a serial connection or a PC phone emulator board. An obstacle to the first-party connection is the physical link that is required between the telephone and desktop PC. This requirement is typically not a problem for stand-alone or small-office environments. However, as the number of integrated PCs and telephones expands, so does the number of additional connections that must be made and maintained. An implementation of one 1000 telephone-PC connections requires managing one 1000 links.

The second type of implementation is the third-party connection, shown in Figure 18-2. In this environment, the desktop computer or "client" is networked to a telephony server that directly connects to the switch via a PBX-to-telephony server connection. The PBX-to-server connection defines a special protocol used by the PBX and server to communicate. Each PBX vendor defines a physical connection using any of several different protocols for the link, including:

- ISDN basic rate interface (BRI)
- TCP/IP
- X.25
- Ethernet

The implementation depends on the individual PBX manufacturer. Figure 18-3 provides a reference for PBX-to-server links.

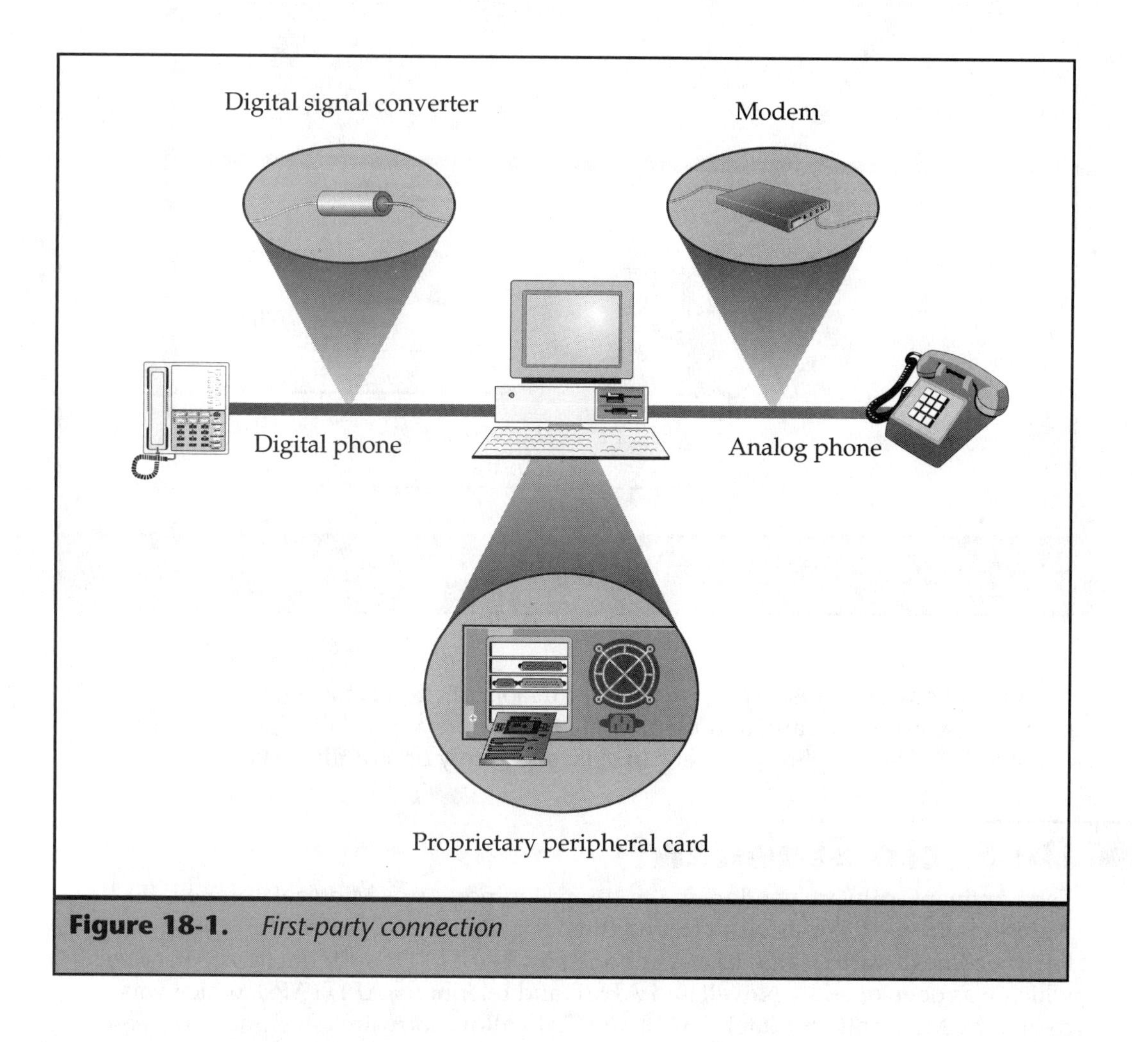

Figure 18-1. *First-party connection*

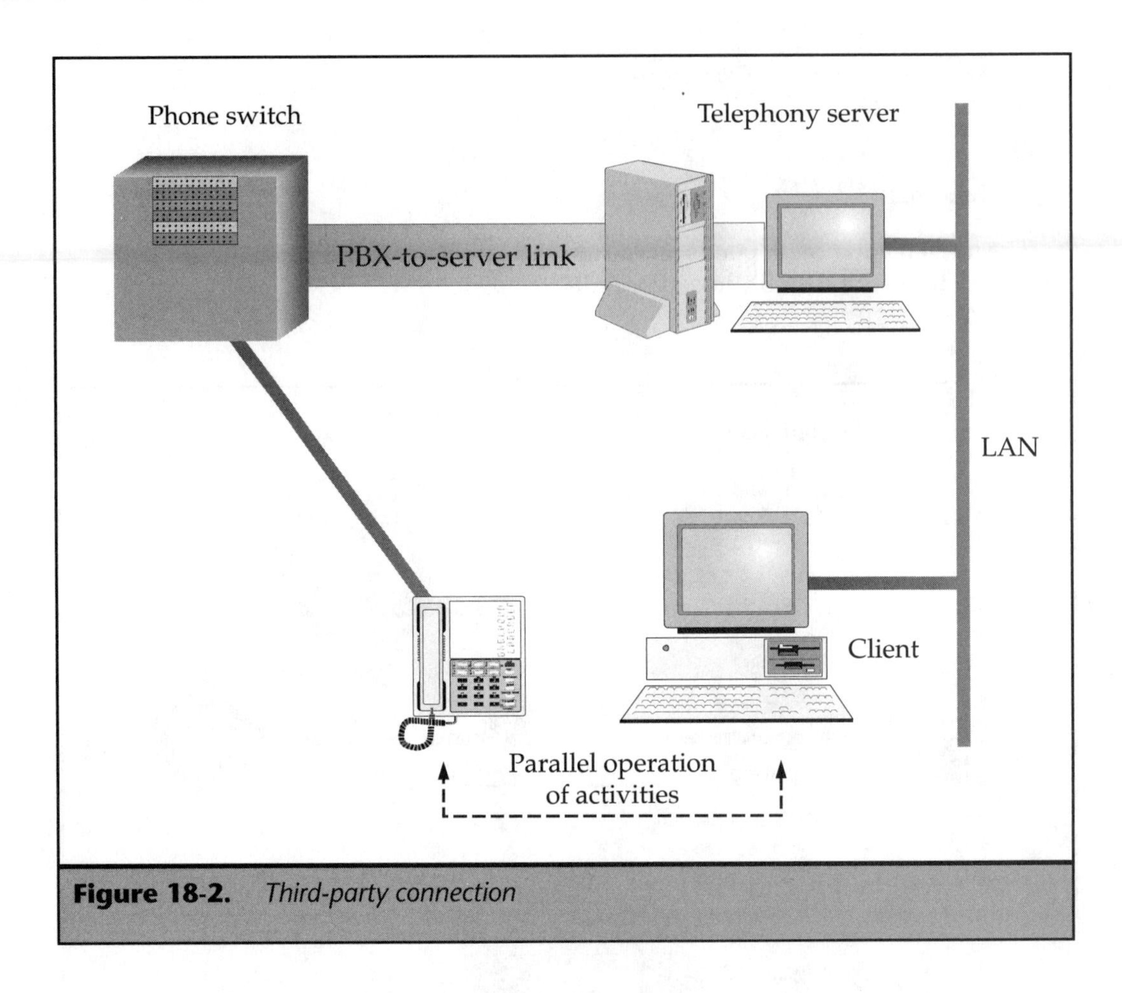

Figure 18-2. *Third-party connection*

A telephony server is a computer whose major function is to control, add intelligence, store, forward, and manipulate the various voice calls flowing into and out of a PBX. The telephony server can exist separately or as a file server.

De Facto Standards

The dominant software vendors in the computer industry—Microsoft, Novell, IBM, Apple, and Sun—have each developed their own APIs for CTI. Two APIs, however, have become de facto standards: telephony services API (TSAPI), mentioned earlier, which was developed by Novell and AT&T, and telephony API (TAPI), which was created by Microsoft and Intel. TSAPI and TAPI allow software developers to create applications that make the link between the telephone and the PC meaningful to a potential user.

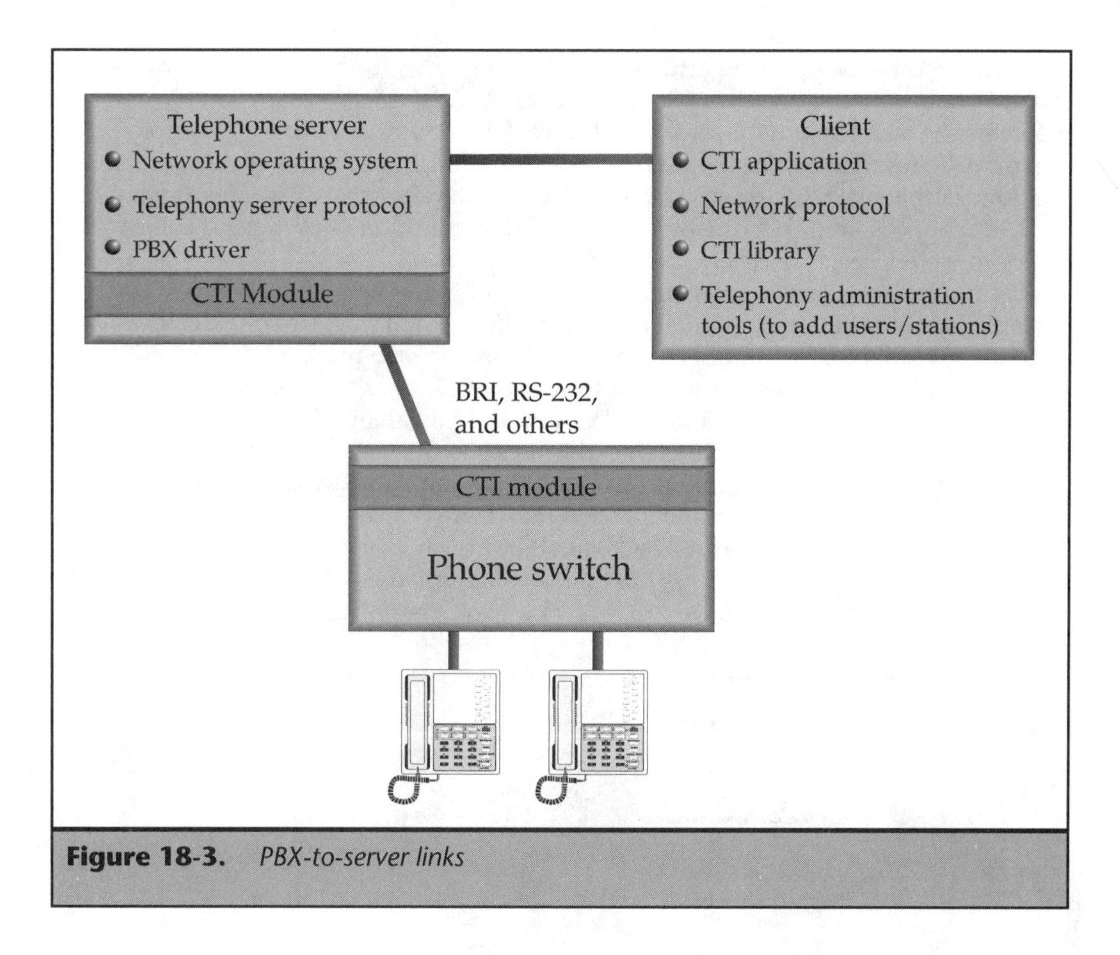

Figure 18-3. *PBX-to-server links*

With the increasing availability of applications based on these APIs, users can take advantage of CTI in almost any environment—from a single-PC/single-phone setup to an enterprise-wide system with thousands of PCs and telephones. The choice of which API is implemented may depend on the type of environment and how that environment evolves.

Windows Telephony Application Programming Interface (TAPI)

The Windows telephony API is called TAPI. Microsoft's TAPI provides a first-party connection. The goal of TAPI is to provide "personal telephony" to the Windows platform. TAPI will be included in Windows 95.

Windows telephony is comprised of a Windows telephony DLL and two standards, shown in Figure 18-4. The first standard is the service provider interface (SPI). The SPI is directed to hardware manufacturers that want to conform to Windows telephony. If a hardware manufacturer's product conforms to the SPI, then it can talk to the Windows telephony DLL. The second standard is called the application programming interface, and it is directed to software developers who write application programs. If those developers' programs adhere to the API, they can take advantage of the Windows telephony DLL to drive whatever telephony devices or services adhere to the SPI.

TAPI supports multiple concurrently running applications sharing and controlling one or more devices.

The applications can be aware of each other and of the state of the devices. For example, one application could dial a number using a digital PBX phone, while a second application could control a data transfer session on a modem, while a third application plays audio recordings over the digital PBX phone to report the status of the transfer process. Of course, one application could do all of this, but TAPI allows multiple applications to cooperate to get the job done.

The TAPI Dynamic Link Library (TAPI.DLL) accomplishes this by playing the role of traffic cop between the applications and the service providers. The service providers are insulated from this complexity, believing they are always servicing one application, and they are not aware of other service providers. Applications, on the other hand, can decide if they want to be aware of other applications and devices.

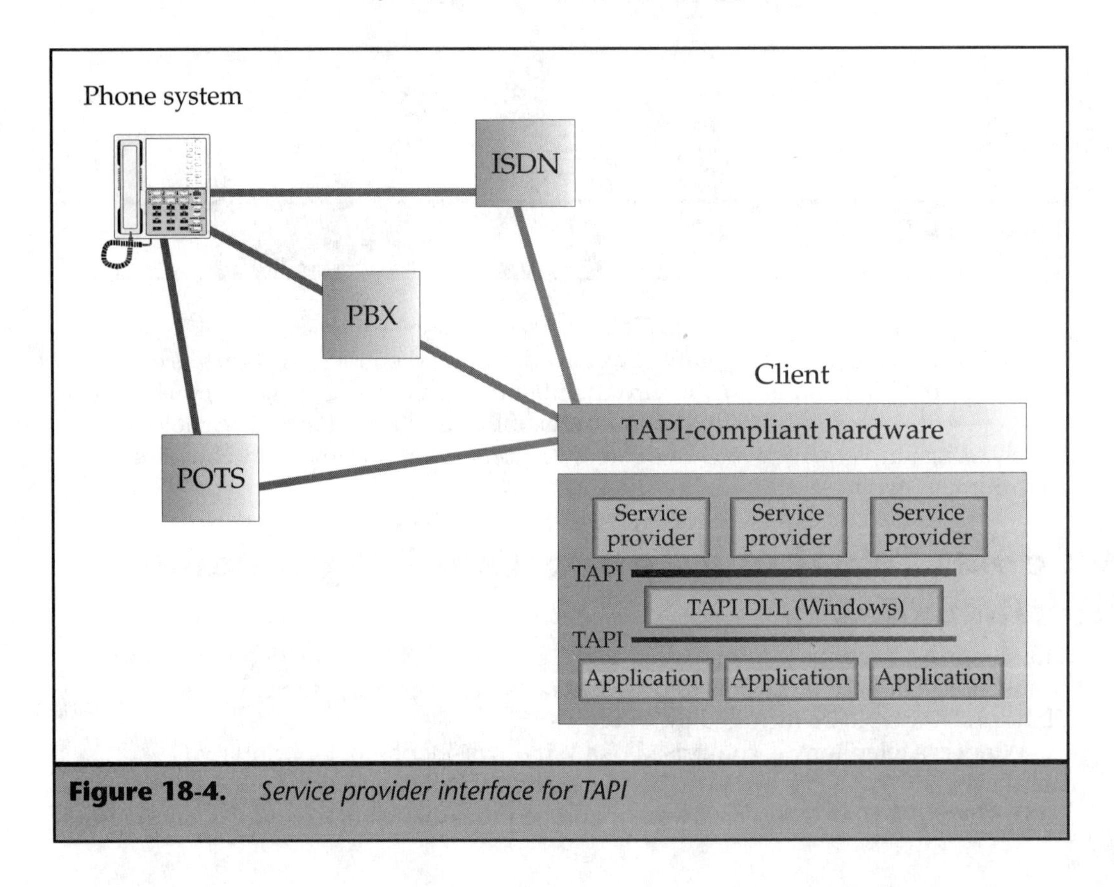

Figure 18-4. *Service provider interface for TAPI*

The power of TAPI is that it allows one application to provide access to a number of devices and media types or cooperate with other applications to provide the necessary access and control.

Telephony services break down into full telephony services and assisted telephony services. Full telephony services are used to build robust telephone control applications. Assisted telephony services are used to add minimal but useful telephone control functionality to nontelephony applications, such as word processors, accounting systems, and personal information managers.

Full telephony defines three levels of service. The lowest level of service is called basic telephony and it provides a guaranteed set of functions that corresponds to the plain old telephone service (POTS)—only make calls and receive calls. Teltone, a company based in Bothel, WA, has a device that hooks up between POTS lines and PC comm ports to provide CTI to the stand-alone PC user. The next service level is supplementary telephone service, which provides advanced switch features, such as hold and transfer. Finally, there is extended telephony services, which enable application developers to access service provider-specific functions not directly defined by the telephony API.

TAPI does not require a LAN to operate, nor does it require a particular telephone system. Under TAPI, a direct connection is made between the PC and the telephone (see Figure 18-5). This can be established in a variety of ways: through an RS-232 cable running from the PC serial port into the telephone, through an add-on card in the PC's expansion slot, or through a modem. In a PC-centric connection, the PC receives the initial telephone service hookup, with the phone attached to the PC. In a phone-centric connection, the phone receives the initial service hookup, with the PC attached to it. The advantage of the phone-centric approach is its simplicity. Function calls from the PC are often based on the Hayes AT command set. This simplicity limits the amount of integration that can be made with other applications residing on the PC. Fax communications and voice-mail, for example, are not easily integrated since the phone still controls the telephone line.

Windows telephony effectively removes earlier barriers to creating PC-driven telephony applications, namely the wide enormity of telephony "network" services—from the many telephone company interfaces (POTS to T-1) to the many more proprietary interfaces behind dozens of proprietary PBXs, key systems, and hybrid phone systems.

Dynamic Link Library (DLL)

A DLL, which is a feature of the Windows operating system, is a collection of executable code that one or more software applications can dynamically call and link at run time. A DLL can be changed without affecting the applications that use it.

Windows Open Services Architecture (WOSA)

Windows telephony is a component of WOSA. WOSA provides a framework for Windows applications to seamlessly access information and network services through a single system-level interface. A front-end application and back-end service don't need to speak each other's languages to communicate as long as they both know how to talk to the WOSA interface.

WOSA defines a system-level Dynamic Link Library (DLL) to provide common procedures and functions for network service providers, i.e., database vendors, e-mail vendors, and PBX vendors, to map their low-level interfaces to. Applications call a common set of system APIs to access computing resources or services such as database

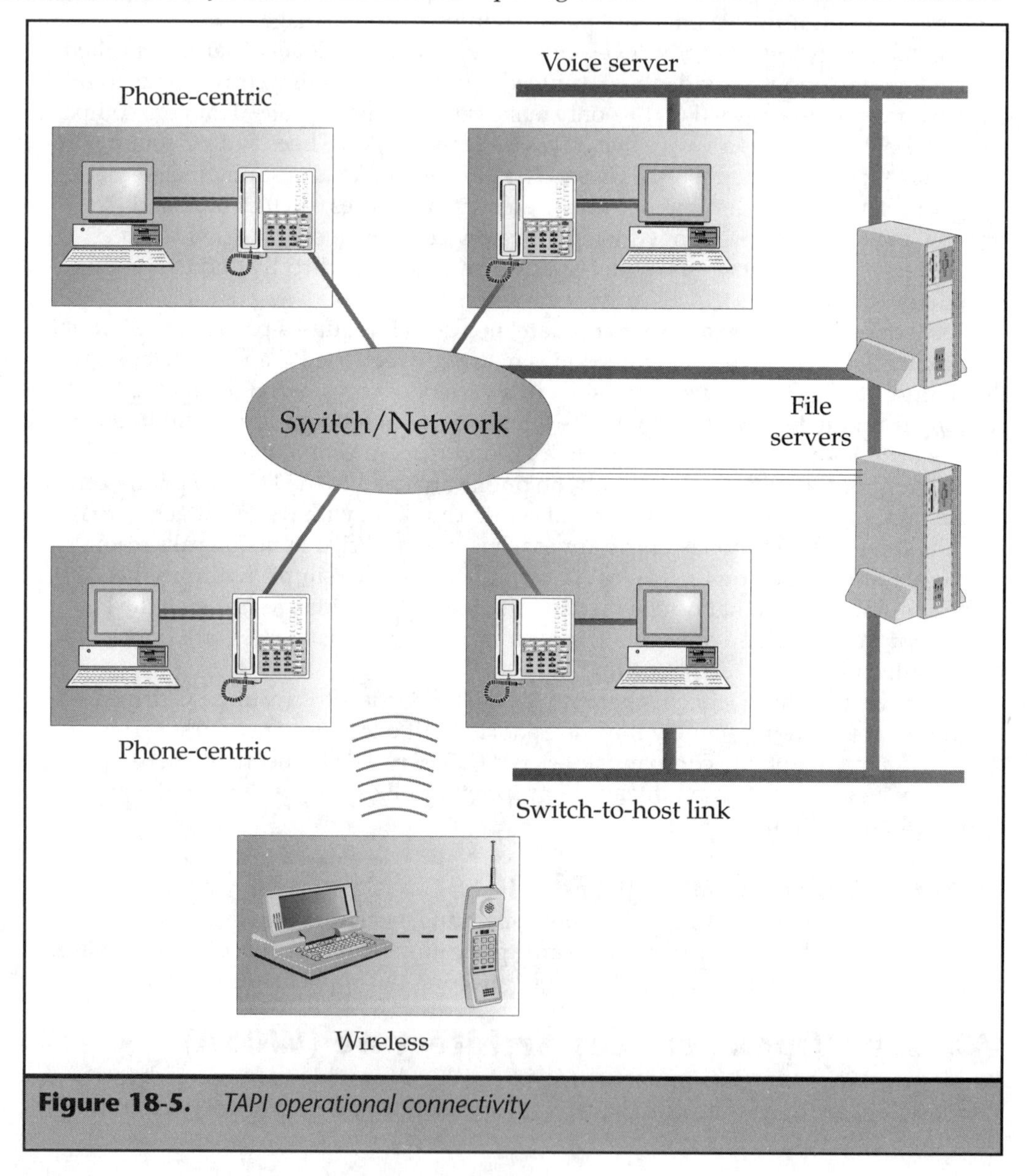

Figure 18-5. *TAPI operational connectivity*

access (ODBC), messaging (MAPI), market data (XRT), and telephony (TAPI). A client application only needs to know how to talk to WOSA and not directly to the low-level protocols and interfaces of each of the service providers. WOSA uses a Windows DLL that allows software components to be linked at run time. This allows applications to connect to services dynamically. Because this set of APIs is extensible, new services and their corresponding APIs can be added as needed.

Client-Server TAPI

Since TAPI is independent of the underlying network and connection model, it supports client-server CTI applications. In this implementation, the service provider communicates to a telephony server on the network while the server talks to the PBX, typically via a CTI link. The application is unaware of how the telephony services are implemented and only knows what functions are supported by the service provider.

The same TAPI application that controls a digital phone using a telephony board on a Windows PC will work unchanged using a network connection to a telephony server, assuming the two service providers support the same functions that the application requires.

One such implementation that uses this approach is Dialogic's CT-Connect product. For example, one configuration may use a Windows NT server to communicate to a Meridian PBX with Windows clients running a TAPI application to dial phone numbers and log calls.

Client-server TAPI is good for applications that provide advanced CTI functionality or organizations that want to implement CTI functionality on hundreds of desktops.

Novell Telephony Services Application Programming Interface (TSAPI)

TSAPI provides a third-party connection between phones and PCs. CTI applications using TSAPI can control any call, even if it comes into a different phone line within the organization. Features such as redirecting calls from one station to another, interactive voice response (IVR), and automatic call distributing (ACD) are possible. Novell and AT&T developed TSAPI to add telephony service to the NetWare network operating system. As a result, telephony services API requires the use of a LAN. Rather than make direct connections between each phone and PC, one connection is made between a server and the phone system. This connection is called the PBX-to-server link, which can be physically established several ways.

Telephony services provide a common interface for front-end applications to talk to back-end switches. Novell's telephony service consists of a telephony server NetWare Loadable Module (NLM), a telephony server API (TSAPI), a Telephony Server Library, and a Telephony Client Library (see Figure 18-6). These components are switch-independent and are supported by any telephony server-compliant switch driver. The switch driver is unique to the specific switch that is being linked to the Novell server.

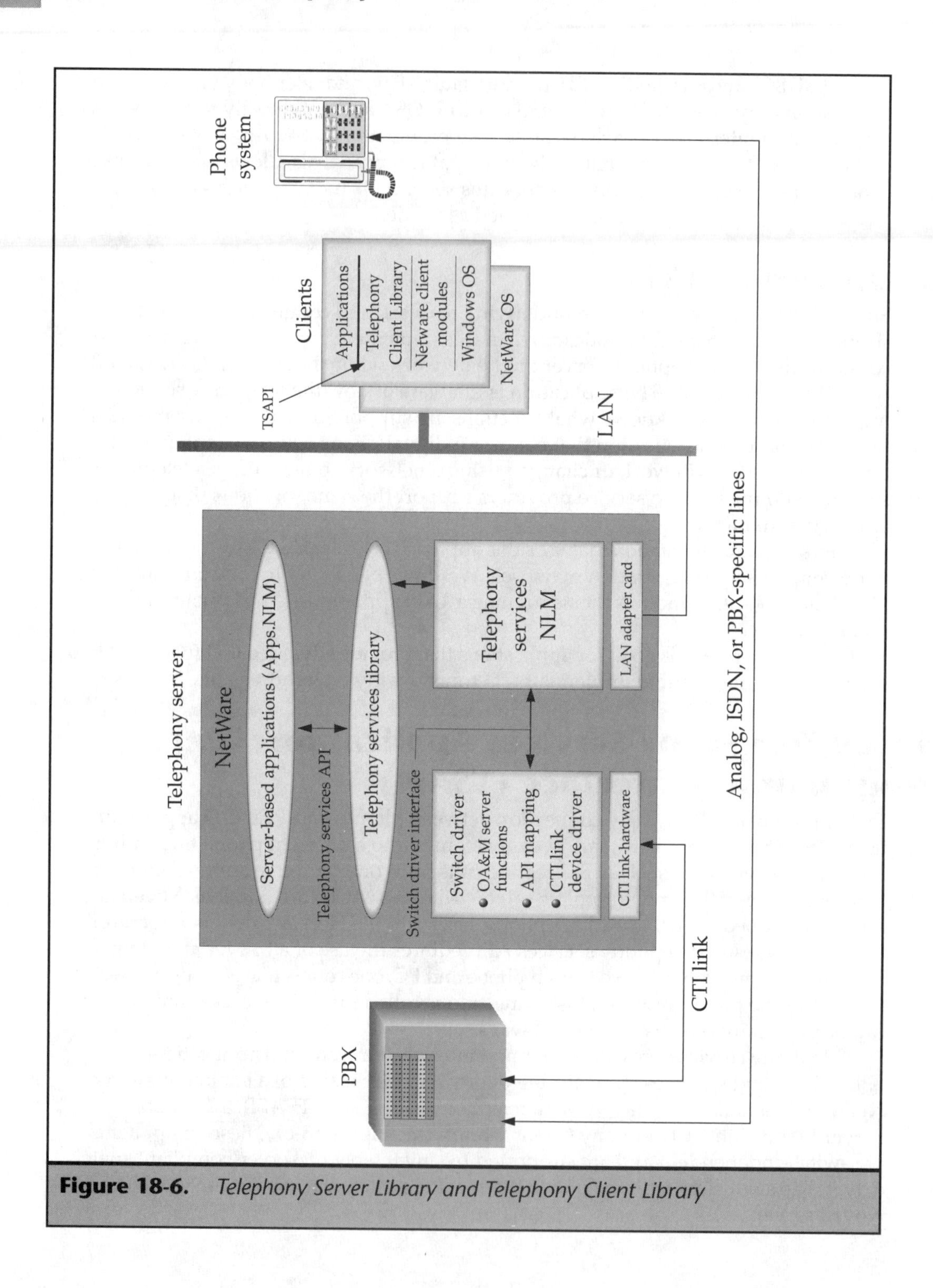

Figure 18-6. *Telephony Server Library and Telephony Client Library*

The telephony server NLM is known as the Tserver. This module acts as a message routing agent, making sure that messages from the switch driver are routed to the clients waiting for telephony events and passing the messages received from the client (API service requests) to the switch driver across the switch driver interface. The telephony server NLM is also responsible for maintaining security according to the user-specific restrictions defined in the telephony server's security database.

The telephony server API is a CSTA-based, C-language definition of the functions (services), data types (parameters and structures), and event messages that will be used by telephony-enabled applications to access telephony services.

The Telephony Server Library is a software module which provides server-based applications with access to the telephony functionality supported by the telephony server API. This module is responsible for presenting the telephony server API to applications running on a server. This includes accepting API service requests and delivering events to applications that are monitoring telephony objects (devices) on the switch. This library can run on the same physical server as the telephony server or on any NetWare server in the network.

The Telephony Client Library is a software module that provides client-based applications with access to the telephony functionality supported by the telephony server API. This module is responsible for presenting the telephony server API to applications running on a client in the network. This includes accepting API service requests and delivering events to applications that are monitoring telephony objects (devices) on the switch. Operating system-specific versions of this library will be provided for all supported client operating systems. All client library modules will support the same functionality unless it is limited by specific operating system limitations.

Third-party connections are much more flexible than first-party connections in client-server networking. Every user on the network immediately has access to the PBX through the existing LAN infrastructure. Accessing other services, such as mainframe or terminal emulation, fax services, and e-mail, is possible through the CTI interface wherever these services are already on the network

A primary obstacle to TSAPI is that the PBX-to-server link between the server and the PBX must be provided by the PBX vendor. Additionally, PBX vendors may not provide a TSAPI driver for all of their PBX models.

NetWare Loadable Module (NLM)

A NetWare Loadable Module is a driver that runs in a server running Novell's NetWare operating system on a local area network, and it can be loaded or unloaded on the fly as it's needed. A telephony NLM might allow a workstation on a LAN to control a telephone system attached to a NetWare file server. It might also allow the workstation to control one or more voice-processing cards sitting in a NetWare server.

The telephony server NLM is the mechanism for passing information between the PBX and the NetWare server. As part of the NLM, an open PBX driver interface allows PBX manufacturers to write drivers that communicate with their respective PBXs. The client-server API provides support across multiple desktop operating systems and allows call control at either the client or server.

TSAPI Versus TAPI Decision Tree

At first glance, the choice seems simple: TAPI for individual Windows PCs and small workgroups, TSAPI for client-server networks in large organizations. But there are several variables. Foremost is compatibility with existing systems within an organization. In almost all instances, telephony services API requires a phone switch upgrade. Telephony API also may require upgrades, depending on the phone system.

Figure 18-7 provides a chart for analyzing the need for a TSAPI or TAPI CTI solution. The factors given relate to the infrastructure and should not be the sole determining factors for deciding whether you should implement TSAPI or TAPI. You should also consider other factors, such as financial issues, when deciding on the CTI solution. To further assist you with the decision-making process, Table 18-1 provides a consolidated look at the major players involved with TSAPI and TAPI integration and a rough estimate of costs to get you started.

Telephony Mapping (T-MAP)

Northern Telecom, an early supporter of TAPI, has introduced T-map, which according to Northern Telecom will translate server-based applications written for TSAPI to the desktop-based TAPI, and vice versa (see Figure 18-8). T-Map interfaces with the TSPI, telephony service provider interface, to facilitate this conversion. Software developers have to write applications to comply with only one of the two APIs, and the applications will work with both. The two de facto standards, therefore, could eventually unite into a single standard with multiple implementation options. If and when that occurs, the greatest beneficiaries will be users. Northern Telecom has stated that this software will be provided at no charge to developers.

CallPath

IBM's CallPath was introduced as a mainframe CTI architecture in the late 1980s. Its intent was to provide an interface for links from PBXs to IBM mainframes in large call-center environments. The software was then expanded to accommodate additional IBM platforms such as OS/2 and AIX. Recently, IBM has announced a client/server version of CallPath, reflecting the importance and prevalence of this environment.

The client-server CallPath is a server-based software package. Rather than focusing on a specific LAN environment (as Novell does with TSAPI for NetWare LANs), IBM has positioned its product as a general CTI server. PCs, workstations, and mainframes can be linked to a CallPath-enabled server. Up to four PBXs can be supported simultaneously, giving IBM leverage in mixed accounts. The software can run on a range of operating systems, including Windows, Solaris, SCO UNIX, IBM's AIX, and OS/2.

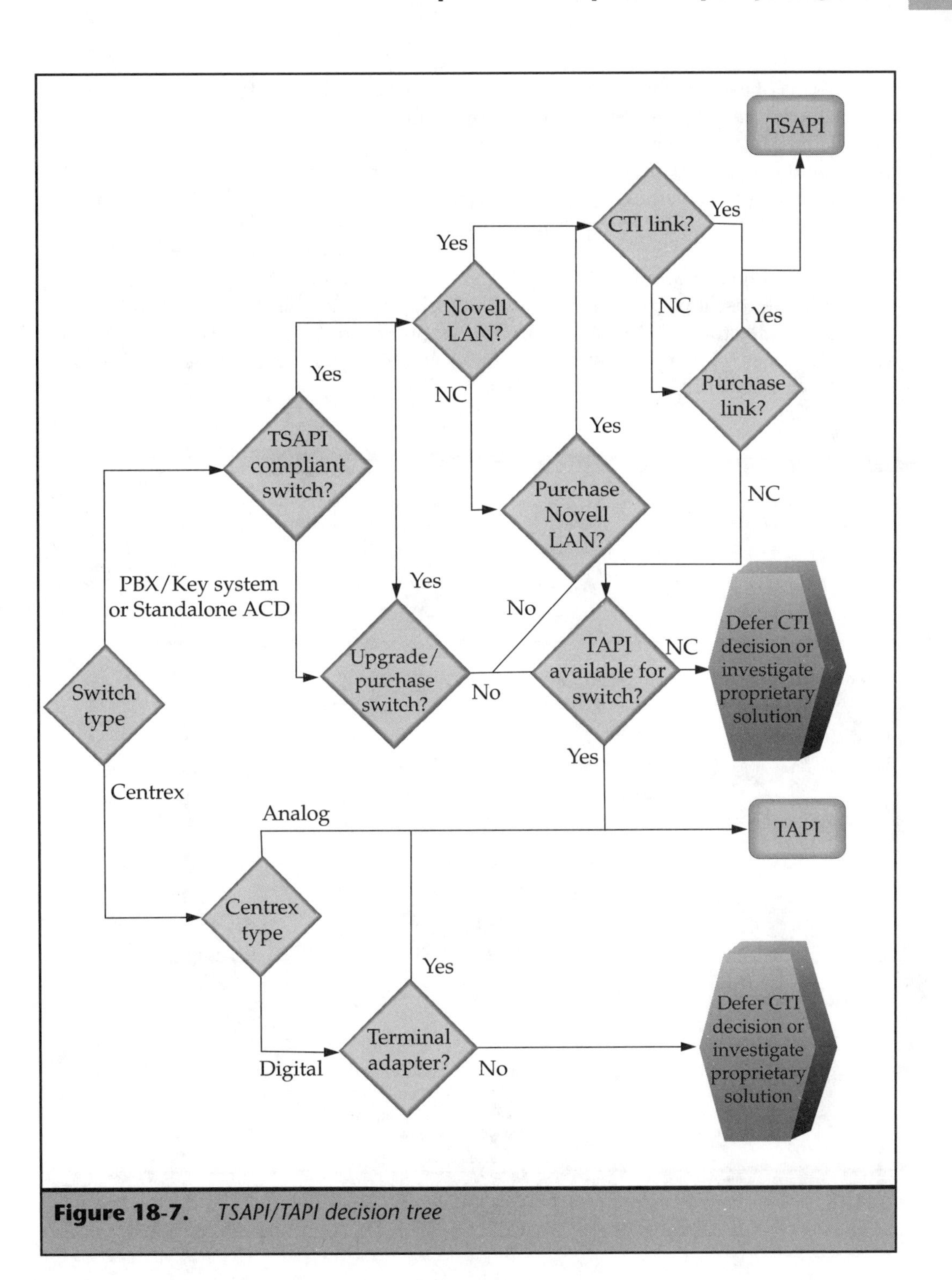

Figure 18-7. *TSAPI/TAPI decision tree*

In terms of applications, IBM is encouraging third-party developers to create CallPath-compliant applications. Their approach is less aggressive than those of Novell and Microsoft. The cost of the software developer kit (SDK) is considerably higher, and IBM is also offering its own client software products.

In addition, IBM is licensing gateway software that converts CallPath messages into the format of a given phone system. This software allows the switches to work with existing CallPath applications and facilitates portability of these applications from one IBM host to another. Aspect Telecommunications, a leading manufacturer of automatic call distributors (ACDs), is the first telephony vendor to license CallPath Services Architecture (CSA) and obtain certification from IBM. Although currently geared to larger call-center environments, this activity is relevant throughout the CTI industry because of the trends it presages.

Company	Links Available	Type of Link	Hardware Costs	Software Costs	Compatibility	Availability	Specific Switches Supported
AT&T	ASAI Gateway	TSAPI	$6,695	$10,000	TSAPI	Now	Definity G3
Comdial	Enterprise for Telephony Services	TSAPI	Included in software price	$67 to $180 per user	TSAPI	Now	DXP
		TAPI	$500	Some apps included in hardware price	TAPI	3Q95	DigiTec, IMPACT and DXP
Fujitsu	TCSI	TSAPI	$10,300	$5,200	TSAPI	Now	F9600
		TAPI	Not announced	Not announced	TAPI	3Q95	F9600
Intecom	TAPI SPI	TAPI	$423/ station	$75 per station or $1,500 per site		2Q95. Currently in beta test	
	PDI-100S Link	Proprietary	$423/ station	$75 per station or $1,500 per site	Proprietary applications	Now	IBX/80 Telari

Table 18-1. *TSAPI/TAPI Provider Specifications*

Company	Links Available	Type of Link	Hardware Costs	Software Costs	Compatibility	Availability	Specific Switches Supported
Mitel	Novell Communications Software	TSAPI	$5,000	$5,000	TSAPI	Now	SX-2000 Light
	MiTai Link	Proprietary	$5,000	$5,000	Proprietary applications	Now	SX-2000 Light
	TAPI	TAPI	$199	Bundled	TAPI	Now	SX-2000 and SX-200
Northern Telecom	Meridian Link	TSAPI	$3,000-$15,000[1]		TSAPI	4Q95	Meridian 1
	Norlink	TSAPI	Not announced	Not announced	TSAPI	4Q95	Norstar Key System
	TAPI Interface	TAPI	$300-$1,200	$0-$300	TAPI	2Q95. Currently in beta test	Meridian 1, Norstar Key System, DMS Centrex
	Meridan Link	Proprietary	$0-$15,000	Varies	Proprietary applications	Now	Norstar Key System
	Norstar Access	Proprietary	Application dependent	$600	Proprietary applications	Now	Norstar Key System
ROLM	CallBridge for Workgroups	TSAPI	$2,300 for two ports	Under $10,000	TSAPI	2Q95	ROLM 9751 CBX
	TAPI	TAPI	Not available	Not available	TAPI	Now	ROLM 9751 CBX
	CallBridge for Workgroups	Proprietary	$2,300 for two ports	$14,000 to $23,000	CallPath, DEC CIT, Tandem CAM	Now	ROLM 9751 CBX
SRX	VisionPath	Proprietary	None	$30,000	VisionPhone PC	Now	Vision LS/MS

[1] *(software included for 20 users)*

Table 18-1. *TSAPI/TAPI Provider Specifications* (continued)

MTA

In 1991, Apple released Macintosh Telephony Architecture (MTA), a desktop-oriented implementation designed to position the Macintosh as a multipurpose computer capable of handling telephony applications. Apple's approach was more broad-based than CTI. It

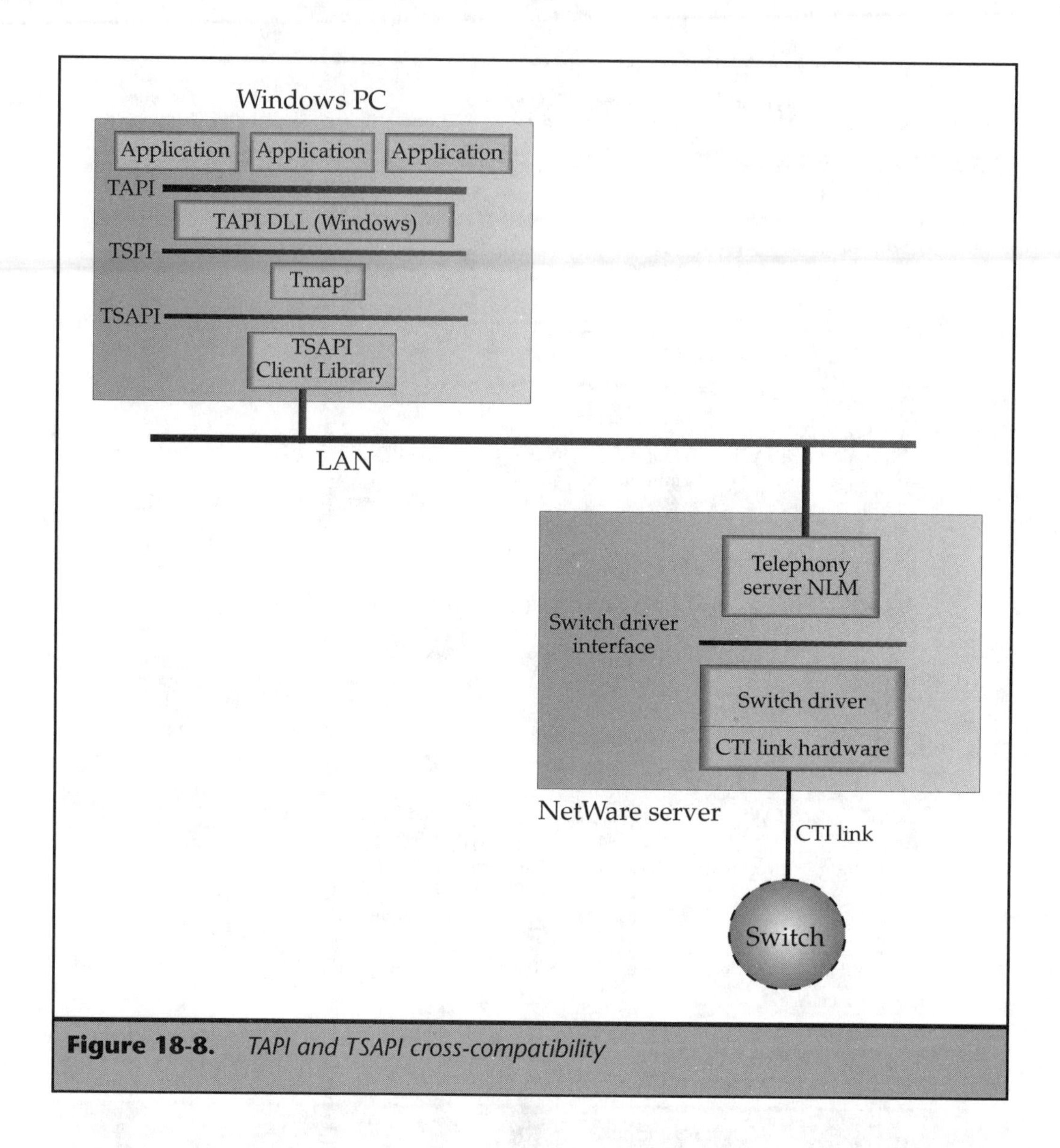

Figure 18-8. *TAPI and TSAPI cross-compatibility*

was intended to incorporate multiple transmission media (data, fax, video, and voice). To support this, Apple developed GeoPort, a high-speed serial interface for Macintosh computers. An adapter, or physical interface box, connects the computer to the phone line. As part of MTA, Apple provided its Telephone Manager application for call-control services and its Telephone Tool software for mapping the Telephone Manager to various telephony platforms.

Apple's concept is excellent, but the company has been slow to implement it for today's more prevalent environments. Specifically, it has no server component, no direct PBX interface, no connection for digital phones (only analog support is provided), and limited applications development. Apple is aware of these limitations and has plans to make MTA more broadly appealing. Industry-wide support is often more important than the concept, the technology, and the actual implementation.

XTL Teleservices

SunSoft has developed the XTL Teleservices platform for Solaris. This platform supports telephony software applications as well as hardware and driver interfaces for third-party device implementations. The XTL architecture allows applications to access telephone technology regardless of the particular topology, telephony interface, or type of phone system being used. It also provides transparent porting between analog, ISDN, ATM, and other technologies.

XTL focuses on desktop telephony services, such as automatic dialing, personal answering machines, voice-mail/e-mail integration, automatic call transfers, and more. The platform can also be used to establish desktop videoconference calls, wide area data communications, multimedia applications, and fax services.

Using SunSoft XTL, multiple desktop applications can share access to workstation telephone interfaces. The applications can manage an individual call without sacrificing data integrity or security. For example, one application can create or receive a call and pass it to another application for voice processing.

SunSoft XTL APIs allow third-party developers to create software applications for telecommunications hardware. At this point, Sun has not encouraged this development as actively as Novell or Microsoft.

SCSA, MVIP

There are two dominant approaches to the interface of PC telephony modules to other devices. One is known as Signal Computing System Architecture (SCSA), the other as Multivendor Integration Protocol (MVIP). Both of these standards are digital telephony buses that allow the transport and switching of telephone data streams within PCs or between PCs and another device. The other device can be any form of preexisting telephone system, including a PBX, key, hybrid, IVR, predictive dialer, ACD, etc. Both standards have similar objectives: allowing integrators to use different PC-based application boards from various vendors to create an end-user solution. Customers, therefore, can benefit from a multiresource system consisting of a voice board, a fax board, a video card, a conferencing board, a voice recognition card, and more—all using the same bus and software interfaces.

MVIP, the first of the two approaches, was spearheaded by a group of telephony board vendors that includes Natural MicroSystems, Rhetorex, and Pika. Dialogic, the market leader in PC-based voice modules, introduced the SCSA bus. The two buses are not compatible, so applications providers and integrators need to choose between the two. Some of their differentiating features are discussed below.

MVIP characterizes itself as being more flexible because of its ability to connect a variety of devices, including networked PCs and WAN equipment, as well as traditional phone systems. SCSA offers a more complete call-processing architecture with a tighter focus on voice and fax processing. It is critical for applications providers to determine carefully which APIs are available on which bus. Since MVIP, for example, has emphasized resource management and connection control, developers have provided tools for connecting to other systems. SCSA has focused more on the specific applications with APIs for speech recognition, fax, and other voice-processing technologies.

If the end user has requirements for a complex system with an array of different connections, then MVIP-compliant products may be a more flexible approach (assuming that the specific applications the customer wants are available in MVIP-compliant form). If the end user is looking for pure voice processing with less of an emphasis on LAN and WAN interconnection, the SCSA product may offer a better solution (with the same caveat as mentioned for MVIP). Since both standards have strong vendor backing, it is safe to assume that continued support will not be a major issue. For more information on SCSA, MVIP, and voice processing, please refer to Chapter 14.

CTI Applications

The early adopters of CTI applications are call centers. A call center is an organization where employees handle incoming and/or outgoing telephone calls to acquire or disseminate information in support of that organization's activities. Call centers can operate business to business, business to consumer or internally only.

CTI can be used to seamlessly interface the caller, the agent, and information on a host computer for a variety of applications. CTI is the "middleware component" that enables a computer platform to talk to a telephone platform. CTI delivers caller ID, automatic number identification (ANI), dialed number identification service (DNIS), and interactive voice response (IVR) dialed digits, such as a customer number, to a software application. CTI also accepts requests from an application, such as "transfer call" or "hold call," and delivers them to the telephone platform. Software applications use the functions of CTI to provide value to the end user.

Software applications can be vertical or horizontal market applications. There is no limit to the number and type of organizations enhanced by CTI-enabled software. Below are some examples of end user software applications typical of CTI-based applications.

911 Emergency Services

Automatic number identification (ANI) is delivered from the central office (CO) to the local switch to enable a software application to automatically look up database information on the caller, such as location, type of building, or disabled or elderly tenant occupants, to dispatch the nearest available mobile unit to the emergency site.

Utilities

IVR digits are collected to determine the customer's account number before the call is being routed to an agent. A software application receives the digit stream as the key for delivering the customer's record to the agent's screen as the call is being delivered to the agent's phone.

Transportation

IVR digits that represent a truck driver's rig number are collected from the truck driver when calling to check in with the dispatcher. The software application receives the caller's telephone number through ANI, and the number is checked against a database of all area codes and exchanges in the United States to determine the location of the calling trucker.

Insurance

A customer who calls the insurance company to see how much was paid on a medical claim enters the customer number on a telephone keypad. After a few additional prompts (e.g., "Press 1 if you have a question about your medical insurance policy"), a software application pulls the account record from the database, and a synthesized voice states how much was paid on the claim—without having to involve an agent.

Desktop/Workgroup

Caller ID, ANI, and IVR digits can provide a software application with the information necessary to provide call screening, call history, call notes, and length of call. In addition, the application can use CTI to provide a graphical user interface to sophisticated phone features such as conferencing and transfer, as shown in Figure 18-9.

Beyond CTI

The primary focus of CTI is on the interplatform information flow between a switch, its connected devices, and the computer. An emerging technology known as business process automation (BPA) expands the focus of information flow to all the communications technologies on the network. It also focuses on integrating these resources with the organization's human components—its employees and customers.

Most organizations have an array of tools for storing, retrieving, and utilizing information. These tools range from computer systems, databases, and application software to communication systems, including switches and their connected devices, fax, e-mail, and paging. However, while productivity tools and applications have proliferated, no one has focused on the core issue: automating the entire business process.

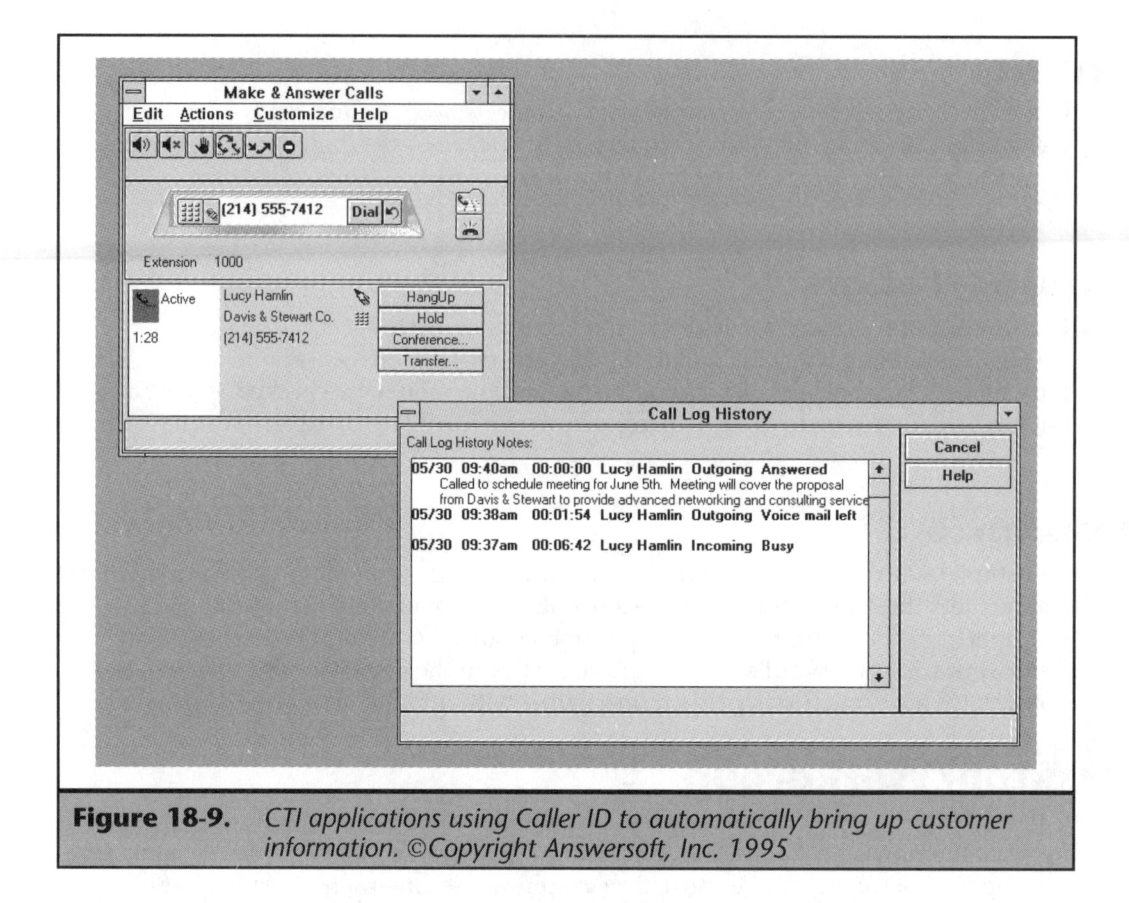

Figure 18-9. *CTI applications using Caller ID to automatically bring up customer information. ©Copyright Answersoft, Inc. 1995*

Business process automation technology generally consists of client-server-based software and systems that enable better decision making and better customer service on the phone.

Business process automation has three objectives:

- Free employees from repetitive tasks, adding value at every stage of customer interaction by automating the front-end, middle, and back-end processes of a phone call
- Allow communications technologies to drive and deliver the dissemination of information
- Provide users with the right information at the right time

Business process automation software provides an umbrella to an organization's databases, applications, and communications systems. It interoperates with these systems, invoking them as needed—it does not replace them. BPA software uses three components. One part monitors events and information flow across computers and networks. Events can be communication-oriented such as phone calls or application-driven such as DDE events. A second part of the automation software applies rules to determine whether or not action is needed. A third initiates any action the "rules engine" calls for. This engine employs rules preestablished by the user.

These three components—dedicated to listening, deciding, and acting—together comprise a remarkably flexible, capable, and almost organic system. The resulting software provides an extremely wide range of behaviors with which to automate business processes.

AnswerSoft, Inc., an organization that pioneered BPA and is one of the industry leaders in computer telephony applications interfacing programming, provides free demonstration diskettes to show the power of computer telephony integration and to help educate people about this incredible technology. Their demo can provide far better insight into this new world of CTI than can be written about in the short space afforded in this book.

Business Process Automation Example

The following example depicts a technical support workgroup that employs various commonly available technologies and applications, including:

- A conventional PBX-type phone system
- A client/server network in which the clients are personal computers
- Server-based software, including:
    - A Lotus Notes-based help desk application
    - An Oracle, Sybase, or Informix database
    - Skyword, a DOS application that automates paging
    - A communications link to the Internet
    - Business process automation software that automates and integrates the support technician's use of the other technologies and systems

The monitor portion of the automation software listens for events and accompanying data such as caller ID on a phone line. When a call comes in, it makes a rules-based decision to check the caller's identity. In this case, it launches, opens a customer profile (Oracle or other relational database) record to perform the ID check. The automation software controls the database, telling the database what to look for and in which field of the records to look. The automation software retrieves data from the record and launches a Lotus Notes help desk application.

Using rules-based intelligence, the automation software determines that the customer profile record doesn't contain any unresolved issues with this caller. It therefore anticipates that the caller will be reporting a new problem. It opens a blank trouble ticket, controlling the help desk application to perform this task. Then it causes the help desk application to prefill the blank form with a serialized reference number and all relevant profile data. All this occurs in seconds. By the time the support technician picks up the phone, the prefilled trouble ticket automatically pops onto the PC screen. As the support technician helps the caller with the problem, the support technician enters a few details into the Lotus Notes trouble ticket. In this case the problem requires immediate on-site repair. With a single mouse-click, the support technician requests a list of local repair centers. The automation software, always listening for events, detects the mouse-click request for repair centers. It responds by:

- Opening a different Oracle database that contains the repair center list
- Using a zip code and phone data from the customer profile to pick out qualifying repair centers
- Popping up on the support technician's screen a Select Repair Center window containing the short list

The support technician clicks on a name. The automation software responds by:

- Launching the Skyword paging application
- Retrieving the repair center's pager number from the Oracle database
- Creating a text message based on entries form the Lotus Notes trouble ticket
- Commanding Skyword to send the page

The support technician quickly completes the call. In the background, transparent to the technician, the automation software:

- Invokes a blank Service Evaluation form
- Prefills it with basic customer information from the profile database, the trouble ticket reference number, and a basic description of the problem from the Lotus Notes application
- Creates a cover note asking the customer to evaluate the service received on this call
- Faxes the form to the customer

The automation software also creates a similar message from the trouble ticket entries, retrieves the account rep's name from the customer profile database, looks up the rep's e-mail address in the company directory, and automatically posts a detailed description to the rep.

Also transparently, the software posts a follow-up callback reminder to the support technician's action list of the next day. Also, it looks up the elapsed time the technician spent on the call and logs the time to the billing department.

Automation scripts tell the automation software what tasks to perform and what rules to apply when performing them—much as a human worker would be similarly instructed. The software thus "knows" what needs to be done, when to do it, what decisions and behaviors will get the job done, and under what conditions.

The next day, when the support technician follows up, the software captures the prompted digits of the outbound call to the customer. Recognizing the phone number, it again launches the Lotus Notes help desk application and displays yesterday's trouble ticket on the screen. Therefore, without even requesting it, the technician has all the information at hand to make an efficient, effective follow-up call to this customer.

This example illustrates how business process automation software gives employees better access to applications, databases, and communications systems in real time and without distraction, even while they are on the phone; how it performs complex actions for users automatically; and how this adds value to the enterprise.

CTI Enables Business Process Automation

A major telecommunications company dedicated to providing premier satellite-based mobile voice and data communications services throughout the United States was searching for a call center solution to provide its customers and associates with exceptional service. The strategy of the company was to present a full-service appearance to the customer with its operator services, directory information, and expert agent resources, and to provide: highly trained, intelligent, and resourceful individuals; information-gathering and dissemination tools; a flexible and fault-tolerant environment; bidirectional access to corporate legacy systems; and effective feedback and escalation procedures.

Implementation of computer telephony integration (CTI) at the application level was a key factor in the success of this program. The ability to make a coordinated presentation of systems and services to the call center agent was paramount. The extent of the applications (i.e., services and programs) requiring support were:

- National accounts: provide premium customer service to corporate customers
- Customer relations: generic customer service to all types of customers
- Installer line: automated assistance for activation of equipment
- Inbound telemarketing: registration of new customers through promotions
- Operator services: provide cellular or satellite callers with the ability to call for operator support, i.e., directory assistance, 911, and customer service
- Order entry/fax: allow new customer registration through the use of facsimile

To support all the application requirements of the call center, a leading BPA systems developer was solicited. The approach taken by the developer to support the "exceptional customer service" requirement of the company focused on providing the call center agents with two abilities: rapid response (initiation of complex actions in real time without distraction) and proactive performance (anticipation of customer needs).

To support call management and control, a client-server application using Novell's telephony services application programming interface (TSAPI) was installed. This provided the agents with dynamic call control and management by integrating enterprise-wide data with sophisticated call functions. Implementation and training cycles on phone and call features were shortened through an intuitive graphical user interface (GUI). And since the physical telephone and agent computer were now integrated, distractions from having to manage two separate devices were gone, and agents could better focus on providing service.

Additionally, the CTI developer added a business process automation (BPA) solution. This provided automation of work and information flow into and out of the call center. Through rules defined by the call center administrator, interaction with and data collection from multiple legacy systems was performed instantaneously and delivered with the call to the agent's station. This intelligent call delivery provided the agent with information about who the caller was and what type of service was being requested. In most cases, this involved launching and controlling (i.e., screen pop) existing enterprise applications running in multiple environments. To remedy the memory constraints caused by running multiple applications simultaneously under Windows, an application memory optimizer was installed. This gave agents the ability to run, concurrently, all the applications necessary to service the customers without experiencing system failures due to insufficient memory.

Conclusion

TAPI and TSAPI are very young technologies that enable programmers to finally combine the strengths of telecommunications with the computer. The two standards are fighting for control in the industry at this time, but one day you might find that they peacefully coexist because of efforts to integrate the two coding methods. The power of CTI is very impressive and must be seen to be truly appreciated. One day, this technology may completely change the face of telephone switching, putting all the power to control one's communication at the desktop. CTI is the wave of the future.

Chapter Nineteen

CATV and Telephony

In today's rapidly changing telecommunications marketplace, a significant number of companies in industries outside of traditional telephone companies have become interested in offering competing telephony services. The cable television (CATV) industry is one such group trying to enter this voice dial tone (telephone) services market. Voice dial tone and other communications services are being added to the traditional CATV television rebroadcast services by upgrading current CATV analog networks with modern telephony technology.

The emergence of CATV as a provider of telephony services is so recent, however, that the supporting technology is still being designed and evaluated; the services are still being structured; and operating procedures are still being developed. Some CATV-based communications service offerings, such as interactive television, are only now being deployed on pilot-and-trial basis in limited areas to test feasibility and market acceptance. Some CATV services, such as local switched voice dial tone, are not available today. This chapter will present an overview of current CATV networks and technology trends to provide a basic understanding of CATV technology, where the CATV industry is going, and where it plans to go. In the next several years, CATV-based services will be evaluated and selected on a comparable basis to traditional telecommunications services.

This chapter will present two aspects of the CATV future: expanding CATV into existing communications services offered by non-CATV companies in a business environment, such as telecommuting, voice telephony, data communications, and Internet access; and new interactive video-based services, such as home shopping, interactive television, and distance learning, which are not typical tools used by business today but will grow to ubiquity in the near future. The CATV systems described in this chapter refer to the United States and Canada, but international CATV developments are very similar to the U.S. experience.

CATV Background

CATV was originally developed as a way to bring broadcast television signals to rural and hard-to-reach areas where television reception was poor. This original deployment of television broadcast-oriented systems (one-way transmission) for improved television signals began in the 1950s. Later, in the 1970s, CATV expanded to carry broadcast television channels outside of a household's normal reception area and to provide access to enhanced specialty channels beamed from satellites, offering programs such as just-released Hollywood movies.

The period from the 1970s to the early 1990s was a time of geographic expansion for the CATV industry. CATV companies' energies were focused mainly on establishing and building new CATV networks in unwired areas. The process for establishing a new CATV network franchise in a municipality basically followed these steps:

1. Negotiate a franchise agreement with the local municipal government, granting either actual or effective exclusive monopoly rights to provide CATV

in the town or city. The actual monopoly rights were contained in the franchise agreement, and the effective monopoly rights were created by the high entry cost and low return to set up a competing (second) CATV system.

2. Build an analog network of coaxial cable, usually in the 350 MHz to 550 MHz capacity range, within the town or city, using existing pole and conduits where possible.
3. Provide cable television service, wire into homes, and provide set-top television converters to subscribers.

It is significant to note that CATV franchises are regulated at the local, state, and federal levels by the local franchise authority, state Public Utilities Commissions (PUCs), and the Federal Communications Commission (FCC), respectively, in contrast to common carriers (telephone companies), which are only regulated by the state PUCs and the FCC.

This licensing at the municipal level has resulted in a patchwork of geographic coverage, with numerous small operations intermixed with large multisystem operators (MSOs). CATV operators range from the over 13,500 individual operator franchises, each with less than 1,000 subscribers, to the top MSO with close to 11 million subscribers. Yet this top MSO commands only 10 percent to 12 percent of the total CATV subscriber base. Individual franchise operations start at about 200 subscribers and can reach up to 1 million subscribers for New York City. CATV operators are usually described according to a penetration ratio, which is the number of subscribers serviced divided by the number of homes passed (subscribers and potential subscribers) by the CATV network. This figure ranges from 50 percent to 99 percent, with 65 percent to 75 percent being the norm.

The CATV industry is now entering a period of consolidation in the mid-1990s with mergers, alliances, and buyouts of various CATV operations. One reason for the consolidation is to provide consistent regional coverage by local CATV networks and operations. Another reason is to build up the financial resources required to meet all the new competition entering the CATV marketplace and to be able to enter new and enhanced services markets.

Emerging in the mid-1990s is another trend, called convergence. *Convergence* denotes the converging and blending of a number of technologies, allowing a range of companies, especially service providers, to reach into new markets and provide network transport, components, and information content in such a way that traditional market labels may not apply. This new market is centered around new technologies of multimedia. Convergence has been extended and forms one of the cornerstones of the much-discussed information superhighway.

The CATV network transport capability focuses on video capability—those services traditionally provided by CATV and television broadcasters—with a view toward expanding into emerging video services, such as distance learning, video-on-demand, and interactive television. In this market, the local regional Bell operating companies (RBOCs) are attempting to compete and offer similar services, while CATV companies are trying to upgrade their networks to offer interactive video capabilities.

For network component capability, the computer, data network, telephone equipment, consumer electronic, and traditional CATV equipment vendors are all striving to participate in this rapidly expanding equipment market. CATV and telephone networks will expand in capacity to handle video streams of data; in addition, new markets for existing devices are opening up (for example, data network routers for CATV networks), and the number of consumer television set-top cable converters (*set-tops*) required for this new network has been projected to be much larger than the current personal computer market. See "Video Servers for Video-on-Demand" later in this chapter for more information.

Concurrent with all of these industry, enhanced services, and network technology changes, the basic CATV foundation service—television rebroadcast—is also undergoing fundamental changes. Interactive television promises to change the structure of television programs from their current serial form, just as hypertext in the World Wide Web (WWW) has changed the structure of written documents. New television programs will have to be created for this new interactive television. The tremendous expansion from less than a half-dozen broadcast TV stations before CATV to 40-60 channels on CATV today, to the digital future of 500-plus CATV channels tomorrow to the home, requires a commensurate increase in the amount of new and unique television programs (content) to fill those channels.

The types of television channels of the future and the consumer response to these new channels can now only be described in theory. The issue and growth in the need for content for this increased network capacity is very similar to the activities occurring on the Internet, in terms of the explosion in access to information-creation tools and information content on the Internet. Both the expected growth in video channel capacity (500-plus channels to the home) and the new presentation mechanisms (personal computers, hypertext, CD-ROM, and multimedia, to name a few) are spurring traditional information and entertainment providers to embrace these new media forms and to face these challenges to traditional product offerings, such as movies, books, and videotape rentals. The companies involved include those in traditional print media (newspapers, book and magazine publishers), movie and television broadcast companies, video game vendors, and emerging multimedia and Internet content providers. The activities involve both taking existing content and modifying it for the new networks and creating entirely new material.

This explosion in the network capacity, and content is all predicated on a digital network, with the content in digital form. Therefore, this chapter will highlight the changes that are being designed into today's analog CATV network (and the RBOCs' competing video network) to bring it into the digital age.

Typical CATV Network Structure

Before examining the changes going into the digital CATV network, let's look at the basic structure of CATV networks and some common terms.

CATV networks are set up as hierarchical tree structures. All signals are broadcast outward from the head end and are transmitted out to all the nodes (households) through the CATV coaxial cables. Since the coaxial network is a system shielded from over-the-air radio transmissions, frequency allocation of television signals is not restricted by FCC frequency assignments. Any return transmissions go back up to the head end on frequencies separate from the broadcast frequencies. Current cable systems use hardly any of the return transmissions capability, except for some interaction with addressable set-tops and early trials of CATV Internet access.

Return transmission becomes much more important in new applications envisioned for the digital CATV network. Some of these interactive applications are explained later in this chapter. The head end communicates with all of the nodes in the network on one set of frequencies (the largest set), and the end nodes (set-top boxes) only transmit directly back to the head end on a different set of frequencies. This is in contrast to Ethernet local area networks, where all devices equally flood the entire network on one baseband frequency. Collision detection and recovery on the return frequencies is important, since the set-top boxes cannot detect simultaneous transmission attempts by other set-top boxes. However, any of a number of local area network protocols can be used to grant unimpeded access to any particular set-top box.

CATV is an entirely contained radio system. All of the components and cables are shielded and are designed to prevent ingress (signals seeping into the cable) or egress (signals seeping out of the cable) of spurious radio signals to cause interference inside or outside the CATV network. The FCC has strict rules covering detecting, preventing, and repairing the "leakage" of CATV broadcast signals.

Refer to Figure 19-1 to see the relationships among the following terms.

ACTIVE DEVICES These are components, such as amplifiers inserted into the CATV network, which require separate line power to operate.

AMPLIFIERS These are inserted in the CATV network wherever the broadcast signal needs to be regenerated in order to be received in the household at an adequate level. Amplifiers are classified by their frequency range (what frequencies they will regenerate) and whether they are one-way or two-way. A one-way amplifier will only regenerate the downstream signal from the head end to the household. It may even prevent return signals from transmitting back to the head end. Two-way amplifiers regenerate signals in both directions.

CABLE This is the distribution mechanism for transporting the television broadcast signals from the head end to the household. It consists of a shielded coaxial cable usually strung along existing pole structures or led through underground conduits.

DOWNSTREAM Main transmission in the direction from the head end to the end nodes (households) in the network. Also called forward signal.

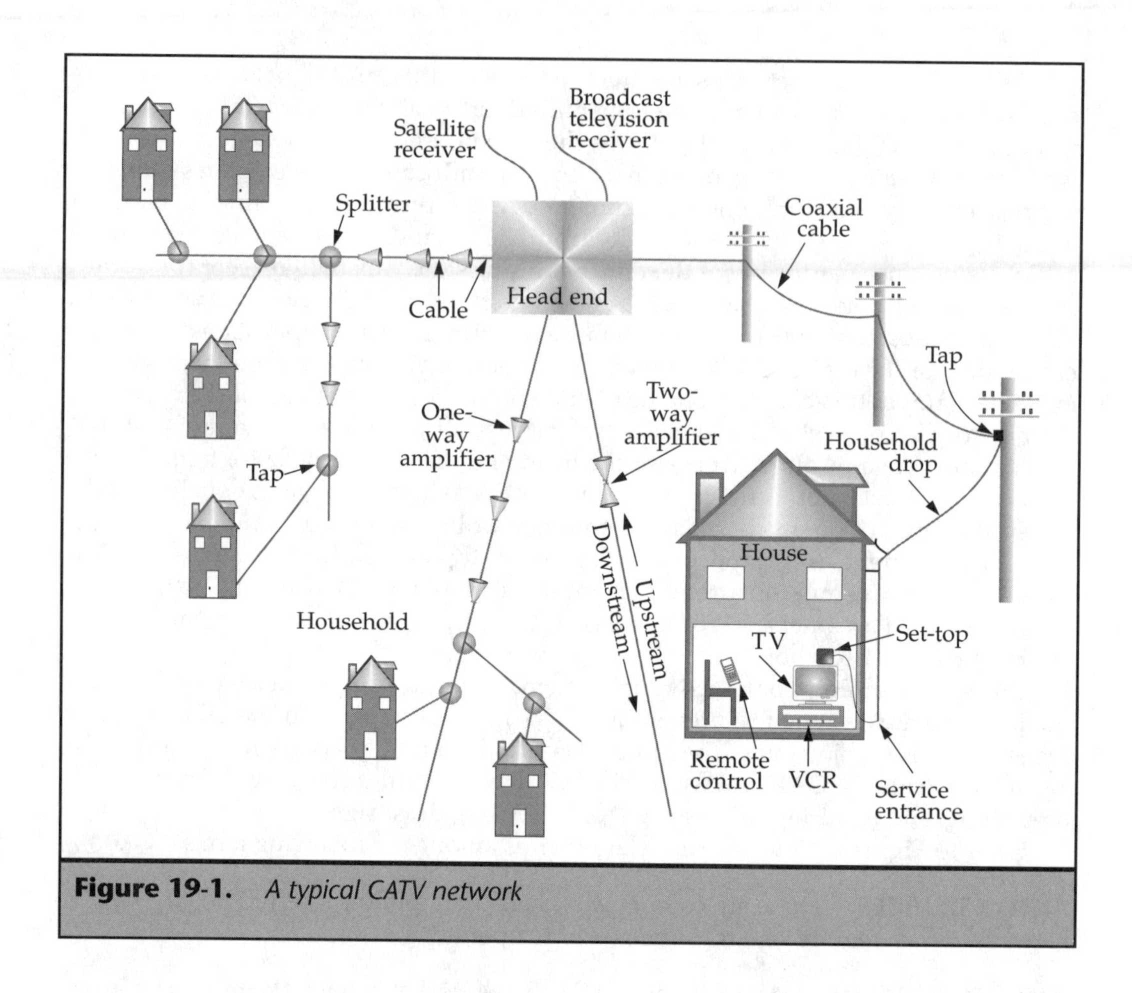

Figure 19-1. *A typical CATV network*

HEAD END This is where all the signals are broadcast into the CATV network and return signals are concentrated for service actions. It usually includes various satellite and television receivers to receive program signals for rebroadcast in the CATV network. The received programming is demodulated to a baseband television signal and then remodulated on the frequency (TV channel) assigned in the cable station lineup. The head end may have some video production capability for local origination programming.

HOUSEHOLD This is the termination or end point of the CATV network. Consumers subscribe to the CATV service, and the CATV network provides a cable *drop* from the network into the house through the *service entrance* to an optional set-top and thence to the television and VCR. The subscriber is then able to view video programming selections they have subscribed to.

PASSIVE DEVICES These are components, such as filters inserted into the CATV network, which do not require line power to operate.

SET-TOP A household-based device that demodulates the CATV broadcast signal and remodulates it to a frequency usable by the subscriber's television and VCR. It's called set-top since it usually sits on top of the television set. The set-top may also be used to control access to additional pay-per-view channels by unscrambling selected subscribed channels. Some CATV networks do not require set-top boxes if they transmit unscrambled signals in the frequencies of over-the-air broadcast or cable-ready television sets. The subscriber's interface to the set-top box is through infrared remote controls. Set-top boxes may provide onscreen program selection guides through the television.

SPLITTER This is a device inserted into the network to create two or more downstream cables from one upstream cable. Any upstream transmissions from the household set-tops are merely combined into the single upstream cable on the same frequency to the head end with no collision control at the splitter. Splitters are used to serve two different neighborhoods, for example, rather than run individual neighborhood cables all the way back to the head end. It splits the radio signals being sent from the head end and routes them down two or more legs. See also *tap*.

TAP Similar to a splitter, though it may not have a signal repeater. This is used to attach the subscriber's cable drop outside the house to the CATV network.

UPSTREAM Transmission in the direction from the cable drops (household wiring) in the network back to the head end. Also called return or backward signal.

Data Communications in Today's CATV Network

It is possible to implement some limited data communications networks over today's analog CATV as an interim step to digital services on a digital CATV network. Modems are available to create dedicated circuits between the home/business drop and the head end by taking over an unused 6 MHz television channel. These modems have capacities up to 43 Mbps over a single channel. These data communications circuits can provide either a metropolitan-level data network or when tied into other wideband modems, a regional or national network. The local CATV network must be investigated to determine whether a spare channel exists and whether the CATV network has the capability to support the frequency allocation.

The New Digital CATV Network

In this section, we discuss the changes being planned, designed, and implemented to bring digital capabilities to the CATV network. You will begin to see the evolutionary path that the CATV industry is currently taking.

Upgrading to a Digital CATV Network

When CATV operators discuss enhancing the CATV network to increased bandwidth and digital capability, they refer to system upgrades or rebuilds. System *upgrades* imply a relatively small-scale effort in which only a portion of the network components are replaced. A system upgrade is employed either when the CATV network already has significant capacity, such as going from a 650 MHz to a 750 MHz bandwidth, or when only a small number of enhanced services are being added, requiring a small increase in network bandwidth. An example of the latter is adding telephone services to an existing CATV network, which may only need 50 MHz of added capacity.

System *rebuilds* imply a major effort where most of the network components and possibly large segments of coaxial cable are replaced. A system rebuild is employed when the CATV network is taking a major step forward into the new digital services, going from a 350 MHz or 450 MHz system up to 750 MHz or 1 GHz capacity. This system capacity for existing and new digital CATV networks is explained in more detail later in this chapter. Each of the major areas of the CATV network will be examined, beginning with fiber deployment.

Fiber in the Network

Fiber-optic transport is an important component in the move to digital capability in CATV networks. The primary question for CATV network upgrades is how close to the home fiber it should run. Fiber costs are now at the level where they can compete with coaxial cable at least for main trunk routes, and they offer several advantages. Fiber offers better signal transmission than coaxial cable, so the need for repeaters and amplifiers is reduced, leading to less signal distortion. Fiber technologies already provide digital transmission capabilities, whereas components for digital transmission over CATV coaxial cables are just coming to market. Fiber also provides a much wider bandwidth capability than coaxial cable. The three main options for fiber deployment are fiber to the home, fiber to the curb, and fiber to the neighborhood (or hybrid). Refer to Figure 19-2 for an illustration of these options.

The fiber to the home option runs fiber all the way into the home to an optical network unit, which then converts light waves into signals for household devices, such as television, telephone, radio, and personal computers. This configuration is currently not cost-competitive with any of the other options built with either fiber/copper combinations or all copper (coaxial cable). As such, it is a future industry goal to offer multimegabit links into each household. Discussions of digital CATV begin with fiber to the household, but none of the current deployment plans utilize this structure.

Fiber to the curb runs fiber to an optical network unit, which serves anywhere from five to 50 households. The optical network unit then distributes signals to households via traditional coaxial cable for video services and one- or two-pair traditional telephone cable for telephony services.

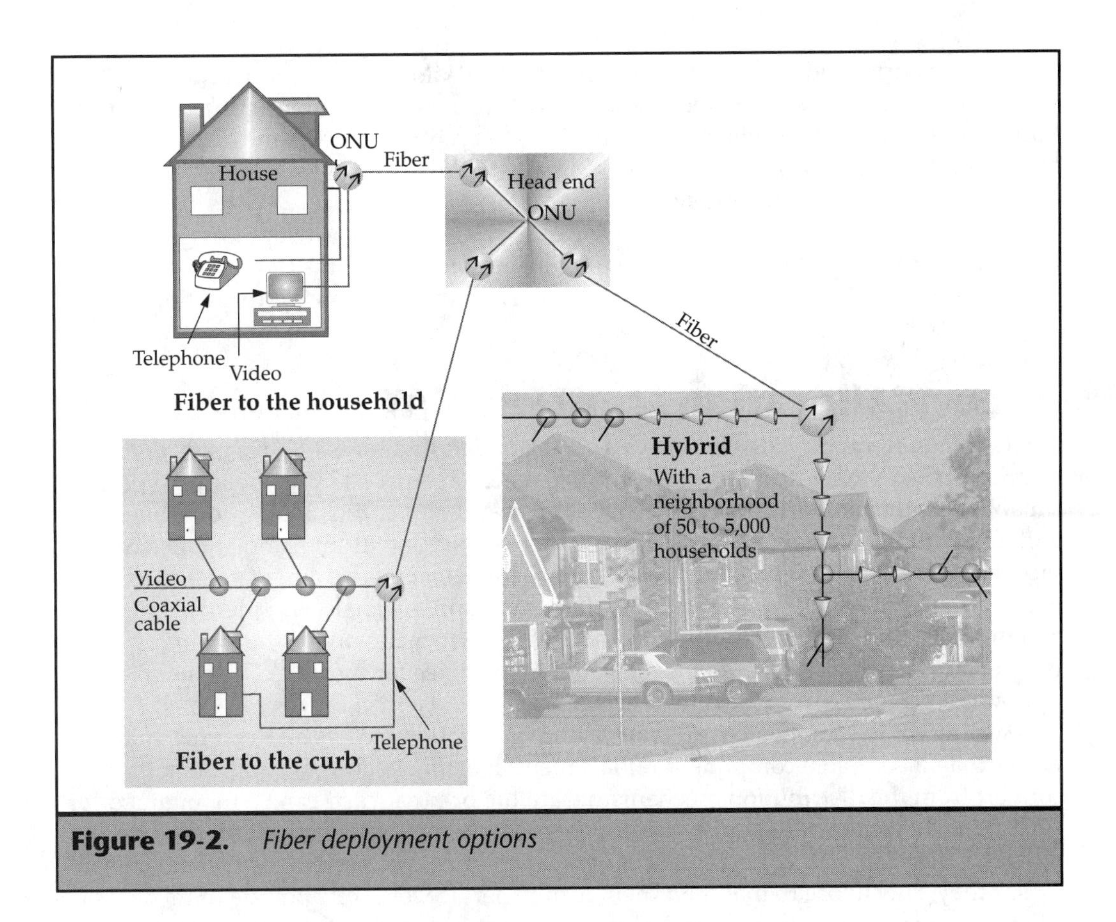

Figure 19-2. *Fiber deployment options*

The most popular CATV configuration is the hybrid network. This setup uses fiber from the head end as main feeder trunks to optical network units serving neighborhoods ranging from 50 to 5,000 households. The distribution of video signals within the neighborhood uses traditional coaxial cable in a tree structure with repeaters, amplifiers, splitters, etc. The choice of the size of the neighborhood is dependent on two factors: cost and return (upstream) frequency utilization. The cost factor is the trade-off of current fiber costs in terms of how many feeder trunks to run versus the cost of coaxial cable. The frequency issue involves the sharing of a common return bandwidth up to the optical network unit by all households in the neighborhood. The questions are how much bandwidth each household will use for the services of the future, and if there is enough bandwidth at the neighborhood level for all the service demanded.

Another distribution option, which will not be explored here, is the asymmetrical digital subscriber line (ADSL) proposed by several RBOCs using the existing public switched telephone network (PSTN). ADSL offers multimegabit digital transport for voice and compressed video to the home and lower-kilobit-speed return transport for voice and interactive video responses over existing copper telephone subscriber lines. Video services would be switched in the telephone central office to an ADSL central office interface, and the households would have an ADSL interface to their television, personal computer, and telephone.

Fiber is also being deployed for back-end CATV use, such as telephony services, program distribution, and advertisement distribution. Most plans call for ATM-based transport over SONET fiber transmission (see Chapter 12 for more information on ATM and SONET).

Frequency Allocations for New CATV Services

The choice of frequency allocation is a significant decision for CATV rebuilds and upgrades. CATV operators have two choices when it comes to implementing digital capability in the network. One choice is to throw out all existing analog components and go all digital for new and existing services, replacing current analog television broadcast frequencies with compressed digital frequencies. The other choice is to expand the frequency capacity of the system, maintain the analog frequency assignments, and add new digital services in the frequencies above the analog frequencies. The latter approach is the solution of choice for most CATV system upgrades and rebuilds, for several reasons.

Adding digital frequencies above existing analog frequencies allows a gradual rather than a complete component replacement. Existing head-end equipment used to convert from the distribution mechanism (satellite or broadcast) can be maintained for existing analog channels. The only additions or replacements are new components that add digital encoding and compression capabilities. The benefit for the subscribers is that they can choose to maintain their current service level or upgrade to the newer enhanced services. This is accomplished in part by choosing to keep the existing analog set-top box to receive only analog television channels or upgrade to a new digital set-top box supporting both existing analog transmissions and new digital transmissions. See Figure 19-3 regarding frequency allocations supported by the new digital set-top boxes.

Both standard analog color television (NTSC) and new digital high-definition television (HDTV) broadcasts utilize 6 MHz frequency bandwidths when not compressed. So, a 650 MHz CATV system can support one hundred analog broadcast channels (the first 50 MHz are not used for television broadcast). System enhancements to digital CATV usually plan for 750 MHz or 1 GHz systems, depending on the expected need for enhanced services beyond compressed video broadcasting. Current upgrade projects are only for 750 MHz upgrades. Figure 19-3 shows representative allocations for both a 750 MHz and a 1 GHz system. The difference between the two systems is that the 750 MHz system only allocates 100 MHz (50 MHz in each direction) for voice, data, and future digital services and utilizes the 0-50 MHz band currently underutilized in analog systems. The 1 GHz systems utilize 250 MHz for telephony and other wideband data services and leave the

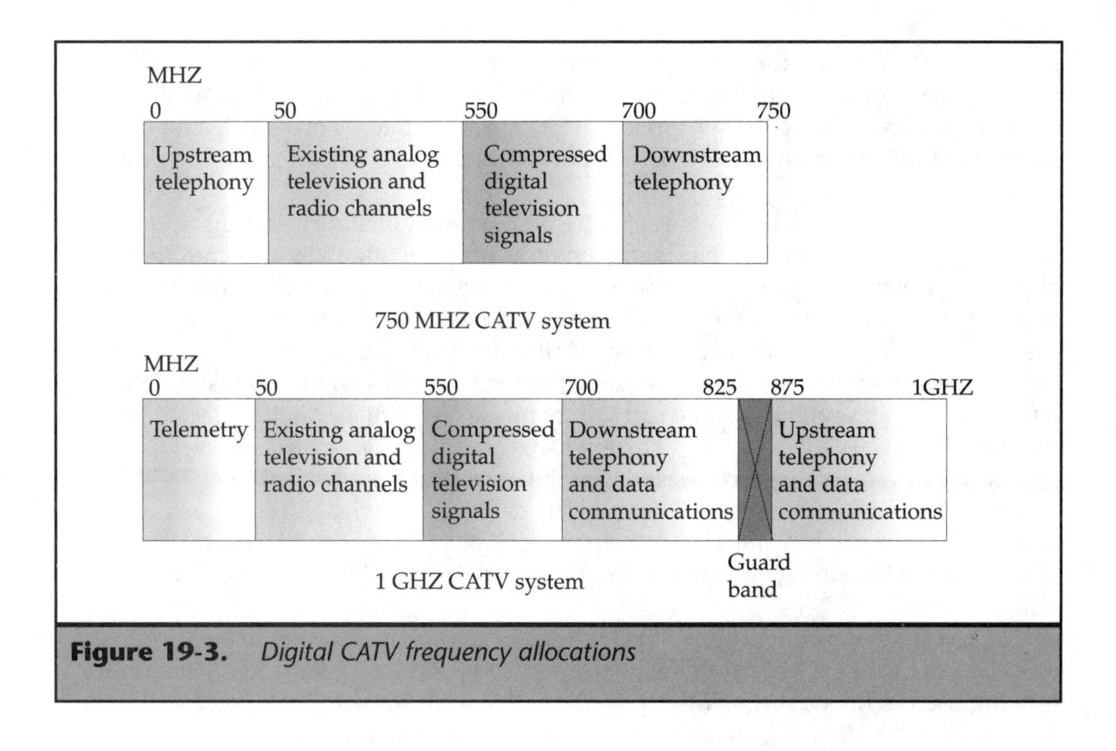

Figure 19-3. *Digital CATV frequency allocations*

low-end 50 MHz open for telemetry communication of system status and services. Current video compression methods being developed and demonstrated can compress about 500-600 standard analog color television (NTSC) channels into the 200 MHz digital video frequency band.

New Digital Set-Tops

Current analog set-tops are limited in functionality. They range in capability from simple frequency conversion (a CATV carrier frequency to a television broadcast frequency—channel 3 or 4) to providing unscrambling of premium channels (extra subscription cost). Some set-top boxes are addressable so that commands to unscramble premium channels can be transmitted to particular set-tops from the head end. This is much more efficient than having to replace passive filters at the pole top cable tap (that little cable TV bubble that you might see in your neighborhood) to the home to change service subscriptions.

Digital set-tops are currently being developed and demonstrated in the prototype stage. The announced digital set-top architectures vary widely, from using personal computers to developing dedicated hardware. All the new set-tops will have the ability to access current analog CATV broadcasts and digital CATV broadcasts, and some will support interfacing to enhanced services through the CATV network, such as home shopping, telephony, Internet networking, interactive video, and video games. Some of the set-tops being promoted are based on graphic workstation architectures, currently costing in the $30,000 to $40,000 range, with the CATV industry goal of offering the set-top in the $90 to $150 range. There is still a lot of work needed to build a digital set-top that supports the new services yet meets the price requirements of the consumer market.

Some of the issues faced in the set-top race to market include selecting a real-time operating system, selecting a hardware architecture, selecting the video compression algorithm, determining what external enhanced service connections need to be supported, selecting the proper interface to the home devices, and having an easy-to-use subscriber interface. This subscriber interface is also called a *navigation system* or *electronic program guide* and uses onscreen displays on the television set. The navigation system provides some form of menu choices, which can be selected by a remote infrared control device. The current focus of set-top user interface development is on how to provide easy service selection and interaction at a usage level comparable to today's infrared television remote controls. The problem of navigation software and hardware design can be compared to trying to control a PC running Microsoft Windows 15 feet away while sitting or lying on a couch and equipped with only a television or VCR remote control.

An example of future in-home CATV devices is a color computer laser printer modified to capture a television image (frame) and print it on hard copy. This printer is being used in the Orlando Full Service Network trials for printing selections from an interactive video shopping television channel for later ordering.

One issue yet to be resolved in the digital set-top unit is industry agreement on the interface between the set-top unit and other consumer devices, such as VCRs and televisions. The arguments range from what the physical interface between the television and the set-top box will be to which functions (such as decompression of digital video, program descrambling, and interactive capability) should exist either in the set-top unit or television. A particularly divisive issue between the set-top manufacturers and the consumer electronics manufacturers is whether the television set should pass through any infrared remote control commands to the set-top unit for set-top execution.

At least two interface standards have been proposed but have not yet achieved widespread support. One interface is the CeBus (consumer electronic bus) for connecting a number of consumer home devices together for the intelligent home. Another possible standard specific to the digital set-top to television/VCR interface is an interface called Multiport—a standard that has been available for several years but has not been implemented. There are at least eight different standards bodies, listed below, attempting to set (sometimes conflicting) standards relating to the set-top box (this list is from "Digital Standards: Looking Beyond MPEG," *Communications Engineering & Design*, July 1994).

- C3AG (Cable/Consumer Electronics/Compatibility Advisory Group): Develop a set-back decoder interface standard; cable-ready receiver specs; third-party module interface

- MPEG (Motion Picture Experts Group): Develop protocols to support various MPEG bit-stream applications in network environments. MPEG is a standard for digital storage of media command and control, a means to compress and store video in plain english
- DAVIC (Digital Audio Video Council): International specifications for open interfaces and protocols for interoperable digital A/V (audio/video) applications and services
- IMA (Interactive Multimedia Association): Facilitate interpretability between set-tops and servers; develop set-top interface specifications
- VESA (Video Electronics Standards Association): Define features and interconnections required for open digital interactive set-top box
- COS (Corporate Open Standards): Open interactive standards for digital set-tops
- ISA (Interactive Services Association): Telecom-based interactive services: audio/video/data
- EIA (Electronics Industry Association): Interface and interoperability standards

This list does not include the proprietary versions that are being proposed. Many hardware vendors in the computer, telecommunications, and CATV arena follow one or more of the standards listed above when developing new product offerings.

Video Compression

Plans for enhanced digital video services call for providing over 500 television channels to the home. This discussion ignores the problems of trying to provide television programs for all 500 channels. However, in order to provide this capacity on CATV, compression of the video signal is required regardless of whether the signal is analog (NTSC) or digital (HDTV), since each channel requires 6 MHz uncompressed. The question then becomes which of several proposed video signal compression algorithms will be deployed in the network.

The video signal compression algorithms being considered are generally JPEG (with enhancements), MPEG2, and proprietary algorithms. JPEG (Joint Photographic Experts Group) is based on digital compression of photographs, while MPEG2 is an extension for motion picture digital compression based on MPEG1, which was developed for digitizing motion pictures on CD-ROMs. Generally, the compression algorithms compare frame-to-frame images and strip out elements that have not changed, then compress the resulting delta (changed) information.

The challenges for any compression technique are: to not introduce any visually noticeable artifacts, such as ghosting, jerky movement, and image slurring, due to losses in the compression technique; to be supportable by hardware coder/decoders (codecs); and to have the compressed signal fit in the selected transmission bandwidth. Another requirement is to be able to support the special effects found on many VCRs, such as slow motion, freeze-frame, and fast-forward, while viewing. In the trade-off between low bandwidth/image quality and high bandwidth/image quality, the compression algorithms for video conferencing such as H.261 are judged as having too much picture loss for broadcast-quality video programming.

Head-End CATV Technologies

There are a number of other communications technologies, many discussed within this book, which will be deployed to interface the CATV network to other communications networks and information sources. Refer to Figure 19-4 during the following discussion of the components going into the next generation head end.

ATM and SONET in the CATV Network

Most of the new enhanced services use the CATV network for final distribution (the "last mile" or local loop equivalent) to the subscriber or destination. To get the desired services delivered to the CATV network, other backbone or distribution networks must be employed. For example, telephone service would require connecting the head

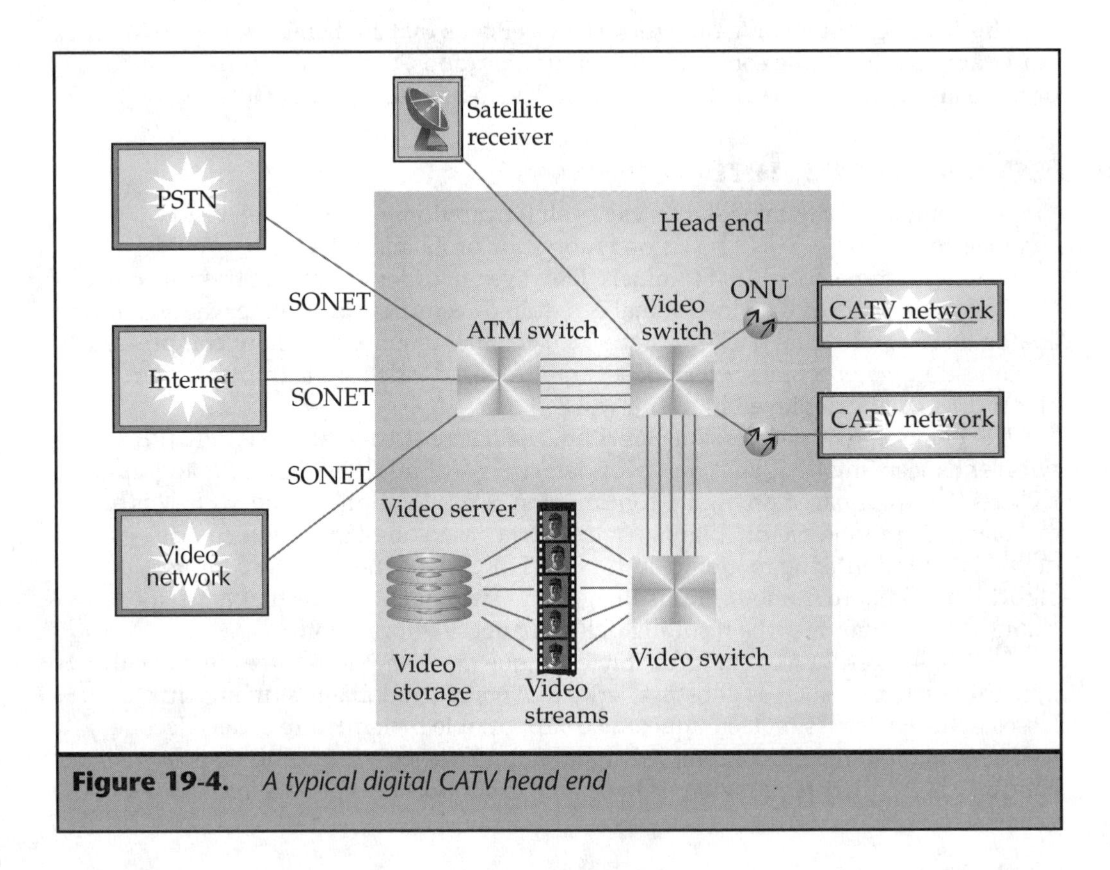

Figure 19-4. *A typical digital CATV head end*

end to existing telephone networks through digital transmission. CATV call switching would be accomplished by installing a central office switch, possibly in the head end. Several new unique video services require potentially new networks, one of which would provide video programming distribution.

Currently, programs are transmitted over satellite links for later broadcast or sent through the mail in the form of videotape. The amount of new digital video programming requiring distribution (some of it real-time broadcast) will exceed current satellite capacity even if the video program is compressed at the point of origination. Video program distribution of the future, whether delivering 500 channels of scheduled material or interactive video, requires the use of wideband fiber-optic networks. In fact, early versions of these video distribution networks have begun to spring up for television advertisement distribution where the local or regional advertisement is inserted into the video programming signal at the head end. These fiber-optic networks allow rapid distribution of advertisement material for last-minute spot advertisement buys, which is easier than the traditional mechanism of mailing copies of videotapes to each head end for local ad insertion.

Most of the plans for upgrading CATV to digital capability call for the use of asynchronous transfer mode (ATM) over synchronous optical network (SONET) transmission to interface to networks such as the Internet, the public switched telephone network (PSTN), and video networks for the transmission of programming content. The advantage of ATM over other data transmission methods is the bandwidth made available to move the signal with ATM and SONET. Another advantage of ATM is that it has the capability to seamlessly support all of the different data types (video, audio, data, and control) and protocols required by the new services envisioned for CATV. This means that the head end can multiplex video, data, audio, and control streams through the ATM switch to the appropriate server. The alternative to ATM is to use dedicated connections to each network in the head end, each with its own type of hardware and software required.

Video Servers for Video-on-Demand

A video broadcast service receiving much attention is video-on-demand, by which consumers can select (via an onscreen navigation interface) movies and television programs to watch whenever they wish. This service is an advance on pay-per-view channels where consumers can view movies released from their movie theater showings starting at regularly scheduled intervals. Video-on-demand also will compete with videotape movie rentals, in that video-on-demand does not require consumers to leave their homes to find the movie of choice. Video-on-demand provides VCR-like capabilities, such as stop, rewind, and fast-forward features, to compete with the video (tape) rental industry.

True video-on-demand services are not currently available but are receiving a significant amount of attention. In the interim, CATV operators are trying to gauge consumer interest in video-on-demand over videotape rental by testing near-video-on-demand (NVOD), which operates as follows: Consumers make movie selections through their onscreen television navigation interface. Periodically (about every 15 minutes), the video server center will collect all of the requests, start the movies that have been selected, and inform the consumer which channel to view the movie on. Commands from the head end to addressable set-tops turn on the descrambling of the selected channel, allowing all consumers who requested movie X to watch it simultaneously. In near-video-on-demand, none of the VCR controls such as stop and rewind are available. Some of these NVOD tests provide the service through traditional linear video components by having a number of VCRs (100 or more) connected to a video stream switch. When a particular movie is requested, the VCR with the movie is started and a video stream switch routes the VCR video signal to the channel selected.

Video servers for true video-on-demand are made up of four components: video stream switch, video storage, video stream sources, and server control. The video stream switch routes and multiplexes the video streams into appropriate time slots for digital transmission or into the appropriate frequency for analog transmission. The video storage provides storage of program material for access by subscribers. The video stream source retrieves the video program when the subscriber makes a selection through the onscreen navigator menu and starts the playing of the program. The server control interacts with consumers for movie selection and provides the VCR control capabilities of stop, rewind, slow motion, and fast-forward. New program material can be downloaded to the video server over the video ATM network described previously.

The architectures of video servers parallel computer servers, especially in the choice of centralized versus distributed architectures. The main difference from computer servers is the wide bandwidth required for the video stream output and some of the behavior of the video storage subsystems. Hard-disk drives used for computer storage have the following feature: When temperature changes affect the tracking of the read-write heads, the drive will momentarily pause and recalibrate its tracking. This recalibration does not impact most computer servers and the applications running on them. However, an interruption in the data delivered to the video stream can impact the video program being viewed. Hard-disk drive vendors have responded to this need with drives that still have the recalibration feature, but implemented in a fashion that does not interrupt access (video stream).

Video program material can be either compressed at the video server when loaded through traditional media such as videotapes or delivered in a compressed form from the point of origination. The point of origination could be the movie studio where the program was made or some centralized distribution center. On the video server, the video program is stored in compressed form and is sent compressed to the set-top devices requesting the program. At the set-top in the home, the program is uncompressed and delivered as a full-bandwidth television channel to the television.

Future CATV Services

The challenge for expanding services (and revenues) in the CATV network is to provide differentiation from existing service providers while allowing potential subscribers to utilize the service on the CATV network in a familiar fashion with existing tools. For this reason, the bulk of the new services use existing interfaces both to consumer devices and to the service backbone network (for example, a standard two- or four-wire modular telephone outlet for the consumer telephone is provided in the set-top box).

Video Services

Enhanced video services is the area where CATV can really stand out and distinguish itself. The uses for some of the new video capabilities outside of the consumer entertainment arena have yet to be thought of. These potentially vast tools open up a wide range of useful applications, including the following:

- Interactive television: At the basic level, the viewer interacts in program selection. At higher interaction levels, the viewer interacts with program material where the interaction is integral to the program content, such as educational material, video shopping catalogs, or even a hypervideo environment involving viewer choice at points in the video program to branch to other video programs.
- Video-on-demand: Providing a library or repository of video programming, which is accessible at each viewer's request.

As can be seen from these service descriptions, CATV video service offerings are still based on a broadcast approach, whereas two-way video such as video conferencing over CATV has not received much attention. Multicast video services also have not added viewer interaction where the return channel is a video stream (video conferencing).

Telephony Services

CATV-based telephony services are implemented the same way as copper local loop replacement (bypass) from a competitive access provider (CAP), also referred to as a long distance telephone provider. The CATV network from a call routing point of view interfaces with the public switched telephone network (PSTN) just as CAP networks do today. These interconnections hold the same CAP issues of local number portability and co-location arrangements. The CATV networks will employ the same end office switches used in telephone company central offices providing all of the familiar features available today, such as call forwarding, call waiting, distinctive ringing, etc. Unlike the CAP or local exchange carrier (LEC), however, the CATV network is not currently engineered with the same robustness as the PSTN. This is the biggest challenge now faced by the CATV industry: to decrease down time to the same level that the PSTN currently provides for telephony services.

The telephone in the home will use the same interface currently used to connect the consumer to the PSTN. This is provided by the hybrid or fiber to the curb CATV network structures by splitting out the telephone signals and running them into the home as a standard copper loop. In the future, it may be possible to have the telephony services provided out of the set-top box in addition to video signal reception.

Data Communications

Just as with adding telephony services onto CATV, data communications on CATV is another type of media to integrate and overlay onto a heterogeneous data network. The head end ties the CATV network into the specified data communications network, such as the Internet. On the subscriber drop side, modems for the selected CATV frequencies are installed and connected to the subscriber's private data network. In the future, it is likely that the data connection will be through the set-top unit. The trials currently in place on CATV networks are providing Internet access via TCP/IP. In the trials, TCP/IP bridges or routers are employed at the head end to tie the CATV nodes into the Internet, and subscribers utilize specialized CATV modems to tie their private TCP/IP networks into the CATV data communications network.

Fax

Proposals have been presented to employ fax capability on CATV to view received faxes on the television. With the video signal printer mentioned earlier, a hard copy of the fax can be produced and, by adding scanning capability to the printer, faxes can also be sent over CATV from the home.

Personal Communications Service (PCS)

The CATV industry has been looking very closely at PCS (i.e., wireless cellular communications) in parallel with telephony services. Several CATV companies have gone so far as to submit bids and win several PCS frequency licenses recently auctioned off by the Federal Communications Commission.

CATV networks can be used to provide the communications links between the cellular or PCS-based station and the mobile switching center (MSC) to support mobile communications. CATV technology claims the advantage of being able to hang cellular telephone or PCS microcells from the pole tops and coaxial cables strung throughout neighborhoods, rather than depending on cellular telephone macrocell deployment involving large tower structures with the attendant community resistance to the large radio towers.

Another benefit is that CATV has a presence in the home, so cellular roaming into the home can be supported integrally by the same network instead of having to use separate land-line PSTN and cellular networks, as was demonstrated by a Bell Atlantic PCS technical trial in Pittsburgh, Pennsylvania. With cellular communications, CATV companies will be making the same choices as the cellular and PCS industries in general—choosing the correct technology for PCS deployment. The choices include what air interfaces to use (CDMA, TDMA, GSM, or new technologies—see Chapter 13 for discussions of these terms), what intersystem interface standards to use, and what kind of geographic coverage can be provided.

CATV's Role in Telecommuting

Telecommuting is where CATV can demonstrate advantages over other service providers. Among these advantages are: CATV provides connections to the home; CATV will have the bandwidth required for true telecommuting; and with PCS, CATV will support the mobility requirements of telecommuting.

Today, in order to support telecommuting from the home, additional telephone lines have to be installed to the home—a business voice line, a fax line, and a data communications line. ISDN basic rate circuits can provide for these requirements for audio and data communications. But as video grows to be an important telecommuting tool, ISDN will not be able to provide the bandwidth that CATV networks can. For more information on ISDN, please refer to Chapter 11.

Competition for Digital CATV

Digital CATV will not be the only way to obtain these new applications and services. There are several alternatives available now or in the near future to provide services similar to the new digital services on CATV networks. It is too early to judge the relative merits or economic advantages of any of these alternatives.

The RBOCs are starting to deploy their own version of CATV services, called Video Dialtone, which is a video transport-only service. This can be thought of as being able to provide the "pipeline" of television channels into the home only, not producing or owning the television shows broadcast on the channels. The RBOCs have been barred by FCC regulations stemming from the 1984 AT&T divestiture from providing or owning video content or programming (movies, television shows, video games, and other consumer entertainment products). The RBOCs have instead partnered with content providers for this programming. Where these RBOC-provided service offerings compete directly with an existing CATV operator in a city or town it is called *overbuild*—where a second video or CATV network is installed to compete directly with the original CATV network franchise.

Finally, another service emerging to compete with CATV for transmission of broadcast video is Direct Broadcast TV (DBTV). With DBTV, the signal is sent via satellite in a digital compressed form to a small (12-18 inches in diameter) receiver at the home. The transmission originates in a centralized control center, then is uplinked to a satellite and rebroadcast simultaneously to all the home receivers. This service is a one-way-only broadcast service but represents significant potential competition to CATV's core business of providing enhanced transmission of consumer video entertainment products. A service that current DBTV cannot offer is local origination programming and advertising that is produced and broadcast at the regional or local level. Another limitation of DBTV is lack of a return channel capability for either data communications or interactive video services.

The Road to Digital CATV Services

The success of the new CATV digital services described is by no means assured. The CATV industry has tremendous potential, built on upgrading CATV networks to digital capability. However, it faces intense regulatory pressures, and CATV competitors are trying to expand into the CATV core business of providing consumer entertainment programming. The following list highlights some of the pressures faced by the CATV industry today:

- Regulation: Regulation of program offerings and rate regulation impact the ability of CATV companies to allocate the capital resources required for network upgrades.
- Competition: Competition from RBOC Video Dialtone services, Direct Broadcast TV satellite television, and wireless cable television will always be an extreme concern of the CATV provider.
- Telephony issues: CATV companies will face the same hurdles faced by CAPs in providing local dial tone telephone service, including number portability and negotiating co-location agreements.
- CATV geographic patchwork coverage: In some cities, multiple CATV operators create a patchwork quilt of coverage. The operators must cooperate and provide compatible services in order to compete successfully against the local telephone company with region-wide coverage.
- Technology: With the large number of equipment vendors developing products for digital CATV networks, the issue of choosing a technology that will develop a dominant or standard position is critical and is similar to technology issues in the computer industry.
- OAM&P: When new services are added to CATV services, new operations, administration, maintenance, and provisioning (OAM&P) systems and procedures have to be integrated into existing CATV OAM&P systems. These new support systems and procedures can be taken from industries already

offering similar services such as the telephone industry. New OAM&P systems will have to be developed to support new technologies and new services. The ability of the CATV companies to develop and integrate successful OAM&P systems will significantly impact new service growth, service acceptance, and customer satisfaction.

- Billing: Billing for new services is just as critical for service acceptance as OAM&P is. Billing methods, such as charging on a per transaction or flat rate (per month) basis, and billing rates have received a lot of discussion, but little consumer acceptance experience. Most of the service trials so far have been technical trials and not market trials. Some extrapolation has occurred with similar services, such as trying to project pay-per-view movie buy rates into full video-on-demand buy rates. CATV billing systems will need to support whatever billing methods are chosen and will require a significant development and integration effort to add the new services into existing CATV billing systems. This includes providing for a clearinghouse function to clear charges between programming content providers and the CATV networks.
- Market orientation: CATV companies currently market to the consumer directly for home entertainment services. It remains to be seen whether CATV companies can shift their marketing and sales to support new distribution channels and new customer requirements such as business data communications needs.

Conclusion

The cable TV industry is destined to expand into other areas of telephony. Remember that telephony is communication, and television is one of the oldest forms of electronic communication. CATV is based on an infrastructure that has a much larger capacity to carry information to our homes than the telephone network currently has in place. The information explosion that we are all experiencing calls for a new way to carry telephony and data services to our homes. The cable owned by the CATV companies looks like a possible solution to the age-old problem of bandwidth and communication.

New alliances between telephone, computer, and CATV companies are being formed slowly but surely. As the technologies discussed in this chapter become reality, look for new and exciting services to be provided to you at your home. Your ablility to communicate will be almost boundless. Remember, the 21st century is just around the corner.

Appendix A

Vendors

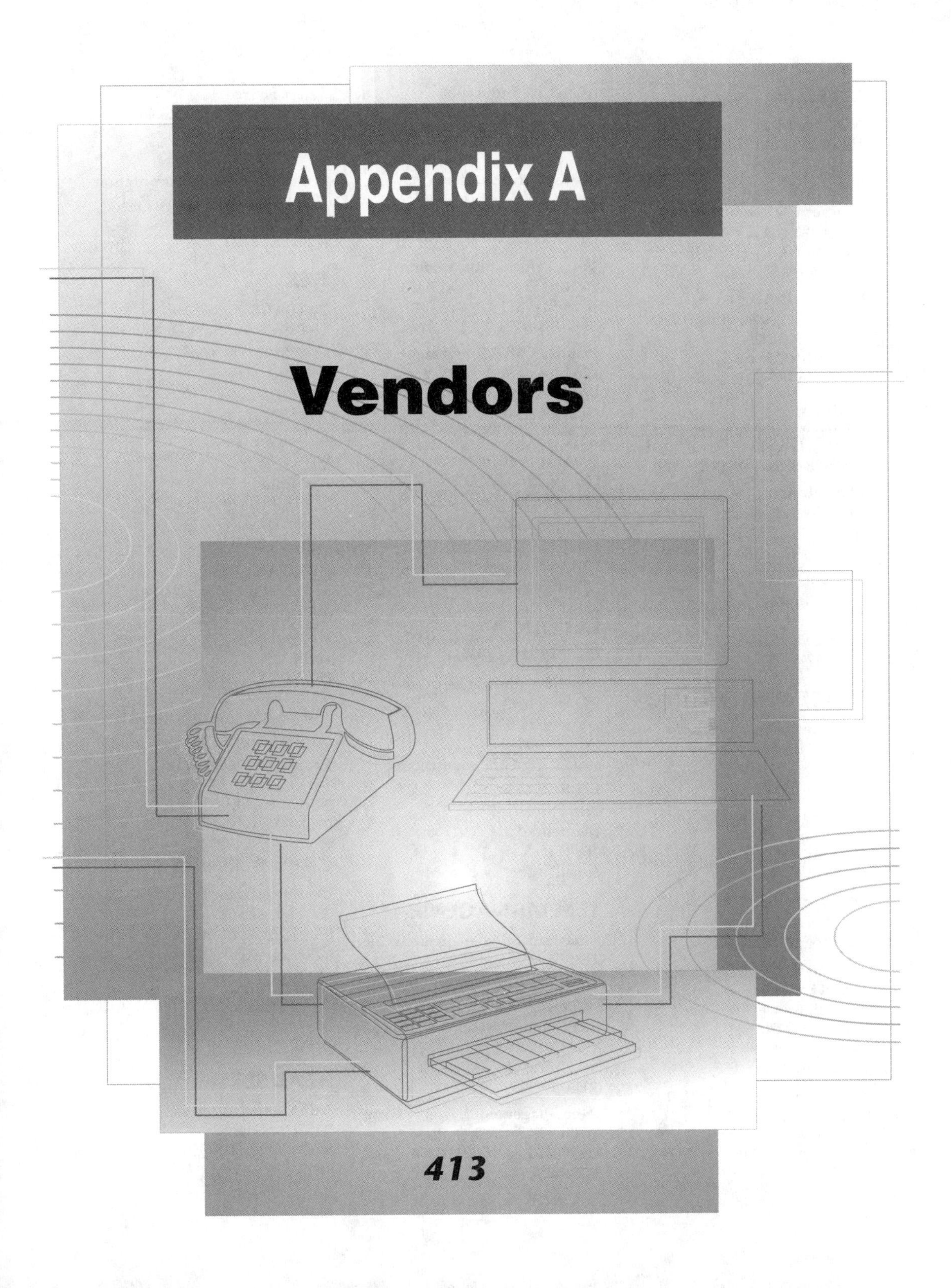

SWITCH

NEC America
1555 W Walnut Hill Lane
Irving, TX 75038
(800)TEAM-NEC or (214)518-4404

Northern Telecom/Norstar
2221 Lakeside Blvd.
Richardson, TX 75082-4399
(214)684-1000

Select Switch Systems, Inc.
8214 Westchester, Suite 907
Dallas, TX 75225
(800)691-9335
e-mail: info@select.switch.com

Harris
300 Bel Marin Keys Blvd.
Novato, CA 94949
(800)888-3763/(415)382-5000

BBS Telecom
3624 North Hills Dr., Ste D205
Austin, TX 78731
(512)502-9220/(800)778-9732

Intecom
Liberty Plaza II
5057 Keller Springs Rd.
Dallas, TX 75248
(214)447-8677

VOICE PROCESSING

Wildfire Communications
20 Maguire Ave.
Lexington, MA 02173
(617)674-1500

Centigram Communications
91 E Tasman Dr.
San Jose, CA 95134
(408)428-3782

Dialogic
1515 Route 10
Parsippany, NJ 07054
(201)993-3000

Expert System
1301 Hightower Trail, Suite 201
Atlanta, GA 30350
(404)642-7575

C3
5255 Triangle Parkway #150
Norcross, GA 30092
(404)453-9686

Symetrics Industries
557 N. Harbor City Blvd.
Melbourne, Florida 32935
(407)254-1500

Natural Microsystem
8 Erie Dr.
Natick, MA 01760-1339
(508)650-1300

Voice Information System
24 North Merion Ave. #353
Bryn Mawr, PA 19010
(800)234-8474

Priority Call Management
226 Lowell St.
Wilmington, MA 01887
(508)694-2688

RHETOREX
200 E. Hacienda Ave.
Campbell, CA 95008
(408)370-0881, ext. 172

Applied Voice Technology
11410 NE 122nd Way
Kirkland, WA 98083
(206)820-4040

INTERNET CONSULTING

Star Communications Group
P.O. Box 1013
Drexel Hill, PA 19026
(610)284-0846
e-mail: info@starcomm.com

Big Byte Ventures
1114 Flintlock Rd.
Diamond Bar, CA 91765
(909)444-3017
e-mail: info@big-byte.com

TELEMANAGEMENT

American Telemanagement, Inc.
7293 Beechmont Ave.
Cincinnati, OH 45230
(513)624-8844

Mer Communications Systems, Inc.
420 Fifth Ave.
New York, NY 10018-2702
(212)719-5959

Telemanagement Systems Group
10700 Jersey Blvd. Suite 190
Rancho Cucamonga, CA 91730
(800) 435-7874

Select Source
8214 Westchester, Suite 907
Dallas, TX 75225
(800) 691-9335

Pinnacle Software Corporation
180 WillowBrook Office Park
Pairport, NY 14450-4282
(716) 381-2750

FAX

FAXBACK
1100 NW Compton Dr. Suite 200
Beaverton, OR 97005
(800)873-8753

FAX Resources
10606 Shady Trail # 20
Dallas, TX 75220
(214)353-0553

Q Comm
1875 S. State St. Suite 2900
Orem, UT 84058
(801)224-5794

RightFAX
4400 East Broadway, Suite 312
Tucson, Arizona 85711
(602)321-7459

Delrina
895 Don Mills Rd.
500-2 Park Centre
Toronto, Ontario Canada M3C1W3
(416)441-4702

Brooktrout Technology
144 Gould St.
Needham, MA 02194
(617)449-4100

CALL BUFFERS

Omnitronix, Inc.
760 Harrison St.
Seattle, WA 98109
(206)624-4985

E-Comms, Inc.
5720 144th St. NW
Gig Harbor, WA 98332
(206)857-3399

HEADSETS

ACS Communications
10 Victor Square
Scotts Valley, CA 95066
(408)438-3883

Plantronics
345 Encinal St.
Santa Cruz, CA 95060
(408)458-4468

MULTI I/O CARDS

Digiboard / Compudata
10501 Drimmond Rd.
Philadelphia, PA 19154
(215) 824-3000

Equinox Systems Inc.
6851 W. Sunrise Blvd.
Ft. Lauderdale, FL 33313

TAPI/TSAPI Vendors

Soft Talk Communications Ltd.
85 Wells Ave. #200
Newton, MA 02159
(617)482-5333

Vodavi Communications Systems
8300 East Raintree Dr.
Scottsdale, AZ 85260
(800)821-2126/(602)443-6000

AT&T
400 A Pierce St.
Somerset, NJ 08873
(908)271-8647

AnswerSoft
3460 Lotus Dr.
Plano, TX 75075
(800)896-ANSR

Teltone Corporation
22121 20th Ave. SE
Bothell, WA 98021-4408
(800)426-3926

Pro CD
222 Rosewood Dr.
P.O. Box 2039
Danvers, MA 01923-5039
(508)750-0000

North American Telecommunications Association (NATA)
2000 M Street, N.W., Suite 550
Washington, D.C. 20036-3367
(202) 296-9800

EUROPEAN VENDORS

Axime Ingénierie
30, chaussée Hôtel de Ville
59650 Villeneuve d'Ascq
France
+33 20 43 88 40

VTCOM
40 rue Gabriel Crié
92240 Malakoff
France
+33 1 46 12 67 89

Newbridge Networks Inc.
593 Herndon Parkway
Herndon, Virginia 22070-5241
(703) 834-3600

Tellabs U.K. Limited
1st Floor, Eton Place
64, High Street
Burnhan, Bucks FL17JT
England
+44 628 66 03 45

EUROPEAN OPERATORS

Austria
OPT: +43 1 515 51 50 40
Mobile:+43 1 515 51 50 40

Belgium
Belgacom: +32 2 202 81 11
Mobile: +32 2 202 81 11

Finland
Telecom Finland: +358 2040 2346
Mobile: +358 2040 2346

France
France Telecom: +33 1 44 76 27 28
Mobile: +33 1 44 76 27 28

Germany
Deutsche Bundespost Telekom: +49 261 123 11
Mobile: +49 511 288 00171

Ireland
Telecom Eircann: +353 1 714444
Mobile: +353 1 6795111

Italy
SIP: +39 02 6211
Mobile: +39 02 6211

Luxembourg
P&T: +352 49 91 55 55
Mobile: +352 49 91 55 55

Netherlands
PTT: +31 70 3434343
Mobile: +31 70 3434343

Portugal
TLP: +351 1 14 32 81
Mobile: +351 1 793 91 68

Spain
Telefonica: +34 1 584 08 44
Mobile: +34 1 584 08 44

Sweden
Telia: +46 8 713 10 00
Mobile: +46 8 707 45 00
Comvik: +46 8 709 12 00
NordicTel: +46 8 626 73 50

Switzerland
Swiss Telecom: +41 313 387 767
Mobile: +41 313 387 767

NETWORK/ISDN/ FRAME RELAY/ATM

The ATM Forum
480 San Antonio Road, Suite 100
Mountain View, CA 94040-1219
(415)962-2585
e-mail: atmforum_info@atmforum.com

Integrated Telecom Technology, Inc.
18310 Montgomery Village Ave.
Suite 300
Gaithersburg, Maryland 20879
(301)990-9890

TranSwitch Corporation
8 Progress Dr.
Shelton, CT 06484
(203)929-8810

The Frame Relay Forum,
North American Office
303 Vintage Park Dr.
Foster City, CA 94404-1138
(415)578-6980
Internet ID: frf@interop.com

Cisco Systems, Inc.
170 West Tasman Dr.
San Jose, CA 95134-1706
(800)553-6387

3Com
5400 Bayfront Plaa
Santa Clara, CA 95052-5166
(408)764-5874

Advanced Computer Communications
10261 Bubb Rd.
Cupertino, CA 95014
(408)366-9649

Alcatel Data Networks
M/S A0223
12502 Sunrise Valley Dr.
Reston, VA 22096
(703)689-5047

Ameritech Services
225 West Randolph HQ12A
Chicago, IL 60606
(312)857-7301

Andrew Corporation
23610 Telo Ave.
Torrance, IL 90505
(310)784-8000

AT&T
5000 Hadley Rd. Room 1B24
South Plainfield, NJ 07080
(908)668-6135

Bell Atlantic
540 Broad St.
Newark, NJ 07102
(201)649-2825

Bellcore
M/S NVC 1-C330
331 Newman Springs Rd.
Red Bank, NJ 07701
(908)758-5743

BellSouth Telecom
675 W Peachtree St. NE
Atlanta, GA 30375
(404) 332-2281

Cable & Wireless
150 N Wacker Dr. Suite 2600
Chicago, IL 60606
(312)899-1792

Cascade Communications
5 Carlisle Rd.
Westford, MA 01886
(508)692-2600, ext. 287

CrossComm
450 Donald Lynch Blvd.
Marlborough, MA 01752-4700
(508)490-5308

Deutsche Bundespost
P. O. Box 100003
64295 Darmstadt
Germany
+49 6151 83 5120

Digital Link Corporation
217 Humboldt Dr.
Sunnyvale, CA 94089
(408)745-6200

Dynatech Communications
991 Annapolis Way
Woodbridge, VA 22191
(703)550-0011

Eicon Technology
2196 32rd Ave.
Montreal PQ, H8T3H7
Canada
(514)631-2592

Ericsson Business
Data Network Division
Esplanaden 3 D
Sundbyberg, S-17293
Sweden
011 46 8 764 0131

FastComm Communications
45472 Holiday Dr. Suite 3
Sterling, VA 20166
(703)318-4362

Fujitsu
3190 Miraloma Ave.
Anaheim, CA 92806
(714)764-2752

Gandalf Systems Corporation
501 Delran Parkway
Delran, NJ 08075
(609)461-8100, ext. 5047

General DataComm, Inc.
1579 Straits Turnpike
Middlebury, CT 06762-1299
(203)574-6263

GTE
600 Hidden Ridge
Irving, TX 75038
(214)718-5078

Helsinki Telephone Co. Ltd
P.O. Box 138
Fin-00381 Helsinki, 00380
Finland
3580-606-4934

Hewlett-Packard
11120 178 St.
Edmonton, AB T5S 1P2
Canada
(403)930-3012

Hughes Network Systems
11717 Exploration Lane
Germantown, MD 20876
(301)601-4000

Hypercom Network Systems
2851 W Kathleen Rd.
Phoenix, AZ 85023
(602)866-5399

IBM
200 Silicon Dr.
Research Triangle Park, NC 27709
(919)254-4141

Infonet
M/S B267
2100 E Grand
El Segundo, CA 90245
(310)335-2693

ITV Telecom Inc
6800 Owensmouth Ave. 2nd floor
Canoga Park, CA 91303
(818)883-6333

Kasten Chase Applied Research
5100 Orbitor Dr.
Mississauga, ON M5V 2Y6
Canada
(905)238-6900, ext. 550

Loral Data Sys
P.O. Box 3041
Sarasota, FL 34230-3041
(813)378-6946

Motorola Codex
20 Cabot Blvd.
Mansfield, MA 02048
(508)261-4754

NETLINK, Inc.
3214 Spring Forest Rd.
Raleigh, NC 27604
(919)878-3569

Netrix
13595 Dulles Technology Dr.
Herndon, VA 22071
(703)742-6000

Network Equipment Tech.
26 Castilian Dr.
Santa Barbara, CA 93117
(805)562-1214

Network Systems Corp.
7600 Boone Ave. N.
Minneapolis, MN 55428
(612)424-1995

Newbridge Networks Inc.
593 Herndon Parkway
Herndon, VA 22070
(703)318-5177

Nokia Telecom
Karaniityntie 1, PO Box 12
Espoo, 02611
Finland
358-0-511-7353

Norwegian Telecom
P.O. Box 6701
Oslo, 0130
Norway
011-47-2-277-8763

Novell
2180 Fortune Dr.
San Jose, CA 95131
(408)577-8341

NTT Corp.
1-1-6 Uchisaiwaicho Chiyodaku
Tokyo, 100
Japan
81 3 3509 5256

NYNEX
125 High St. Room 481
Boston, MA 02110
(617)743-1756

OKI America
666 Fifth Ave. 12th Floor
New York, NY 10103
(212)489-8873

Pacific Bell
2600 Camino Ramon
San Ramon, CA 94583
(510)901-6498

Primary Rate Incorporated
5 Manor Parkway
Salem, NH 03079
(603)898-1800

Siemens Stromberg-Carlson
900 Broken Sound Parkway
Boca Raton, FL 33487
(407)955-5000

SITA
Sophia Antipolis
1041 route des Dolines
Valbonne, 06560
France
011 33 92 96 65 57

Sprint
M/S Varesao115
12490 Sunrise Valley Dr.
Reston, VA 22096
(703)689-5488

Stentor Resource Center
160 Elgon St. Suite 1040
Ottawa, ON K1G 3J4
Canada
(613)781-8972

StrataCom
1400 Parkmoor Ave.
San Jose, CA 95126
(408)494-2140

Sync Research
7 Studebaker
Irvine, CA 92718-2013
(714)588-2070

Tekelec
26580 W Agoura Rd.
Calabasas, CA 91302
(818)880-7875

Toshiba America Info
9740 Irvine Blvd.
Irvine, CA 92713
(714)587-6928

Transpac
85 rue du Gouverneur Général Eboue
Issy-Les-Moulineaux, 92141
France
33 1 46 48 17 27

U S West
150 S 5th St. Suite 3200
Minneapolis, MN 55402
(612)663-5060

Unitel Communications
200 Wellington St. W
Toronto, ON M5V 3C2
Canada
(416)345-2810

Wandel & Goltermann
1030 Swabia Ct.
Research Triangle Park, NC 27709-3585
(919)941-5730

Wellfleet Communications
2 Federal St., M/S 2025
Billerica, MA 01821
(508)436-3835

WilTel
8665 New Trails Dr.
Woodlands, TX 77381-4254
(713)364-4174

AT&T GIS
2700 Snelling Ave. N, M/S S065
St Paul, MN 55113
(612)638-7543

BBN Communications
10 Moulton St., M/S 3C
Cambridge, MA 02138
(617)873-2924

GTE Laboratories
40 Sylvan Rd., M/S 28
Waltham, MA 02254
(617)466-2585

TTC
20410 Observation Dr.
Germantown, MD 20876-4023
(301)353-1550

ATM VENDORS

Adaptive Corporation
200 Penobscot Dr.
Redwood City, CA 94063
(415)366-9500

Cellware Breitband Technologie GmbH
Gustav Meyer Allee 25
D-13355 Berlin 65
Germany
49(0)30/46 70 82 0
e-mail: info@cellware.de

Fore Systems, Inc.
174 Thorn Hill Rd.
Warrendale, PA 15086-7535
(412)772 6600
e-mail: info@fore.com

GTE Government Systems
Building 7, 77 A St.
Needham Heights, MA 02194
(617)455-5182

MPR Teltech Ltd.
8999 Nelson Way
Burnaby, BC V5A 4B5
Canada
(604) 294-1471

Motorola Codex
20 Cabot Blvd.
Mansfield, MA 02048-1193
(508)261-4000

PMC-Sierra, Inc.
8999 Nelson Way
Burnaby, BC V5A 4B5
Canada
(604)293-5755

Telecommunications Techniques Corporation
20410 Observation Dr.
Germantown, MD 20876
(800)638-2049

Trillium Digital Systems, Inc.
2001 S. Barrington Ave., Suite 215
Los Angeles, CA 90025
(310)479-0500

Wandel & Goltermann Inc.
2200 Gateway Centre Boulevard
Morrisville, NC 27560-9228
(919)460-3300

DESKTOP VIDEO CONFERENCING

AT&T Global Information Solutions
211 Mount Airy Rd.
Basking Ridge, NJ 07920
(800)225-5627/(513)445-5000

British Telecom
2100 Reston Parkway
Reston, VA 22091
(800)778-6288/(703)715-4231

Compression Labs Inc.
2860 Junction Ave.
San Jose, CA 95134-1900
(800)538-7542/(408)435-3000

Creative Labs
1523 Cimarron Plaza
Stillwater, OK 74075
(800)998-5227

Datapoint Corp.
8400 Datapoint Dr.
San Antonio, TX 78229-8530
(800)378-6469/(210)593-7660

IBM
3039 Cornwallis Rd.
Research Trinagle Park, NC 27709
(800)342-6672

ImageLink, Inc.
300 Mt. LeBanon Blvd., Suite 2201
Pittsburgh, PA 15234
(412)344-7511

InSoft, Inc.
4718 Old Gettysburg Rd. Suite 307
Mechanicsburg, PA 17055
(717)730-9501

Intel Corp.
5200 N.E. Elam Young Parkway
Hillsboro, OR 97124
(800)538-3373/(503)629-7354

Invision Systems Corporation
8500 Leesburg Pike, Suite 300
Vienna, VA 22182
(800)847-1662/(918)584-7772

Northern Telecom Inc.
P.O. Box 833858
Richardson, TX 75083-3858
(800)667-8437/(214)684-9414

PictureTel Corp.
222 Rosewood Dr.
Danvers, MA 01923
(800)761-6000/(508)762-5245

Target Technologies
6714 Netherlands Dr.
Wilmington, NC 28405
(800)666-2496/(910)395-6100

Vivo Software Inc.
411 Waverly Oaks Rd.
Waltham, MA 02154-8414
(800)848-6411/(617)899-8900

VTEL
108 Wild Basin Rd.
Austin, TX 78746
(800)284-8871/(512)314-2700

C - Phone / Target Technologies
6714 Netherlands Dr.
Wilmington, NC 28405
(910)-395-6108

WIRELESS

Motorola Corporate Communications
1303 E. Algonquin Rd.
Schaumburg, IL 60196
(708)576-5000

RAM Mobile Data
10 Woodbridge Center Dr.
Woodbridge, NJ 07095
(908)602-5484

RadioMail Corporation
2600 Campus Dr.
San Mateo, CA 94403
(800)597-MAIL

Motorola (Wireless Data Group)
50 E. Commerce Dr. Suite M1
Schaumburg, IL 60173
(800)233-0877

Ericsson Mobile Communications AB
45C Commerce Way
Totawah, NJ 07512
(201)890-3600

Motorola, Inc. (Land Mobile Products Sector)
1301 E. Algonquin Rd.
Schaumburg, IL 60196
(800)247-2346

Cylink Corp.
310 N. Mary Ave.
Sunnyvale, CA 94086
(408)735-5800

Laser Communications Inc.
1848 Charter Lane Suite F
Lancaster, PA 17605-0066
(800)527-3740

Persoft, Inc.
465 Science Dr.
Madison, WI 53744-4953
(800)368-5283

Microware Radio Corporation
20 Alpha Rd.
Chelmsford, MA 01824-4168
(508)-250-1110

Motorola (Wireless Data Group)
50 E. Commerce Dr. Suite M1
Schaumburg, IL 60173
(800)233-0877

AT&T Global Information Systems
WWIS Network Systems Solutions
1700 S. Patterson Blvd. PCD/6
Dayton, OH 45479-0001
(513)445-2468

AIRONET Wireless Communications
3330 W. Market St.
Akron, OH 44334-0292
(800)800-8016

Black Box Corporation
P.O. Box 12800
Pittsburgh, PA 15241
(800)355-8001

INFRALAN Technologies Inc.
12 Craig Rd.
Acton, MA 01720
(508)635-0806

Proxim, Inc.
296 N. Bernardo Ave.
Mountain View, CA 94043
(415)960-1630

Appendix B

Glossary

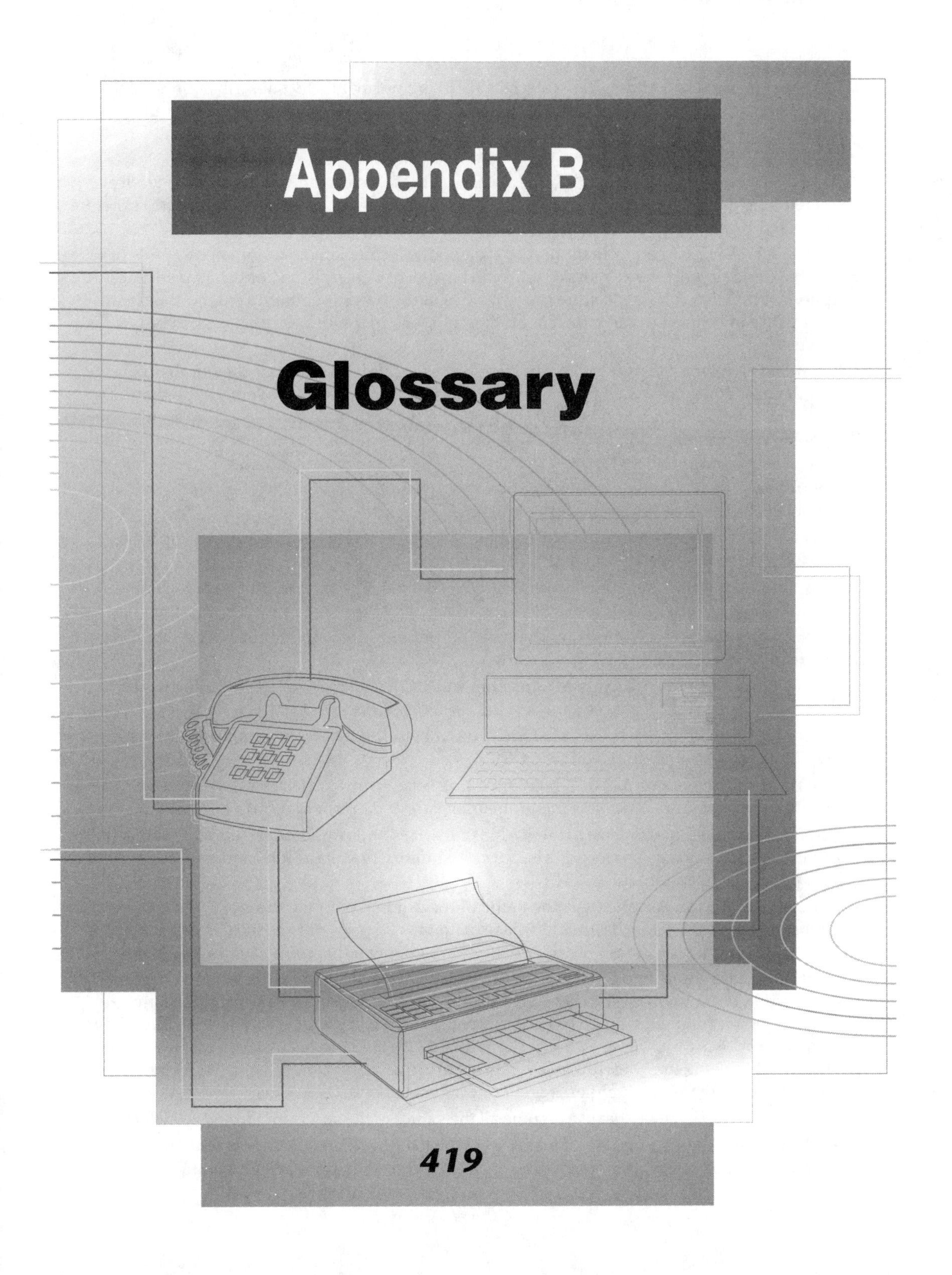

2B+D: Also referred to as Basic Rate Interface (BRI) is a normal ISDN subscriber circuit. It contains two B channels that are capable of transmitting 64Kbps each and one D channel capable of transmitting 16Kbps of packet switched (usually X.25) data yielding a total throughput of up to 144Kbps.

23B+D: Also referred to as Primary Rate Interface (PRI) which is the standard ISDN service delivered over T-1 circuits. PRI can supply up to 1.544 Mbps transmission rate using 23 separate or combined B channels at 64Kbps per channel and and one 16Kbps D channel.

50 Pair Block: A piece of equipment used by telephone technicians to terminate the ends of telephone cable. Twenty-five+ pairs of wire can be attached to each side of the 50 pair block. A telephone technician can then bridge the two sides of the block together to extend the circuit. Station wire can be used to carry the circuit to another 50 pair block or the pairs can be left for future use.

800 Number: A telephone number that is "free" to the caller. The owner of the number pays for all the charges, which are billed by the long-distance carrier. Due to an oversight by the FCC, some 800 number are not free. These 800 number are typically linked to services provided over the phone (sex lines) that don't bill for the telephone call, but bill for the service. Look for the FCC to close this loop hole in the near future.

900 Number: A pay-per-call telephone number with national accessibility. Callers are billed for the charges of the call on their monthly phone bill.

Access Line: A communications line (e.g. circuit) interconnecting a frame-relay-compatible device(DTE) to a frame-relay switch (DCE). *See also* Trunk.

Access Rate (AR): The data rate of the user access channel. The speed of the access channel determines how rapidly (maximum rate) the end user can inject data into a frame relay network.

Active Devices: These are components, such as amplifiers, inserted into the CATV network which require separate line power to operate.

Adaptive Differential Pulse Code Modulation (ADPCM): encodes only the value of the difference between adjacent voice samples. *See also* Pulse Code Modulation.

Advanced Mobile Phone Service (AMPS): The analog cellular service currently deployed in the United States and several other countries.

Alternate Mark Inversion (AMI): A bipolar coding scheme in which successive ones (marks) must alternate in polarity. The ones mentioned here are as in zeros and ones of binary.

American National Standards Institute (ANSI): Devises and proposes recommendations for international communications standards. *See also* Comite Consultatif International Telegraphique et Telephonique (CCITT).

Amphenol Connector: Amphenol is the name of the original manufacturer of these connectors. The most famous connector they made was used to take 25 pairs of wire into 50 pair blocks, Centrex, PBX, and key systems. The telephone companies call this connector an RJ-21.

Amplifiers: These are inserted in the CATV network wherever the broadcast signal needs to be regenerated in order to be received in the household at an adequate level. Amplifiers are classified by their frequency range (what frequencies they will re-generate) and whether they are one-way or two-way. One-way amplifier will only regenerate the downstream signal from the head end to the household. It may even prevent return signals from being transmitted back to the head end. Two-way amplifiers regenerate signals in both directions.

Analog: A continuous signal that can assume all the values of the spectrum through which it passes.

Asynchronous: A method of data transmission which allows characters to be sent at non-predetermined intervals by preceding each character with a start bit and ending each character with a stop bit. These extra bits slow down data transmission, but also make the communication more forgiving of errors, because the bits help keep the transmission accurate even when poor line quality exists.

Asynchronous Transfer Mode (ATM): A data transmission technology developed for Broadband ISDN service. It is considered to be a fast, cell-switched technology and uses a fixed 53 byte cell that contain 48 bytes of data.

ATM Adaptation Layer (AAL): The ATM adaptation layer sits on top of ATM to convert data into segments of 48 bytes (also known as an octet) for transmission using ATM. It also handles the reassembly of the octets into the original data.

Attenuation: Signal power loss through equipment, lines or other transmission devices measured in decibels.

Authorization Code: Authorization codes are used on PBX and Centrex systems to change the class of service of a telephone so that a telephone call can be placed. They are similar in functionality to calling credit cards . The codes are typically between 1 and 18 digits in length, although seven digits are the most common in business. Authorization codes are usually assigned to people, extensions, departments, projects or general ledger numbers. The code appears on the call record generated by the telephone system. A telemanagement system collects the call records and assigns the call back to the responsible party based upon the authorization code on the call and found in the telemanagement system.

Automatic Number Identification (ANI): Refers to the process that identifies the phone number of callers provided by service bureaus or carriers. The public switched telephone network (PSTN) can send the telephone number of the calling party directly to the destination telephone in either dual tone multifrequency (DTMF) or dialed number identification service (DNIS) signaling.

Backward Explicit Congestion Notification (BECN): A bit set by a frame relay network to notify an interface device(DTE) that congestion avoidance procedures should be initiated by the sending device.

Bandwidth: A term defining the information carrying capacity of a channel—its throughput. In analog systems, it is the difference between the highest frequency that a channel can carry minus the lowest, measured in hertz. In digital systems, the unit of measure of bandwidth is bits per second (bps). The bandwidth determines the rate at which information can be sent through a channel —the greater the bandwidth, the more information that can be sent in a given amount of time.

Bi-Bop: Commercial name of the CT2 service by France Tin Paris, Strasbourg and Lilles.

Bipolar: A signaling method that represents a binary one by alternating positive and negative pulses and a binary zero by the absence of pulses.

Bridge: A place where two or more locations are joined. In telephony, a bridge is where a telephone circuit terminates and another circuit begins and ends in another location. A bridge is made between the two circuits to extend the telephone circuit to the distant location. Bridges can also direct the telephone circuit in more than one direction. The telephones in your home are all bridged to a single telephone line. In videoconferencing —a bridge connects three or more conference sites so that they can simultaneously communicate. They are often referred to as Multipoint Control Units.

Bridge Clips: A small metal clip that is used to connect two adjacent sides of a 50 pair block together.

Broadband Integrated Services Digital Network (BISDN): 100Mbps or greater data transmission speeds based on basic ISDN architecture. *See* ISDN.

Building Distribution Frame: A location in a building where all telephony wiring for the building is located. Typically outside feeder cable is located in one section and then there are several other sections of the frame that contain station wire, riser cable, and inside feeder cable.

Butt Set: A one-piece telephone used by telephone technicians. It is called a butt set because a technician can "butt in" on telephone conversations from any point in the telephone circuit where the wires are accessible. The butt set is usually equiped with wire clips and a standard RJ plug for interfacing with the telephone circuit. Many models also have speakers built in.

Cable: In telecommunications, this is a group of telephony wires combined into one larger cable. Telephony cable ranges in size from as small as two pairs of wire to over 1,000 pairs of wire. In cable television, this is the distribution mechanism for transporting the television broadcast signals from the head end to the household. It consists of shielded coaxial cable usually strung along existing pole structures or led through underground conduits.

Cable/Plant Management (CPM): One of the functions of a telemanagement system. It is the mapping, monitoring, tracking, and managing of the entire telephony infrastructure. Every pair of wire, or any other medium such as fiber optics, coax, twinax, ethernet, 10base-T, etc. can be logically connected and monitored. Pairs can be identified to the precise user. CPM can also identify in-use, bad, terminated, and available circuits.

Call Pickup Group (CPUG): A CPUG is when multiple telephone extensions ring when the same phone number is dialed and the calls are evenly distributed between the appearances of the telephone extension. The most common example of CPUGs are operator stations that answer all calls made to the main telephone number or answer all internal calls directed to the operator.

Call Resale: The process of buying local and long-distance telephone usage at one rate and then reselling the usage to individuals at higher rates. This process is made possible through a telemanagement system that collects the telephone call data, then prices and posts the telephone charge to the responsible party. From this point on standard accounts receivable practices are followed.

Calling Pattern: If you take a close look at the telephone numbers that an organization makes over a period of time you will see that a pattern develops that tells the educated telecommunications person valuable information about the way the organization does business. The calling pattern can be used to help shop for long-distance, cut back on toll fraud, and it can also be used for network optimization.

Carrier: A continuous signal at a fixed frequency that is capable of being modulated with a second (information carrying) signal.

Centi Call Seconds (CCS): A unit of measure used for calculating call traffic through a network. A centi call second is 100 seconds of traffic through the network.

Central Office (CO): This is where the telephone company houses the switching system that provides telephone service throughout a neighborhood. A Centrex switch is what is housed in a CO location.

Centrex: Any one of a number of large telephone switching systems that the telephone company might deploy to provide telephone service to an area. DMS100, DMS200, 5AESS, DEX 600, and 400 are among the more common modern centrex switches used today.

Channel: A path for electrical transmission between two or more points. Also called a line, link, circuit, or facility.

Channel Service Unit (CSU): A device used to connect a digital connection, such as a T-1, being delivered from the phone company to either a multiplexor, channel bank, or other device that accepts digital signals. The CSU also regenerates the digital signal to boost clarity and remove background noise.

Circuit-switching: The process of establishing a connection for the purpose of communication in which the full use of the circuit is guaranteed to the parties or devices exhanging information.

Class of Service (COS): In order to let one telephone be able to access certain outbound facilities, features, and functionality, while another phone is given a different group of facilities, features, and functionality, a class of service is assigned which defines these criteria. An example of a COS difference is a hall phone that can only make calls within the organization, and the phone of the president of the company, which can make calls anywhere in the world and has voice mail, call forwarding, and a host of other features. COS can also be assigned to authorization codes to temporarily increase the COS of a telephone for a single call.

Code-Division Multiple Access (CDMA): A digital mobile communications technology which competes with GSM in the U.S.A.

Codec: Codec is a contraction for coder/decoder or compress/decompress. It is a sophisticated digital signal processing unit that takes analog input and converts it to digital on the sending end. At the receiving end, the process is reversed to take the digital signal and reconvert it to analog.

Comite Consultatif International Telegraphique et Telephonique (CCITT): In the English language this translates to International Telegraph and Telephone Consultative Committee, a committee of the International Telecommunications Union (ITU), and an agency of the United Nations. With representatives from many governments, the committee is organized into 15 study groups that makes recommendations for telecommunications standardization on a global basis.

Common Carrier Bureau (CCB): This bureau, established by the FCC, specifically regulates interstate telephone systems. Its responsibilities include licensing, rate monitoring, and working with state public utilities commissions.

Common Channel Signaling (CCS): A method of signaling information relating to circuits or functions of network management, and conveyed over a sigle channel by addressed messages.

Common Intermediate Format (CIF): Part of the ITU's H.261 standards, which specifies 288 non-interlaced luminance lines, each containing 352 pixels, that are then sent at a rate of 30 frames per second.

Compression: Any of several techniques that reduce the number of bits required to present information in data transmission.

Computer Telephony: A term that describes the industries that concentrate on integrating computer intelligence to telecommunications devices. Computer telephony covers everything from modems, to voice processing, and a host of other integrated hardware/software applications available now and in the future. Almost all subjects in this book are computer telephony related.

Concert: British Telecom-MCI alliance offering IVPN in Europe.

Corporate Open Standards: Open interactive standards for digital set-tops.

Credit Card Processing: This is a feature on a fax-on-demand service that can take the credit card number and expiration date from a caller over the phone (using touch-tone input) and immediately process a charge to that account.

Crosstalk: Unwanted transfer of energy from one circuit to another.

CT2 (Cordless Telephone 2): A second-generation digital cordless telephone standard.

Cyclic Redundancy Check (CRC): A confirmation performed on data transmission to see if an error has occurred in the transmission, reading, or writing of the data being transferred. A CRC character is calculated from the data being transferred. The software compares the character stored at the beginning on the transmitting end to the character calculated and returned from the receiving end. If the characters are the same, the data is assumed to be accurate.

Data Communications Equipment (DCE): The equipment providing functions that establish, maintain, and terminate a data transmission connection (such as a modem or printer).

Data Service Unit (DSU): A device used to connect a computing device (Data Terminal Equipment: DTE) to a digital phone line to allow for fully digital communications.

Data Terminal Equipment (DTE): A device transmitting data to, and/or receiving data from a piece of data communications equipment (such as a modem). Your computer, particularly the communications port, is a DTE device.

DCS 1800 (Digital Communications System): A variant of GSM in the 1.8 GHz band.

Decibel (db): Unit for measuring relative strength (ratio) of two signals.

DECT (Digital European Cordless Telephone): The European digital cordless telephone standard, more advanced than CT2.

Demarcation Point: The location where two separate ends of telephone cable or wiring meet. Typically used to mean the place where the telephone company's wiring and the customer premises wiring come together. This is where the responsibility of the local telephone company ends and the customer's (your) responsibility begins.

Dialed Number Identification Service (DNIS): A more advanced method of delivery of automatic number identification that uses out-of-band signaling (signals that can not be heard by the human ear). DNIS also provides the capability to T-1 telecommunications equipment to accommodate simultaneous calls to independent phone numbers connected through computer technology. This feature prevents callers from getting busy signals when they call. DNIS can include the calling parties telephone number within the signal and can also tell the receiving system what telephone number was dialed so that many calls coming over the same telephone lines can be broken apart to separtate destinations, such as an operator, IVR (interactive voice response) system, or voice mail. DNIS can also pass a host of other information in a similar fashion to telephony devices to gather further information.

Digital: A discrete signal that can assume only certain specific values within its range. The use of a series of binary codes (zeros and ones) to represent information.

Digital Audio Video Council (DAVIC): International specifications for open interfaces and protocols for interoperable digital A/V applications and services.

Digital Loopback: A technique for testing the digital processing of a communications device.

Digitized Speech: The process of converting speech to digital format for computer storage.

Direct Inward System Access (DISA): A method for dialing into a telephone system. DISA can be used for everything from accessing one's voice mail to connecting to an outbound trunk to make long distance calls. DISA can be the cause of toll fraud if not properly implemented.

Distortion: The unwanted change in a signal's waveform occurring between two points in a transmission system.

Domain Name Server (DNS): The part of the distributed database system for resolving a fully qualified domain name into the four-part IP (Internet Protocol) number used to route communications across the Internet. When a domain name is not recognized by the first DNS encountered, the DNS will forward the request up through a hierarchy of domain name servers in order to make every attempt at resolving the domain name into an IP address.

Downstream: Main transmission in the direction from the head end to the end nodes (household) in the network. Also called forward signal.

Dual Tone Multi Frequency (DTMF): Simply put, the tones generated by your touch tone telephone to denote the ten numbers, star, and pound keys on your telephone. There are four other less common DTMF tones: A, B, C, and D. Each of these tones are generated in hertz cycles in predetermined ranges between 697Hz and 1633Hz. *See also* Automatic Number Identification.

E&M Signaling: Uses separate paths for signaling and voice signals. The E lead (derived from the ear) receives incoming signals as either ground or open condition. The M lead (derived from the mouth) transmits ground or battery towards the switch.

E1: European equivilant of a T-1 circuit. Transmission rate of 2.048 Mbps on E1 communications lines. An E1 facility carries a 2.048 Mbps digital signal over 32 channels. *See also* T1 and Channel.

Echo-Signal: Distortion occurring when a transmitted signal is echoed back (reflected) to the originating station.

Electronics Industry Association (EIA): An interface and interoperabiltiy standards organization.

Electronic Serial Number (ESN): Part of the data transmission registration signal that is sent from cellular telephones to identify the cellular phone and its location. The manufacturer is supposed to make this number impossible for the user to change, but unfortunately thieves have found a way to change a cellular telephone's ESN and steal cellular service.

Equipment Bill Reconciliation: The process of downloading your equipment charges from your local telephone company in some electronic form and producing discrepancy reports from a facilities management system database. The information produced is used to get refunds or reductions in phone company charges.

Equipment/Feature Inventory: The process of tracking, monitoring, allocating, and accounting for all equipment and telephony features being managed by a facilities management system.

Erlangs: A measure of telephone traffic over a network. One Erlang is equal to one full hour of use of a telephone circuit, i.e. 3600 seconds of communications.

Ermes: The European radio messaging system. The standard European paging system.

Extended Super Frame (ESF): A T-1 framing format that uses the framing bit to provide maintenance and diagnostic functions.

Facilities Management: The process of managing all telephony related products and services provided by a telecommunications department. Facilities management is one of several functions that make up a telemanagement system. Facilities management is made up of three major components: cable/plant management, equipment/feature inventory, and work order processing.

Fax-on-Demand: Also known as interactive fax or fax retrieval. This is a new technology where a recipient requests a document and it is delivered by fax instantly and automatically.

Fax Retrieval: *See* Fax-on-Demand

Fax Vault: A computing device used to store fax documents is a high quality format. A fax-on-demand system will draw on the fax pages stored in a fax vault for transmission to users requesting information.

Federal Communications Commission (FCC): Government agency established in 1934 to regulate interstate and foreign communication over wire, radio, and television. The FCC plays an important role in shaping what we see and hear.

File Server: In the context of frame relay network supporting LAN-to-LAN communications, a device connecting a series of workstations within a given LAN. The device performs error recovery and flow control functions as well as end-to-end acknowledgment of data during data transfer, thereby significantly reducing overhead within the frame relay network.

File Transfer Protocol (FTP): The set of standards that allow you to exchange complete files across different computer hosts. Using an FTP client, you can search for files and retrieve them from software archives on the Internet.

Firewall: A means of securing an internal network from unwelcome intrusions from an external network, typically in the case of a business or research computer network by protecting its data and operations from hackers. Firewall protection is usually accomplished by a router configured to exclude traffic of certain types (such as Telnet, FTP, etc.) or from certain sources (as determined from domain names or IP addresses), but it can also be accomplished by the server or by having separate networks for services, such as World Wide Web, which are not connected to other internal computers or networks.

Forced Authorization Code (FAC): *See* Authorization Code.

Forward Explicit Congestion Notification (FECN): A bit set by a frame relay network to notify an interface device (DTE) that congestion avoidance procedures should be initiated by the receiving device. *See also* Backward Explicit Congestion Notification (BECN).

Frame: An individual television, film, or video image. Television sends images at 30 frames-per-second (fps), while film (movies) uses 24 fps. Videoconferencing systems typically send between 8 and 30 fps, depending on the transmission bandwidth offered. A sequence of time slots, with each slot containing a sample from one of the channels available on the circuit.

Frame Check Sequence (FCS): The standard 16-bit cyclic redundancy check used for HDLC and frame relay frames. The FCS detects bit errors occurring in the bits of the frame between the opening flag and the FCS, and is only effective in detecting errors in frames no larger than 4096 octets. *See also* Cyclic Redundancy Check (CRC).

Frame Relay Frame: A variable-length unit of data, in frame-relay format that is transmitted through a frame relay network as pure data. *Contrast with* Packet. *See also* Q.922A.Frame Relay

Frame Relay Network: A network interface providing high speed frame or packet transmission, with minimum delay. It has less protocol overhead than X.25.

Frame Relay-Capable Interface Device: A communications device that performs encapsulation. Frame-relay-capable routers and bridges are examples of interface devices used to interface the customer's equipment to a frame relay network.

Franchise: The grant or agreement by a municipality with a CATV company for the right to build a CATV network, sell television re-broadcast services to residents, and charge subscribers for the service. The franchise agreement may be exclusive.

Frequency: The number of repetitions of a signal within a given time.

Full-Duplex Audio: An audio channel which allows conversation to take place interactively and simultaneously between the various parties, without electronically cutting off one or more participant if someone else is speaking. With half-duplex, only one party can speak at a time without cutting the other end off.

FX (Foreign Exchange) line: Telephone service provided from a central office (CO) other than the local central office that would normally service that location. FX lines are used to make telephone calls to and from the FX area local, thus reducing overall costs. FX lines cost more than regular CO lines, but after a certain amount of call traffic, cost savings are realized.

Geosynchronous Orbit: The positioning of a satellite over the earth in a specific location. Satellites that are designed for communications between one or more points attempt to stay in a specific location in the heavens above the earth. A ground station transmits its signals towards the satellite's location so the signal can be relayed to the destination location.

Global Positioning System (GPS): A series of satellites combined with specialized devices that allow you to pinpoint exactly where you are on the earth, typically in longitude and latitude.

Global System for Mobile Communications(GSM): The European digital mobile communications standard used throughout Europe and in many other countries.

Ground Station: A telecommunications location used to send signals from the earth to satellites in geosynchronous orbit about the earth and to receive signals relayed from the same satellite back to earth. The nonmobile equipment is distributed at various sites so that the mobile equipment can communicate with it.

H.261: The ITU's standard that allows dissimilar video codecs to interpret how a signal has been encoded and compressed, and to decode and decompress that signal. Also referred to as PX64, it incorporates multiplexing, demultiplexing, and framing of multimedia data as well as transmission protocols, call setup, and teardown.

H.320: Representing a family of standards for videoconferencing, H.320 includes a number of individual recommendations for coding, framing, signaling, and establishing connections. It can be a video system's sole compression method or a supplementary algorithm to a proprietary one so dissimilar codecs can interoperate.

Hand-Off: The process by which a mobile telephone switching system transfers calls from one cell location to another when the signal grows weak from the first switching system and another switching system can better support the communications.

Head End: This is where all the signals are broadcast into the CATV network and return signals concentrated for service actions. It usually includes various satellite and television receivers to receive program signals for rebroadcast in the CATV network. The received programming is demodulated to a baseband television signal and then remodulated on the frequency (TV channel) assigned in the cable station lineup. The head end may have some video production capability for local origination programming.

High-Level Data Link control (HDLC): A generic link-level communications protocol developed by the International Organization for Standardization (ISO). HDLC manages synchronous, code-transparent, and serial information transfer over a link connection. *See also* Synchronous Data Link Control (SDLC).

Hop: A single trunk line between two switches in a frame relay network. An established PVC consists of a certain number of hops, spanning the distance from the ingress access interface to the egress access interface within the network.

Household: This is the termination or end point of the CATV network. Consumers subscribe to the CATV service and the CATV network provides a cable drop from the network into the house through the service entrance to an optional set-top and thence to the television and VCR. The subscriber is then able to view video programming selections they have subscribed to.

Hub: A central point of a network, usually providing some level of network-wide coordination. A hub can also be a single point of failure. Often, it is efficient to locate the fileserver or connection to a wider-area network at the hub.

HyperText Markup Language (HTML): The programming language used to create web pages that contains a system of notification, called tags, text formatting, and hypertext links that can be compiled and run by users of the Internet with browsers, such as Netscape or Mosaic.

HyperText Transport Protocol (HTTP): The set of standards that defines World Wide Web services, allowing the web browser or client to access, link to, and retrieve information and graphics from different Web servers.

Impedance: The combined effect of resistance, inductance, and capacitance on a transmitted signal.

In-Band Signaling: Signaling utilizing frequencies within the information band of the channel.

Industrial, Scientific and Medical (ISM): Radio frequencies reserved for radiation by various equipment. Available for use by communications equipment without license if that equipment complies with maximum emitted power limits. Equipment which uses the ISM Band must tolerate interference from other such equipment.

Information Industry Association (IIA): A national association that dedicates a portion of its services to the pay-per-call industry.

Information Superhighway: The ultimate interaction of computer networks. A new evolving communications medium which promises to link the entire world through video, sound, and digital communications. The Internet is also known as the Information Superhighway.

Inside Feeder Cable: Telephone cable used to provide large amounts of telephony services from one side of a building to another location on the same floor. Inside feeder cable usually begins in a main distribution frame (MDF), building distribution frame (BDF), or intermediate distribution frame (IDF), and ends in another IDF.

Integrated Services Digital Network (ISDN): A service that allows a variety of switched digital data and voice transmission to be accommodated simultaneously. It is a networking concept that provides subscribers with end-to-end fully digital communications.

Interactive fax: *See* Fax-on-Demand.

Interactive MultiMedia Association (IMA): Facilitate interpretability between set-tops and servers; develop set-top interface specifications

Interactive Services Association (ISA): Telecom-based Interactive Services—audio/video/data.

Interactive Voice Response (IVR): An interactive computing device that combines prerecorded voice information with information gathered from conventional computing devices. *See also* Voice Processing.

Interface: A shared boundary, defined by common physical interconnection characteristics, signal characteristics, and meanings of exchanged signals.

Intermediate Distribution Frame (IDF): A location where telephone cabling is accessible by a person to create telephony circuits. IDFs may contain station wire, riser cable, inside feeder cable, blocks, and a host of other telephony related components.

International Standards Organization (ISO): An international standards group that is concerned with defining standards for data communication and network management. Committees throughout the world contribute to the ISO standards set forth.

International Telecommunications Union (ITU): This international organization is a specialized agency composed of telecommunications administrators whose focus is on developing and improving standards.

Internet: The global system of networks interconnected by TCP/IP (and IP-related protocols), which include over 30 million users from the private sector, educational instututions, government, nonprofits, and individuals. Internet users gain access to e-mail, file transfer, remote login, gopher, news, World Wide Web, and other related services.

Internet Access Provider (IAP): An organization that provides connectivity to the Internet from a cental location or locations. This organization usually markets their services to end users in a specific community.

Internet Protocol (IP): Part of the TCP/IP familiy of communications protocols that are designed to track an Internet address of nodes, routes outgoing messages and recognizes incoming messages.

Internet Service Provider (ISP): An organization that offers access to the Internet through dial-up or dedicated lines. Supports customers to various degrees in the areas of special services, such as domain name registration, listservers, FTP site creation and maintenance, and Web page creation and maintenance.

Intraframe Coding: A compression technique where redundancy within a video frame is eliminated.

Inverse Multiplexer: Equipment that receives a high-speed input and breaks it up for the network into multiple 56 or 64 Kbps signals so that it can be carried over switched digital services. It also provides the synchronization necessary to recombine the multiple channels into a single integrated transmission at the receiving end.

Isochronous: Signals which are dependant on some uniform timing or carry their own timing information embedded as part of the signal.

Jack: Any one of a number of pieces of equipment that are designed to provide quick (tool-less) wiring connectivity, as in a telephone jack. RJ-11, RJ-22,RJ-45 and RJ-48 are the most common examples of jacks—there are dozens more.

Jitter: The deviation of a transmission signal in time or phase.

Key System: A telephone system that can allow users to access multiple telephone lines on a telephone set which may contain multiple telephone line appearances. Key systems can supply a wide variety of advanced telephone functions such as speed dialing, conference calling, authorization code verification, and a host of other functions, but they typically can not switch or transfer a call from one telephone line to another.

LAN Protocols: A range of LAN protocols supported by a frame relay network, including Transmission Control Protocol/Internet Protocol (TCP/IP), Apple Talk, Xerox Network System (XNS), Internetwork Packet Exchange (IPX), and Common Operating System used by DOS-based PCs.

LAN Segment: In the context of a frame relay network supporting LAN-to-LAN communications, a LAN linked to another LAN by a bridge. Bridges enable two LANs to function like a single, large LAN by passing data from one LAN segment to another. To communicate with each other, the bridged LAN segments must use the same native protocol.

Land Line: A term used for wired telephony, especially in contrast with wireless. Telephone circuitry that travels on land.

Lata: Any of 161 areas of the country where a LEC has been awarded the exclusive rights to provide telephone service to that area.

Lead Retrieval: A system that will capture the caller ANI (automatic number identification), their fax number, name of caller, and company name.

Line Consolidation: The process of reducing the number of incomming and outgoing telephone lines from a location, but not reducing the accessibility to telephony service. Example: A company has 100 employees with telephones on their desk. That same company might have 15 two-way telephone lines that service the location. Sixteen employees would have to attempt to make or receive a telephone call simultaneously before the sixteenth person would experience call blockage. Line consolidation is heavily based on statistical data that determines the optimal number of lines the serve a given population and prevent against call blockage.

Link Access Procedure Balanced (LAPB): The balanced-mode, enhanced, version of HDLC. Used in X.25 packet-switching networks. *Contrast with* LAPD.

Link Access Procedure on the D-channel (LAPD): A protocol that operates at the data link layer (layer 2) of the OSI architecture. LAPD is used to convey information between layer 3 entities across the frame relay network. The D-channel carries signaling information for circuit switching. *Contrast with* LAPB.

Loading: The addition of inductance to a line in order to reduce amplitude distortion.

Local Area Network (LAN): A privately owned network that offers high-speed communications channels to connect information processing equipment in a limited geographic area.

Local Exchange Carrier (LEC) lines: Local Exchange Carrier or local phone company telephone lines used to supply telephone service to a location. LEC lines are needed by all organizations to handle local telephone traffic. Local telephone traffic is defined as telephone calls that do not leave the lata (intralata).

Local Loop: The transmission path between a user's premises and a central office. The connection between the local telephone company's network and the customer premises equipment is formally called the network interface or the point of termination.

Main Distribution Frame (MDF): A wall of telephone cable, typically connected to blocks, that contains all wiring in and out of an installation. The MDF is usually located in the telephone switch room. An MDF might also be a terminal (junction box) that simply provides connectivity access.

Menu-Driven Line: An interactive voice response application where callers choose from specific options and then receive customized information.

Minitel: Generic name of a very popular Videotex terminal from France Télécom.

Mobile Telephone Switching Office (MTSO): Sometimes called mobile switching center. The central office for a number of cellular base station sites. Makes the interface between the cellular system and the PSTN.

Modem: A type of computer equipment that links computers via the telephone for the transmission of computer files, letters, invoices, etc. Modems are considered to be DCE devices.

Modulation: The alteration of a carrier wave in relation to the value or samples of the data being transferred.

Motion Compensation: In a motion sequence, each pixel can have a distinct displacement from one frame to the next. Motion compensation is a compression technique that exploits similarities in neighboring pixels, and devides the image up into rectangular blocks and uses a displacement vector to describe movement of all pixels within the block.

Multidrop: A communications configuration in which multiple devices share a common transmission facility.

Multipoint Control Unit (MCU): A device that bridges together multiple inputs so that more than three parties can participate in a videoconference.

Multi System Operator (MSO): These are CATV companies owning more than one franchise to provide CATV services. The franchises may or may not be geographically adjoining. This geographic distribution results in operating multiple cable systems or networks to provide television signals to subscribers.

National Association for Information Services (NAIS): A national organization for pay-per-call services which provides industry and government standards, public relations, and comprehensive information about regulations, resources, and education for members involved in all facets of the industry.

Narrow-Band ISDN: Includes basic interface (2B + D or BRI) and primary rate interface (23B + D or PRI). Copper-based at speed up to 1.544 Mbps.

Network: A telecommunications network based on frame relay technology. Data is multiplexed. *Contrast with* Packet-Switching Network.

Network Optimization: The process of monitoring all traffic over a given network and then analyzing the data using traffic management techniques in order to reconfigure the network components for optimal performance.

Network Terminating Equipment: The part of a digital communications circuit that completes the final leg of the circuit to the data terminal equipment on the customer side of the connection.

Network Terminating Equipment Type 1 (NT-1 or NT1): The first customer premise device on a two-wire ISDN circuit coming in from the telephone company. It, among other things, converts the two-wire signal called a "U" interface to four-wires, so multiple devices can be attached to the ISDN circuit.

Network Tie-Line Reconciliation: When calls are routed through multiple switching network locations, a call detail record (CDR) is generated at each switch. A telemanagement system will collect the call data from each location and then combine the call records down to a single consolidated record that reflects the beginning and termination points, as well as calculating the costs and identifying the party responsible fo the call in the first place.

NNX/NXX: Also referred to as the exchange. The three digit prefix that follows the area code of a telephone number which identifies a specific central office that a call should be routed to.

Nordic Mobile Telephone (NMT): An analog mobile phone technology initially developed for Scandinavian countries, and eventually used in other European countries.

Open Systems Interconnect (OSI): The OSI is a reference model that serves as a framework for the development of communications protocols. The model is intended to provide guidance for the establishment of common standards to facilitate communications between different vendors' products.

Outside Feeder Cable: Telephone cable that is placed in the ground either buried, in conduit, on telephone poles, or in tunnels, and used to create circuits between two locations.

Packet: Typically refers to sizes of from 8 to 8K bytes. Some systems will always transfer a fixed size packet, while others will transfer from one byte up to the maximum packet size. The data, call control signals, and possible error control information are arranged in a predetermined format. Packets do not always travel the same pathway but are arranged in proper sequence at the destination side before forwarding the complete message to an addressee. *Contrast with* Frame Relay Frame.

Packet-Switching Network: A telecommunications network based on packet-switching technology, wherein a transmission channel is occupied only for the duration of the transmission of the packet. *Contrast with* Frame Relay Network.

Parameter: A numerical code that controls an aspect of terminal and/or network operation. Parameters control such aspects as page size, data transmission speed, and timing options.

Passive Devices: These are components, such as filters, inserted into the CATV network which do not require line power to operate.

Pay-per-Call: Service where caller is billed for the charge of a telephone call by the phone company.

PBX (Private Broadcast Exchange): Any privately owned switching system. A PBX gives a company control over their own telephone system. In contrast to a Centrex system that is run by the local telephone company and all control remains in the hands of the phone company that serves you. Modern PBX systems provide a great deal of flexibility and manageability to an organization, with regards to telecommunications, that can not be acheived by any other means.

PCMCIA: People Can't Memorize Complex Industry Acronyms. Actually, it stands for Personal Computer Memory Card International Association, which is unfortunate, since the standard has grown to include almost any kind of computer peripheral in a credit-card sized package. The association has trademarked the term PCCard for the format, but the term PCMCIA card will live on. The standard includes not only the electrical connector interface, but also the package size, so that properly designed card can fit conveniently just within the slot designed to hold them. The standard includes three types, distinguished by increasing thickness: Type 1 (very thin memory cards), Type 2 (most modems and interfaces) and Type 3 (twice as thick as Type 2, intended for disk drives and other complex interfaces). Many modern laptops provide two Type 2 slots which can also be used as a single Type 3 slot. Adding the term "Extended" implies that the card is extra long, and will stick out of the side of the laptop. On some standard-violating "extended" cards, the extension can grow to a large bump holding oversized connectors, batteries, antennas, and/or displays.

PCS 1900 (Personal Communications System): A variant of GSM in the 1.9 GHz band specifically designed for the American market.

Peer-to-Peer: In networking, communications between equals with no central system or device providing coordination.

Permanent Virtual Circuit (PVC): A frame relay logical link, whose endpoints and class of service are defined by network management. Analogous to an X.25 permanent virtual circuit, a PVC (often referred to as a PVC) consists of the originating frame relay network element address, originating data link control identifier, terminating frame relay network element address, and termination data link control identifier. *Originating* refers to the access interface from which the PVC is initiated. *Terminating* refers to the access interface at which the PVC stops. Many data network customers require a PVC between two points. Data terminating equipment, with a need for continuous communication, use PVCs.

Personal Authorization Code (PAC): *See* Authorization Code

Personal Communications Service (PCS): The generic name for the services which will be provided in a new set of frequencies allocated by the government. Cellular telephone is likely to be the first use for these frequencies.

Personal Identification Number (PIN): A number assigned to an individual which accesses exclusive information housed in a telecommunications system. *See also* Authorization Code.

Plain Old Telephone Service (POTS): The standard telephone service that we are all familiar with. The service is provided on two wires and uses analog signaling to provide single line telephone service. No additional features such as call waiting, call forwarding, conference calling, voice mail, etc. All you can do is make or receive calls.

Plesiochronous: Signals which are arbitrarily close in frequency to some defined precision. They are not sourced from the same clock and so, over the long term, will be skewed from each other. Their relative closeness of frequency allows a switch to cross connect, switch, or in some way, processs them. That same inaccuracy of timing will force a switch, over time, to repeat or delete frames (called frame slips) in order to handle buffer underflow or overflow.

Point-to-Point: Any connection between two locations. A point-to-point connection can be either dial-up (the connection is established by making a telephone call and the public switched telephone network creates the connection), or the connection is dedicated (the two points on the connection are fixed). A dedicated point-to-point connection is also called a leased line or a tie-line.

Postalized: To charge a fixed amount per minute for a call instead of charging a variable rate based upon the mileage the call travels from origination to termination. Just like the post office charges the same amount for a letter, whether it travels across the country or across the street.

Point-to-Point Protocol (PPP): A protocol that allows a computer to use TCP/IP over a standard telephone line with a high-speed modem to become a fully participating member of the Internet. PPP is favored as a replacement for SLIP, an older protocol that accomplished the same function.

Predictive Codebook Linear Predictive Coding: A modern algorithm for voice compression which imitates the human vocal tract, utilizing codebooks of common phonetic sounds.

Private Broadcast Exchange (PBX): A telephone system that is capable of performing the switching function and is owned by you. In effect, the organization is its own customer, as opposed to having the organization be the customer of the local phone company. The PBX still requires that the telephone lines be supplied by the local exchange carrier (LEC) or by a long distance vendor through the LEC, but the power of control, line consolidation, and routing is in the hands of the owner. The features and functionality of modern telephone system are yours to command.

Protocol: A formal set of conventions governing the formatting and relative timing of message exchange between two communication systems.

Public Switched Telephone Network (PSTN): The telecommunications network commonly accessed by all of us virtually every day of our lives when we make a telephone call. The PSTN is a gigantic maze of switching computers that can connect any two (sometimes more) telephony points in potentially hundreds of different ways.

Pulse Code Modulation (PCM): The speech signal is sampled at 8khz, and each sample is encoded using an 8-bit code, 8,000 samplings x 8 bits per sample = a 64 Kbps bit stream.

Q.922 Annex A (Q.922A): The international draft standard that defines the structure of frame relay frames. Based on the Q.922A frame format developed by the CCITT. All frame relay frames entering a frame relay network automatically conform to this structure. *Contrast with* Link Access Procedure Balanced (LAPB).

Q.922A Frame: A variable-length unit of data, formatted in frame relay (Q.922A) format, that is transmitted through a frame relay network as pure data (i.e., it contains no flow control information.) *Contrast with* Packet. *See also* Frame Relay Frame.

QCIF: Quarter Common Intermediate Format. A mandatory part of the ITU-T standard which requires that non-interlaced video frames be sent with 144 luminance lines and 176 pixels.

Quantizing: The process by which the amplitude is measured of each sample.

Radiocom 2000: An analog mobile phone technology used mainly in France.

Request For Proposal (RFP): The process of compiling a set of specifications for any procurement (telephone system, telemanagment system, voice mail, etc.) to clearly define all of the requirements of said procurement. The RFP is then distributed to all prospective vendors, so they may respond in detail to the requirements and provide a quotation for the project as defined within the RFP. The goal of the RFP process is to procure the item(s) that are detailed within the RFP.

Revenue Generating Number: A telephone number which is designed to yield profits from the information provided on the program.

Riser Cable: Telephony cable that travels from one floor of a building to one or more floors to provide telephony services. Riser cables typically start in either MDFs, BDFs, or IDFs.

RJ-11(Registered Jack): A telephone jack usually wired with four conductors (although it could be wired with six), although POTS telephone lines only use two. RJ11 jacks are the most common jacks used in the industry today.

RJ-21: Another name for an Amphenol connector. *See* Amphenol Connector.

RJ-45: An eight wire jack that is typically used for data transmision over regular telephone wires. If serial data is being transmitted, flat wire (wire that does not twist around itself) can be used. If network connections such as 10BaseT are used, twisted pair wire must be used.

Route: Also called a trunk group. A group of telephone lines that serve the same purpose. Your AT&T long distance lines, your tie-lines over to corporate headquarters, your local telephone lines, your foreign exchange lines are all examples of routes.

Router: A device that supports LAN-to-LAN communications. Routers may be equipped to provide frame relay support to the LAN devices they serve. A frame-relay-capable router encapsulates LAN frames in frame relay frames and feeds those frame relay frames to a frame relay switch for transmission across the network. A frame-relay-capable router also receives frame relay frames from the network, strips the frame relay frame off each frame to product the original LAN frame, and passes the LAN frame on to the end device. Routers connect multiple LAN segments to each other or to a WAN. Routers route traffic on the Level 3 LAN protocol (e.g., the Internet Protocol address).

Routing: The process of choosing a specific route to send telephone calls out on based upon the telephone number dialed. On many telephone switching systems there is a table called an automatic route selection (ARS) that contains all the area codes(NPAs) and many exchanges(NNXs) that can be dialed. The ARS tells the switch which route to send a call over. Second and third choices are also sometimes listed in the ARS table. Access codes can also be used to select the route to use, such as "9" to select a long distance line or "8" to select a local line.

Service Bureau: An independent company that provides technical support, computer hardware, marketing services, and other telephony related services.

Service Profile Identifier (SPID): A code used to identify a specific ISDN set on a given ISDN circuit when more that one ISDN device is attached to the same circuit.

Set-Top: A household based device which demodulates CATV broadcast signal and remodulates it to a frequency usable by the subscriber's television and VCR. Called set-top since it usually sits on top of the television set. The set-top may also be used to control access to additional pay-per-view channels by unscrambling selected subscribed channels. Some CATV networks do not require set-top boxes if they transmit unscrambled signals in the frequencies of over-the-air broadcast or cable-ready television sets. The subscriber's interface to set-top box is through infrared remote controls. Set-top boxes may provide on-screen program selection guides through the television.

Signaling: Transmission of information for the purpose of directing voice traffic over a telecommunications network.

Signaling System 7 (SS7): A common channel signaling protocol used between switching systems.

Simple Mail Transport Protocol (SMTP): Defines the standard for naming and transferring mail across a network. Addresses in SMTP are in the form of: user@host.network.

Space: In telecommunications, the absence of a signal.

Spans: Module which interoperates signalling and routing with Fore Systems ASX switch and various host interfaces. SPANS is a trademark of Fore Systems, Inc.

Splice: To join two or more pieces of wire in a perminant fashion so the length of the circuit created is increased, usually to provide telephony service to remote locations.

Splitter: This is a device inserted into the network to create two or more downstream cables from one upstream cable. Any upstream transmission from the households set-tops are merely combined into the single upstream cable on the same frequency to the headend with no collision control at the splitter. Splitters are used to serve two different neighborhoods, for example, rather than run individual neighborhood cables all the way back to the headend. It splits the radio signals being sent from the headend and routes them down two or more legs. *See also* Tap.

Station Wire: The final leg of the telephone circuit to the station. Two or more pairs of wire that typically connect a distribution frame to a telephone jack. Station wire is also used to jumper two or more blocks in a distribution frame.

Statistical Multiplexing: Interleaving the data input of two or more devices on a single channel or access line for transmission through a frame relay network. Interleaving of data is accomplished using the DLCI.

Subscriber Identity Module (SIM): A credit-card sized card that holds all subscription information for a GSM user, letting him plug his or her SIM in any GSM handset and use it as if it were his or her own.

Switch: Another word for a PBX or a Centrex system. *See* Centrex and PBX.

Synchronous: Signals that are sourced from the same timing reference. These have the same frequency. (*Contrast with* Plesiochronous) Asynchronous means signals that are sourced from independent clocks.

Synchronous Data Link Control (SDLC): A link-level communications protocol used in International Business Machines (IBM) Systems Network Architecture (SNA) network that manages synchronous, code-transparent, serial information transfer over a link connection. SDLC is a subset of the more generic High-Level Data Link Control (HDLC) protocol developed by the International Organization for Standardization (ISO).

T-1 (or T1): Transmission rate of 1.544 Mbps on T-1 communications lines. A T-1 facility carriers a 1.544 Mbps digital signal over 24 channels. Also referred to as digital signal level 1 (DS-1). *See also* E1 and Channel.

T.120: This ITU standard for audiographics exchange specifies how to efficiently and reliably distribute data in a multipoint, multimedia meeting using a set of infrastructure protocols.

Tap: Similar to a splitter, though may not have a signal repeater. This is used to attach the subscriber's cable drop outside the house to the CATV network.

Tariff: Charge, cost, price, or fare required for service.

Telemanagement System (TMS): A group of software modules specifically designed to manage telecommunications and telephony related operations. A TMS can be responsible for: call accounting, telephone/long-distance resale, facilities management, network management, labor, systems integration, and just about anything else you can think of pertaining to the telecommunications industry.

Telephone System: Any one of three types of equipment: Centrex, PBX, or Key system. All telephone systems provide basically the same function—telephone service in one form or another to their users. Telephone systems at the very least usually provide the owner of the telephone system with the ability to perform line consolidation.

Telephony Application Programming Interface (TAPI): Software jointly introduced by Microsoft and Intel. The software has two main specifications—service provider interface which defines hardware specification for interfacing with a Windows dynamic link library (DLL), and the applications programming interface which is geared towards software developers so they may write applications that will integrate with the hardware that meets the service provider interface requirements. TAPI lets a PC communicate directly with telephone systems.

Telephony Server Application Programming Interface (TSAPI): Software primarily introduced by Novel and AT&T, also referred to as telephony service for NetWare. TSAPI is a similar, but competing, standard to TAPI. TSAPI centers the control of the telephone from a computer on the LAN server instead of each individual PC.

Tie-Line: A telephone circuit or group of circuits that connect one private telephone system to another telephone system. These lines are typically leased from the phone company to reduce the costs of communications between the two locations.

Time Division Multiplexing (TDM): The process of dividing the capacity of the transmission facility into discrete time slots, with each individual channel assigned a specific time slot.

Toll Call: Any communications made over telephone lines that the caller pays a charge for placing the call. A toll call can be charged at a fixed rate for placing the call, a per minute rate, or a postalized rate.

Toll Fraud: When someone fraudulently uses (steals) telephone toll call service.

Toner: A device used by telephone technicians that can be used to test circuit continuity, or it can send electronic tones through a telephone circuit that can be detected by either a butt set or a probe.

Trunk: A single circuit between two points, both of which are switching centers or individual distribution points.

Trunk Group: A set of telephone circuits provided by any number of telecommunications firms for a specific purpose, such as to place long distance toll calls, to provide local calling capability, to provide WATS service, to provide inbound calling, etc.

U Interface: The two-wire ISDN circuit provided by the phone company. It is capable of supporting the 2B+D signaling (144kbps) and an additional 16kbps for network signaling between the NT1, or other customer premise device, and the central office switching system for a total of 160Kbps of data transmission.

Uniform Resource Locator (URL): A consistent format by which Web browsers, such as Netscape, locate and access information in a networked environment. A URL consists of a service type (http, FTP, or gopher), a host name (www.apple.com, or www.eg.bucknell.edu, for instance), and a pathname (/docs/info/plan.html/) to a particular source document. Special URL service types can be used to evoke e-mail, Netnews, and telnet sessions.

Unisource: Group from various European operators (PTT Telecom Netherlands, Swiss PTT, Telefonica, and Telia) offering IVPN in Europe.

Upstream: Transmission in the direction from the cable drops (household wiring) in the network back to the headend. Also called return or backward signal.

Vendor Bill Reconciliation: The process of downloading your telephone bill electronically from the phone company to a telemanagement system (TMS). The TMS then compares the phone company's billing to the call accounting information that is created by the TMS to create a reconciled descrepancy report that can be submitted to obtain a refund from the telephone company.

VESA (Video Electronics Standards Association): Define features and interconnections required for open digital interactive set-top box.

Voice Capture: An audio text feature that records the callers name, address, and telephone number, and/or message at the end of the program.

Voice Processing: Any one of several industries that combine computers with the telephone system to provide automated services. Voice processing systems can understand touch tone digits entered by a user at their telephone. Some can even understand the spoken word and can speak themselves. Some forms of voice processing that you might be familiar with are voice mail and banking by phone.

VSAT (Very Small Aperture Terminal): Small terrestrial station providing bi-directional communications with a remote site thanks to a link with a satellite.

Wide Area Telecommunications Service (WATS): A discounted telephone toll service provided by many of the telephone companies and long distance vendors. Inbound and outbound WATS service can be provided. 800 telephone service is often supplied on WATS lines. The lines can be configured for one-way or two-way telephone traffic. The main thing to remember about WATS is that it is cheaper!

Web Page: Refers to either the HTML source document or the hypermedia image that is interpreted by a Web browser such as Netscape or Mosaic.

Wire Wrap Tool: A device that is used to strip and coil a single strand of wire around a terminal peg to create a very reliable connection between two pieces of telephony equipment. A wire wrap tool is used on wire wrap terminal blocks, which are a substitute for 50 pair punch down blocks.

Wire Snake: A long rigid metal wire used by telephone technicians to put telephone station wire in a wall or ceiling after a building is constructed to add telephony service to a location.

World Wide Web (WWW or Web): A graphical, interactive, hypertext information system that is cross-platform and can be run locally or over the global Internet. The Web consist of Web servers offering pages of information to Web browsers who view and interact with the pages. Pages can contain formatted text, background colors, graphics, as well as audio and video clips. Simple links in a Web page can cause the browser to jump to a different part of the same page or to a page on a Web halfway around the world. Web pages can be used to send mail, read news, and download files. A Web address is called a URL (Uniform Resource Locator). *See also* HTML and HTTP.

Index

A

D

E

I

Q

R

T

U

V

LAN TIMES Free Subscription Form

❍ Yes, I want to receive (continue to receive) LAN TIMES free of charge. ❍ No.

I am ❍ a new subscriber ❍ renewing my subscription ❍ changing my address

Signature required ____________________ Date ____________________

Name ____

Title ____________________ Telephone ____________________

Company ____

Address ____

City ____

State/County ____________________ Zip/Postal Code ____________________

Free in the United States to qualified subscribers only

International Prices (Airmail Delivery)

Canada: $65 Elsewhere: $150

❍ Payment enclosed ❍ Bill me later

Charge my: ❍ Visa ❍ Mastercard ❍ Amer. Exp

Card number ____________________

Exp. Date ____________________

All questions must be completed to qualify for a subscription to LAN TIMES. Publisher reserves the right to serve only those individuals who meet publication criteria.

1. Which of the following best describe your organization? (Check only one)
- ❍ A. Agriculture/Mining/Construction/Oil/Petrochemical/Environmental
- ❍ B. Manufacturer (non-computer)
- ❍ C. Government/Military/Public Adm.
- ❍ D. Education
- ❍ E. Research/Development
- ❍ F. Engineering/Architecture
- ❍ G. Finance/Banking/Accounting/Insurance/Real Estate
- ❍ H. Health/Medical/Legal
- ❍ I. VAR/VAD Systems House
- ❍ J. Manufacturer Computer Hardware/Software
- ❍ K. Aerospace
- ❍ L. Retailer/Distributor/Wholesaler (non-computer)
- ❍ M. Computer Retailer/Distributor/Sales
- ❍ N. Transportation
- ❍ O. Media/Marketing/Advertising/Publishing/Broadcasting
- ❍ P. Utilities/Telecommunications/VAN
- ❍ Q. Entertainment/Recreation/Hospitality/Non-profit/Trade Association
- ❍ R. Consultant
- ❍ S. Systems Integrator
- ❍ T. Computer/LAN Leasing/Training
- ❍ U. Information/Data Services
- ❍ V. Computer/Communications Services: Outsourcing/3rd Party
- ❍ W. All Other Business Services
- ❍ X. Other ____________________

2. Which best describes your title? (Check only one)
- ❍ A. Network/LAN Manager
- ❍ B. MIS/DP/IS Manager
- ❍ C. Owner/President/CEO/Partner
- ❍ D. Data Communications Manager
- ❍ E. Engineer/CNE/Technician
- ❍ F. Consultant/Analyst
- ❍ G. Micro Manager/Specialist/Coordinator
- ❍ H. Vice President
- ❍ I. All other Dept. Heads, Directors and Managers
- ❍ J. Educator
- ❍ K. Programmer/Systems Analyst
- ❍ L. Professional
- ❍ M. Other ____________________

3. Which of the following best describes your job function? (Check only one)
- ❍ A. Network/LAN Management
- ❍ B. MIS/DP/IS Management
- ❍ C. Systems Engineering/Integration
- ❍ D. Administration/Management
- ❍ E. Technical Services
- ❍ F. Consulting
- ❍ G. Research/Development
- ❍ H. Sales/Marketing
- ❍ I. Accounting/Finance
- ❍ J. Education/Training
- ❍ K. Office Automation
- ❍ L. Manufacturing/Operations/Production
- ❍ M. Personnel
- ❍ N. Technology Assessment
- ❍ O. Other ____________________

4. How many employees work in your entire ORGANIZATION? (Check only one)
- ❍ A. Under 25
- ❍ B. 25-100
- ❍ C. 101-500
- ❍ D. 501-1,000
- ❍ E. 1,001-5,000
- ❍ F. 5,001-9,999
- ❍ G. 10,000 and over

5. Which of the following are you or your clients currently using, or planning to purchase in the next 12 months? (1–Own; 2–Plan to purchase in next 12 months) (Check all that apply)

Topologies	1	2
A. Ethernet	❍	❍
B. Token Ring	❍	❍
C. Arcnet	❍	❍
D. LocalTalk	❍	❍
E. FDDI	❍	❍
F. Starlan	❍	❍
G. Other	❍	❍

Network Operating System	1	2
A. Novell Netware	❍	❍
B. Novell Netware Lite	❍	❍
C. Banyan VINES	❍	❍
D. Digital Pathworks	❍	❍
E. IBM LAN Server	❍	❍
F. Microsoft LAN Manager	❍	❍
G. Microsoft Windows for Workgroups	❍	❍
H. Artisoft LANtastic	❍	❍
I. Sitka TOPS	❍	❍
J. 10NET	❍	❍
K. AppleTalk	❍	❍

Client/Workstation Operating Sys.	1	2
A. DOS	❍	❍
B. DR-DOS	❍	❍
C. Windows	❍	❍
D. Windows NT	❍	❍
E. UNIX	❍	❍
F. UnixWare	❍	❍
G. OS/2	❍	❍
H. Mac System 6	❍	❍
I. Mac System 7	❍	❍

Protocols/Standards	1	2
A. IPX	❍	❍
B. TCP/IP	❍	❍
C. X.25	❍	❍
D. XNS	❍	❍
E. OSI	❍	❍
F. SAA/SNA	❍	❍
G. NFS	❍	❍
H. MHS	❍	❍

6. Is your Organization/Clients network...(Check all that apply)

- ❍ A. International
- ❍ B. National
- ❍ C. Regional
- ❍ D. Metropolitan
- ❍ E. Local
- ❍ F. Other ______________

7. What hardware does your department/client base own/plan to purchase. (Check all that apply)

	Owns	Plan to purchase in next 12 months
A. Bridges	❍	❍
B. Diskless Workstations	❍	❍
C. Cabling System	❍	❍
D. Printers	❍	❍
E. Disk Drive	❍	❍
F. Optical Storage	❍	❍
G. Tape Backup System	❍	❍
H. Optical Storage	❍	❍
I. Application Servers	❍	❍
J. Communication Servers	❍	❍
K. Fax Servers	❍	❍
L. Mainframe	❍	❍
M. Network Adapter Cards	❍	❍
N. Wireless Adapters/Bridges	❍	❍
O. Power Conditioners/UPSs	❍	❍
P. Hubs/Concentrators	❍	❍
Q. Minicomputers	❍	❍
R. Modems	❍	❍
S. 386-based computers	❍	❍
T. 486-based computers	❍	❍
U. Pentium-based computers	❍	❍
V. Macintosh computers	❍	❍
W. RISC-based workstations	❍	❍
X. Routers	❍	❍
Y. Multimedia Cards	❍	❍
Z. Network Test/Diagnostic Equipment	❍	❍
1. Notebooks/Laptops	❍	❍
2. DSU/CSU	❍	❍
99. None of the Above	❍	❍

8. What network software/applications do you/your clients own/plan to purchase in the next 12 months? (Check all that apply)

- ❍ A. Network Management
- ❍ B. Software Metering
- ❍ C. Network Inventory
- ❍ D. Virus Protection
- ❍ E. Menuing
- ❍ F. E-mail
- ❍ G. Word Processing
- ❍ H. Spreadsheet
- ❍ I. Database
- ❍ J. Accounting
- ❍ K. Document Management
- ❍ L. Graphics
- ❍ M. Communications
- ❍ N. Application Development Tools
- ❍ O. Desktop Publishing
- ❍ P. Integrated Business Applications
- ❍ Q. Multimedia
- ❍ R. Document Imaging
- ❍ S. Groupware
- ❍ Z. None of the above

9. What is the annual revenue of your entire organization or budget if non-profit (Check only one)

- ❍ A. Under $10 million
- ❍ B. $10-$50 million
- ❍ C. $50-$100 million
- ❍ D. $100-$500 million
- ❍ E. $500 million-$1 billion
- ❍ F. Over $1 billion

10. How much does your organization (if reseller, your largest client's company) plan to spend on computer products in the next 12 months? (Check only one)

- ❍ A. Under $25,000
- ❍ B. $25,000-$99,999
- ❍ C. $100,000-$499,999
- ❍ D. $500,000-$999,999
- ❍ E. $1 billion

11. Where do you purchase computer products? (Check all that apply)

- ❍ A. Manufacturer
- ❍ B. Distributor
- ❍ C. Reseller
- ❍ D. VAR
- ❍ E. System Integrator
- ❍ F. Consultant
- ❍ G. Other ______________

12. In which ways are you involved in acquiring computer products and services? (Check all that apply)

- ❍ A. Determine the need
- ❍ B. Define product specifications/features
- ❍ C. Select brand
- ❍ D. Evaluate the supplier
- ❍ E. Select vendor/source
- ❍ F. Approve the acquisition
- ❍ G. None of the above

ICS1639

fold here

Place Stamp Here

LAN TIMES

McGraw–Hill, INC.

P.O. Box 652

Hightstown NJ 08520-0652